# Porous Metals with Directional Pores

Hideo Nakajima

# Porous Metals with Directional Pores

Springer

Hideo Nakajima
Director, The Wakasa Wan Energy Research Center
Tsuruga, Japan

Emeritus Professor, Osaka University
Suita, Japan

ISBN 978-4-431-54016-8 ISBN 978-4-431-54017-5 (eBook)
DOI 10.1007/978-4-431-54017-5
Springer Tokyo Heidelberg New York Dordrecht London

Library of Congress Control Number: 2013942871



Printed on acid-free paper

Springer is part of Springer Science+Business Media (www.springer.com)

# Preface

Porous metallic materials such as foamed metals, sponge-like metals, structural cellular metals, metals with directional pores, and sintered metals are increasingly looked upon as potential light-weight structural and functional materials with, for example, superior sound absorption, damping, and filtering properties. The porous metals are a new category of promising engineering materials from the point of view of both fundamental science and industrial applications. To realize such applications, various problems in fabrication such as uniformity in pore size and porosity controlled with variable factors must be solved; furthermore, their properties must be sufficiently understood and elucidated. Toward that end, investigations into the science and technology of porous and foamed metallic materials definitely have been expanded recently.

It is well understood that porous and foamed metals should be useful in solving some major issues of the twenty-first century such as environmental preservation, aging societies, and energy problems. For example, foamed aluminum is expected to be used for crash absorbers in automobiles and for sound absorption, while porous metals with elongated directional pores may be useful for medical devices, machine tools, heat sinks, and in other ways. There are various fabrication methods, classified by power sintering, foaming, and casting techniques. Porous metals are further grouped into porous and cellular metals depending upon the magnitude of their porosity.

Among these porous and foamed metals, the porous metals with directional pores, the so-called lotus and gasar metals, have been attracting attention owing to their long cylindrical pores aligned in one direction. These are considered new types of porous metals. Recently, methods for their fabrication have almost been established at the mass production level. Various unique physical, chemical, and mechanical properties have been discovered and already have been fairly well understood. In addition, several applications are proceeding at the industrial manufacturing level. Thus, I consider that the present is a good time to organize and present the science and relevant technology of porous metals with directional pores. I hope that readers of this book can understand the present status of research and development of porous metals with directional pores for the benefit of progress in their research.

Tsuruga, Japan<br>Suita, Japan

Hideo Nakajima

# Acknowledgements

The author would like to acknowledge the collaborations with Prof. S.K. Hyun of Inha University in Korea, Prof. T. Ikeda of Ibaraki University, Prof. S. Suzuki of Waseda University, Dr. M. Tane, Dr. T. Ide, Prof. K. Nakata, Prof. M. Hirao, Prof. H. Utsunomiya, Prof. T. Nakano, Prof. S. Fujimoto of Osaka University, Prof. O. Yoshinari of Nagoya Institute of Technology, Dr. S. Ueno of Nippon University, Dr. T. Murakami of Tohoku University, Prof. B.-Y. Hur of Gyeongsang National University in Korea, Dr. V. Shapovalov of MER Corporation, Ltd. of USA, Dr. Y. Higuchi of Osaka Dental University, Prof. T. Ogushi of Hiroshima International University and Dr. H. Chiba of Mitsubishi Electric Corporation, Ltd.

The author also expresses his appreciation to Prof. G. Stephani of Fraunhofer Institute and Prof. J. Banhart of Helmholtz Center Berlin of Germany, and Prof. D. Dunand of Northwestern University of USA.

# Contents

# Chapter 1
# Introduction

**Abstract** In natural world, a number of porous materials are observed, for example, woods, animal bones, leaves, and stalks. These holes or pores carry out the functions of supply, lightweight property, fluid permeability, keeping warm ability, masticability, etc. Wood, trabecular bone, and bamboo are all anisotropic; their mechanical properties depend on the direction of loading. Natural cellular materials exploit anisotropy to increase their mechanical efficiency, placing material where it is most needed to resist the applied loads. Throughout this book, we shall see the way in which porous materials exploit anisotropy to give unique functional performance.

**Keywords** Bamboo • Cancellous bone • Cellular materials • Natural biomaterials • Solidification defects

## 1.1 Porous Materials Widespread in Natural World

When we look around the natural world, we notice that a number of porous materials are existent such as woods, animal bones, leaves, and stalks. Besides, artificial materials such as foods, clothes, and buildings are not immaculate, but porous. Figure 1.1 shows the microstructure of several cellular materials [1]. Wood and cork are honeycomb-like cellular materials with prismatic cells like the hexagonal cells in a bee's honeycomb. Sponge and cancellous tissue consist of connected ligaments, while a coral and a bone of cuttlefish consist of stacked rectangular cells. On the other hand, leaves are network of round cells and stems of plants have also similar shape, but these are cross section of bundles of many tubes. As shown in Fig. 1.2, the bones are much familiar to our human body [1]. Most bones are an elaborate construction, made up of an outer shell of dense, compact bone, enclosing a core of porous cellular, cancellous, or trabecular bone. Most of the configuration minimizes the weight of bone while still providing a large bearing area, a design which reduces the bearing stresses at the joint [1]. In a sense these bones are

H. Nakajima, *Porous Metals with Directional Pores*,
DOI 10.1007/978-4-431-54017-5_1, 

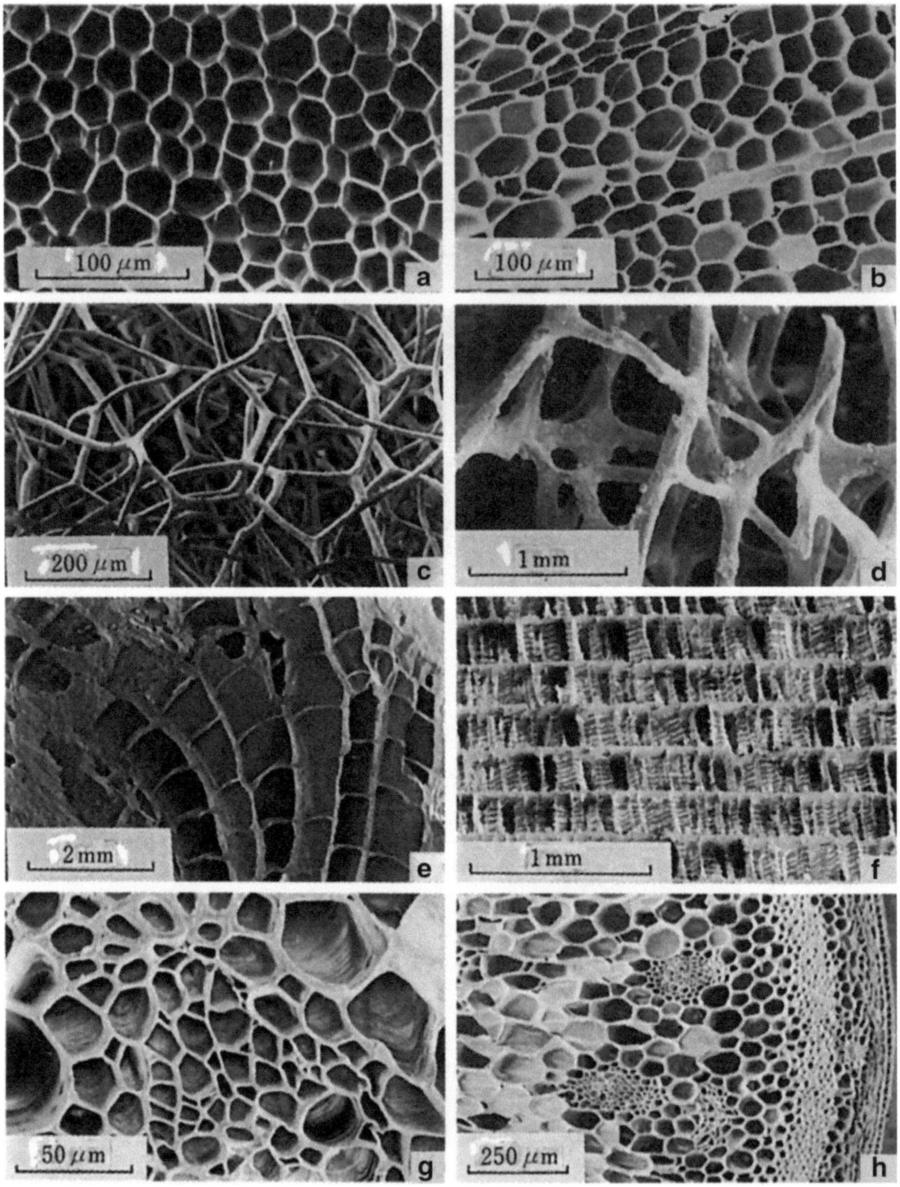


**Fig. 1.1** Natural cellular materials: (**a**) cork, (**b**) balsa, (**c**) sponge, (**d**) cancellous bone, (**e**) coral, (**f**) cuttlefish bone, (**g**) iris leaf, and (**h**) stalk of a plant (Reprinted with permission from [1] © 1997 Cambridge University Press)

considered as functionally gradient materials; the outer immaculate skin layer holds strength and the porosity of the bone increases as the depth from the top surface is inward. It is more interesting that the bones of birds that fly in sky are much lighter weight, just like a tube, and more porous than those of animals that crawled on

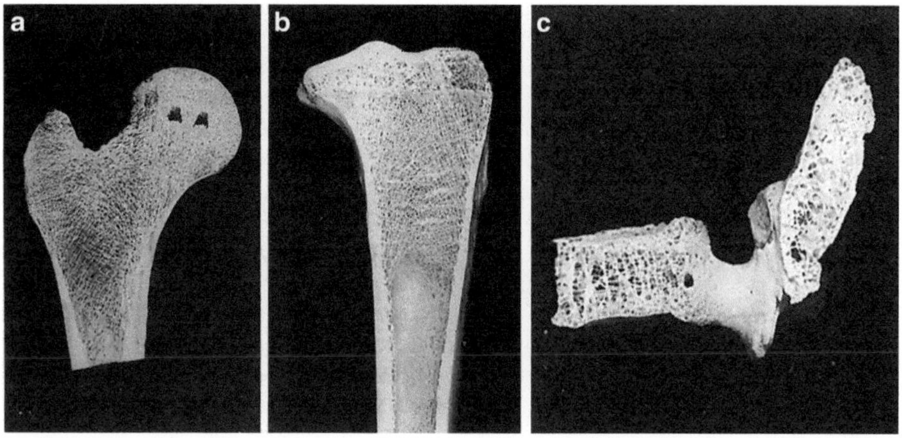


**Fig. 1.2** Cross-sectional views of (**a**) the head of a femur, (**b**) the tibia, and (**c**) a lumbar vertebra. In each case, there is an outer shell of almost fully dense compact bone surrounding a core of porous, low-density, cancellous bone (Reprinted with permission from [1] © 1997 Cambridge University Press)

the ground as shown in Fig. 1.3 [http://www3.famille.ne.jp/~ochi/kaisetsu-01/05-te-ashi.html]. Thus, it may be considered that such natural biomaterials are one example to be designed as advanced materials. We, researchers who are investigating metallic and inorganic advanced materials, should learn various ideas from natural environment, in particular, the biomaterials in nature. These holes or pores carry out the functions of channels of supply, lightweight property, fluid permeability, keeping warm ability, masticability etc. Superficially, bones look fairly solid, but deceptive.

It is no exaggeration to say that most of various parts of industrial products are manufactured by casting or powder sintering techniques. In these processes, casting defects and sintering defects such as gas pores are usually considered as harmful defects to impede an efficiency or functional properties of the manufactured products. Therefore, it is said to be indispensable for high performance of the products to manufacture high-density materials with a porosity as small as possible. Figure 1.4 shows cross-sectional view of rimmed steel ingot [2]. The top part of the ingot has large shrinkage cavity, which is caused by volume reduction during solidification (not appear in this photo). Inside the ingot, there are a number of cylindrical pores which are called casting (solidification) defects. Such defects are known to be formed by evolution of hydrogen bubbles, carbon monoxide gas bubbles, etc. In particular, in the bottom part, some longer elongated pores are observed. Such long pores are grown and elongated in the direction of solidification. This book focuses attention to formation of the elongated solidification defects, properties of metals with the elongated defects, and utilization of such elongated defects in metals.

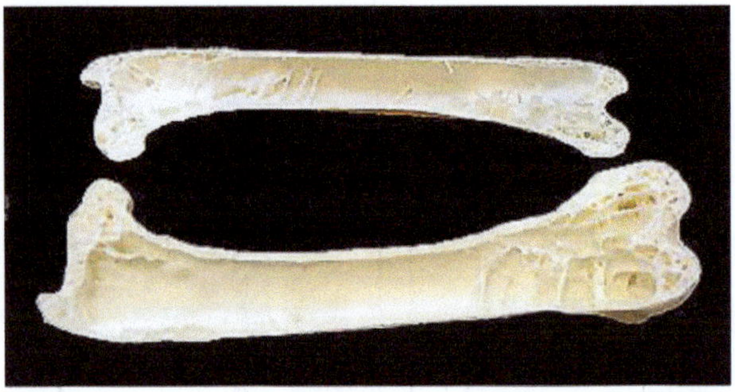

Fig. 1.3 Cross sections of a bone of a bird that flies in sky (Reprinted with permission from [http://www3.famille.ne.jp/~ochi/kaisetsu-01/05-te-ashi.html] © Nature Photo Gallery, Shinji Ochi)

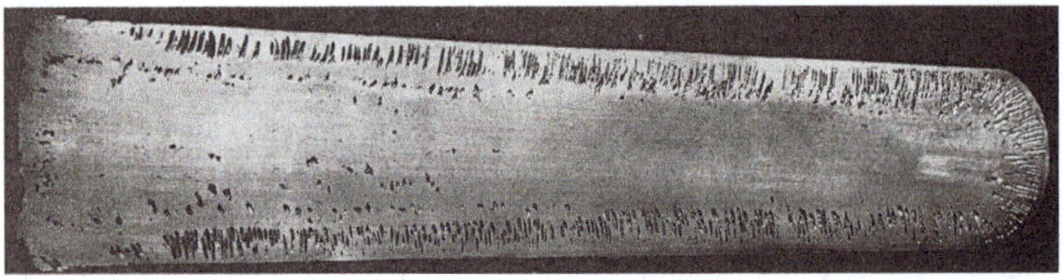

Fig. 1.4 Cross-sectional view of rimmed steel ingot. Two typical solidification defects are observed; top part is shrinkage cavity and surroundings are gas pores [2]

Up to the present, for example, penetrated pores can be used for filtering materials and large surface area resulted from high porosity can be utilized as electrode materials. However, if the porous materials whose mechanical strength does not become inferior significantly can be produced, wide and various applications to lightweight structural, functional materials, transportation materials, etc., could be possible prospectively. The porous materials with directional pores may meet these demands.

Natural structures such as bones have a gradient in density, rather than two distinct solid and cellular components. For example, bamboo epitomizes this (Fig. 1.5); the volume fraction of dense fibers increases radially toward the periphery of the stem. Bamboo is also tubular, again increasing the bending stiffness of its cross section. Wood, trabecular bone and bamboo are all anisotropic; their mechanical properties depend on the direction of loading. Natural cellular materials exploit anisotropy to increase their mechanical efficiency, placing material where it is most needed to resist the applied loads. In tree, for instance, the highest stresses, resulting from bending in the wind, act along the length of the trunk and branches. Wood is much stiffer and stronger in this direction, along the grain, than across it, as a result of its honeycomb-like cellular microstructure as well as the composite nature of the

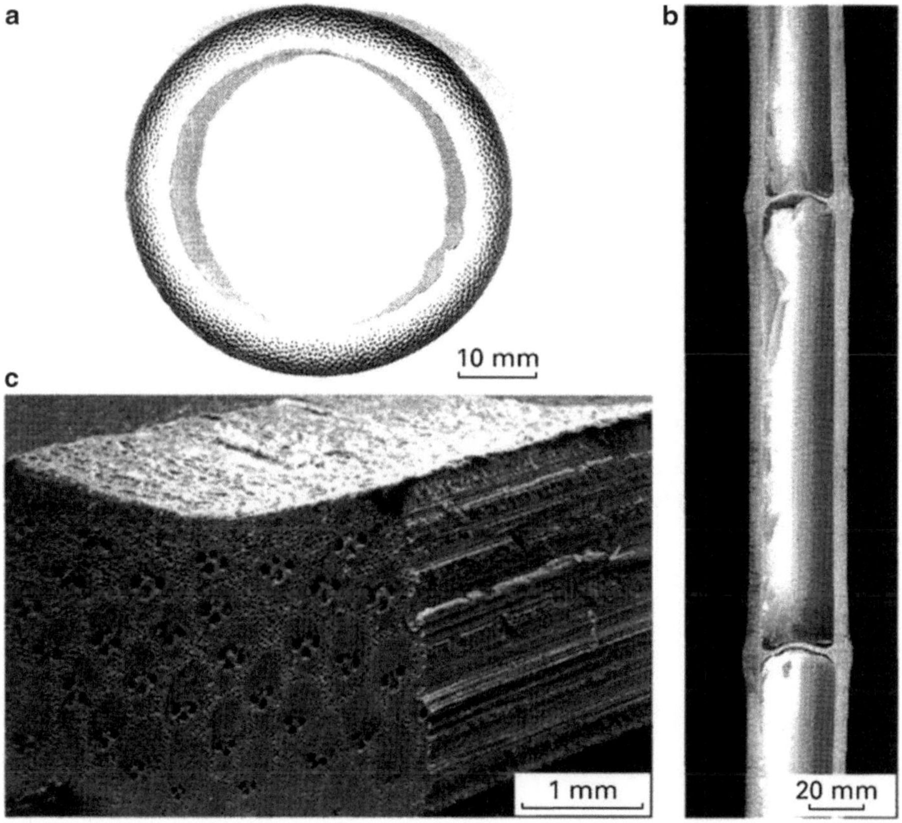


**Fig. 1.5** (**a**) A cross-section of bamboo, showing the tubular (**b**) structure. (**b**) A longitudinal section of bamboo, showing the more or less evenly spaced diaphragms. (**c**) Scanning electron micrograph of a cross-section of bamboo, showing the radial density gradient. (Reprinted with permission from [3] © 2010 Cambridge University Press

solid cell wall material. Bone grows in response to applied loads; the trabeculae in human vertebrae, for instance, which are subjected primarily to compressive loading from the weight of the body, align in the vertical direction, increasing the stiffness and strength in that direction. Throughout this book, we shall see the way in which porous materials exploit anisotropy to give exceptional mechanical performance.

This book is divided into three parts. The first part summarized the various fabrication methods of cellular and foamed metals (Chap. 2), fabrication methods of porous metals with directional pores, in particular explaining various casting techniques and various gas-supplying techniques including the historical background (Chap. 3). Furthermore, nucleation and growth mechanism of pores in metals in Chap. 4 and control methods of pore size and porosity in metals in

Chap. 5 were described. Then we described the details of fabrication techniques of various materials through various fabrication methods in Chap. 6. In the second part, mechanical properties (Chap. 7) and various physical and chemical properties (Chap. 8) and processing techniques (Chap. 9) are explained. Finally, in the third part, various applications to heat sinks, vibration-damping materials, golf putter, and medical devices are described somehow in details in Chap. 10. In Chap. 11, the conclusions are summarized.

## References

1. Gibson LJ, Ashby MF (1997) Cellular solids. Cambridge University Press, Cambridge
2. Hultgren A, Phragmen G (1939) Solidification of rimming-steel ingots. J Iron Steel Inst 139:133–244
3. Gibson LJ, Ashby MF, Harley BA (2010) Cellular materials in nature and medicine. Cambridge University Press, Cambridge

# Chapter 2
# Various Fabrication Methods of Cellular Metals and Foamed Metals

**Abstract** This book concerns about porous metals with directional pores. In general, porous materials are defined as holey materials, which are mainly characterized by porosity. The porous materials with high porosity and low density are called as foamed materials, which resemble bubbling materials such as soap and beer. Cellular materials look like honeycomb, whose porosity is also high and the density is low. The holey materials, whose porosity is less than 70 %, are usually called porous materials. Depending upon the shape of pores, isotropic and anisotropic porous materials are classified. The former has rather isotropic spherical pores, while the latter has cylindrically elongated pores aligned in one direction. In this chapter, various fabrication techniques of foamed metals and cellular metals are presented.

**Keywords** Closed pores • Hollow spheres • Metal foams • Open pores • Porosity

## 2.1 Materials Definition

There are several nomenclatures to call metallic materials which possess high porosity. Usually, it is accustomed to classify these materials by their relative density, $\rho^*/\rho_s$, the density of the high-porosity metal $\rho^*$ divided by that of the solid metal (nonporous metal) $\rho_s$. Special ultra-low-density foams can be made with a relative density as low as 0.001. Polymeric foams used for cushioning, packing, and insulation have relative densities which are usually between 0.05 and 0.2. As the relative density of the foamed metals increases, the cell walls thicken and the pore space shrinks; above about 0.3 there is a transition from a cellular structure to porous structure which is better thought of as a solid containing isolated pores. Thus, the porous materials are defined as the materials whose porosity ranges less than 70 %.

The shape of pores in most of foamed and sintered materials is nearly spherical and isotropic, while the shape in lotus and gasar-type porous materials is

H. Nakajima, *Porous Metals with Directional Pores*,
DOI 10.1007/978-4-431-54017-5_2, 

anisotropic and long cylindrical, which is evolved unidirectionally during the solidification. Thus, the pore morphology is defined with the following parameters:

Pore diameter (or pore size)
Pore length
Pore aspect ratio: the ratio of pore length against the diameter
Pore orientation
Porosity: pore volume fraction

Regardless of porosity, pore size, and pore shape, two kinds of pores are existent depending upon the fabrication methods: closed pores and open pores. "Closed pores" are surrounded by material wall and each pore is isolated, which are usually filled with gas, while "open pores" are pores connected to each other in space; the pores are held by struts. The closed pores are not permeable to gas or fluid, but the open pores are usually permeable.

## 2.2 Various Fabrication Methods

A metallic foam was first tried to be produced by Sosnik in 1948 [1]. In order to create pores, he put mercury into molten aluminum. In 1956 Elliot replaced mercury with foaming agents generating gas by thermal decomposition [2], and until now, many researchers have developed metallic foam using this technique without the toxicity of mercury. Then in 1963, Allen invented a powder-compact foaming technique to manufacture metallic foams, and the basic processing techniques for such metallic foams were almost completed [3].

A variety of cellular and foamed metals have been fabricated by researchers and industries. Metal foams and porous metals are made by the following processes, which are summarized from the book [4].

### *2.2.1 Melt Gas Injection (Air Bubbling) [4]*

Pure liquid metals cannot easily be caused to foam by bubbling a gas into them. Drainage of liquid down the walls of the bubbles usually occurs too quickly to create a foam that remains stable long enough to solidify. However, 10–30 % of small, insoluble, or slowly dissolving particles, such as aluminum oxide or silicon carbide, raise the viscosity of the aluminum melt and impede drainage in the bubble membrane, stabilizing the foam. Gas-injection processes are easiest to implement with aluminum alloys because they have a low density and do not excessively oxidize when the melt is exposed to air or other gases containing oxygen. A variety of gases can be used to create bubbles within liquid aluminum. Bubbles formed by this process float to the melt surface, drain, and then begin to solidify. Low relative density, closed-cell foams can be produced by carefully controlling the gas-injection process and the cooling rate of the foam. Various techniques can be used to draw off the foam and create large (up to 1 m wide and 0.2 m thick) slabs containing closed cell pores with diameters between 5 and 20 mm. This technique is the least costly to implement and results in a foam with relative densities in the range 0.03–0.1.

### *2.2.2 Gas-Releasing Particle Decomposition in the Melt [4]*

Metal alloys can be foamed by mixing into them a foaming agent that releases gas when heated. The widely used foaming agent titanium hydride ($TiH_2$) begins to decompose into Ti and gaseous $H_2$ when heated above about 738 K. By adding titanium hydride particles to aluminum melt, large volumes of hydrogen gas are rapidly produced, creating bubbles that can lead to a closed-cell foam, provided foam drainage is sufficiently slow, which requires a high-melt viscosity. The Shinko Wire Company has developed an aluminum foam trade named Alporas using this technique. The process begins by melting aluminum and stabilizing the melt temperature between 943 and 963 K. Its viscosity is then raised by adding 1–2 % of calcium which rapidly oxidized and forms finely dispersed CaO and $CaAl_2O_4$ particles. The melt is then aggressively stirred and 1–2 % of $TiH_2$ is added in the form of 5–20 μm diameter particles. As soon as these are dispersed in the melt, the stirring system is withdrawn, and a foam is allowed to form above the melt. Control of the process is achieved by adjusting the overpressure, temperature, and time. When foaming is complete, the melt is cooled to solidify the foam before the hydrogen escapes and the bubbles coalesce or collapse. The cell size can be varied from 0.5 to 5 mm by changing the $TiH_2$ content, and the foaming and cooling conditions. Relative densities from 0.2 to as low as 0.07 can be obtained. As produced, the Alporas foam has predominantly closed cells, though a subsequent rolling treatment can be used to fracture many of the cell walls in order to increase their acoustic damping.

### *2.2.3 Gas-Releasing Particle Decomposition in Semisolids [4]*

Foaming agents can be introduced into metals in the solid state by mixing and consolidating powders. Titanium hydride, a widely used foaming agent, begins to decompose at about 738 K, which is well below the melting point of pure aluminum and of its alloys. This raises the possibility of creating a foam by dispersing the foaming agent in solid aluminum using powder metallurgy processes and then raising the temperature sufficiently to cause gas release and partial or full melting of the metal, allowing bubble growth. Cooling then stabilizes the foam. It begins by combining particles of a foaming agent with an aluminum alloy powder. After the ingredients are thoroughly mixed, the powder is cold compacted and then extruded into a bar or plate of near theoretical density. This precursor material is chopped into small pieces, placed inside a sealed split mold, and heated to a little above the solidus temperature of the alloy. The titanium hydride then decomposes, creating voids with a high internal pressure. These expand by semisolid flow and the aluminum swells, creating foam that fills the mold. The process results in components with the same shape as the container and relative densities as low as 0.08. The foam has closed cells with diameters that range from 1 to 5 mm in diameter.

### 2.2.4 *Casting Using a Polymer or Wax Precursor as Template [4]*

An open-cell polymer foam mold template with the desired cell size and relative density is first selected. This can be coated with a mold casting slurry which is then dried and embedded in casting sand. The mold is then baked both to harden the casting material and to decompose the polymer template, leaving behind a negative image of the foam. This mold is subsequently filled with a metal alloy and allowed to cool. The use of a moderate pressure during melt infiltration can overcome the resistance to flow of some liquid alloys. After directional solidification and cooling, the mold materials are removed, leaving behind the metal equivalent of the original polymer foam. Metal powder slurries can also be used instead of liquid metals. These are subsequently sintered. The method gives open-cell foams with pore sizes of 1–5 mm and relative densities as low as 0.05. The process can be used to manufacture foams from almost any metal that can be investment cast.

### 2.2.5 *Metal Deposition on Cellular Preforms [4]*

Open-cell polymer foams can serve as templates upon which metals are deposited by chemical vapor decomposition (CVD), by evaporation, or by electrodeposition. Nickel is deposited by the decomposition of nickel carbonyl, $Ni(CO)_4$. An open-cell polymer is placed in a CVD reactor and nickel carbonyl is introduced. This gas decomposes to nickel and carbon monoxide at a temperature of about 373 K and coats all the exposed heated surfaces within the reactor. Infrared or RF heating can be used to heat only the polymer foam. After several tens of micrometers of the metal have been deposited, the metal-coated polymer foam is removed from the CVD reactor and the polymer is burnt out by heating in air. This results in a cellular metal structure with hollow ligaments. A subsequent sintering step is used to densify the ligaments. The foam with open sizes in the 100–300 μm diameter range is available. It gives the lowest relative density (0.02–0.05) foams available today.

### 2.2.6 *Entrapped Gas Expansion [4]*

In the process Ti-6Al-4V powder is sealed in a canister of the same alloy. The canister is evacuated to remove any oxygen and then backfilled with between 0.3 and 0.5 MPa of argon. The canister is then sealed and consolidated to a high relative density (0.9–0.98) by HIPing causing an eightfold increase in void pressure. This is too low to cause expansion of Ti-6Al-4V at room temperature. The number of pores present in the consolidated sample is relatively low, so a rolling step is introduced to refine the structure and create a more uniform distribution of small pores. In titanium alloys, rolling at 1,173–1,213 K results in void flattening and elongation in the rolling direction. As the voids flatten, void faces come into contact and diffusion bond, creating strings of smaller gas-filled pores. Cross rolling improves the uniformity of their distribution. Various cold sheet forming processes can then be used to shape the as-rolled plates. The final step in the process sequence is expansion by heating at 1,173 K for 20–30 h. The high temperature raises the internal pore pressure by the ratio of the absolute temperature of the furnace to that of the ambient, i.e., to between 10 and 16 MPa, causing creep dilation and a reduction in the overall density of the sample. This process results in shaped Ti alloy sandwich construction components with a core containing a closed-cell void fraction of up to 0.5 and a void size of 10–300 μm.

### 2.2.7 *Hollow Sphere Structures [4]*

Several approaches have recently emerged for synthesizing hollow metal spheres. For one example, hollow spheres are formed from a slurry composed of a decomposable precursor such as $TiH_2$, together with organic binders and solvents. The spheres are hardened by evaporation during their flight in a tall drop tower, heated to drive off the solvents and to volatilize the binder. A final heat treatment decomposes the metal hydride, leaving hollow metal spheres. The approach can be applied to many materials and is not limited to hydrides. As an example, an oxide mixture such as $Fe_2O_3$ plus $Cr_2O_3$ can be reduced to create stainless steel.

In another method developed at IFAM, Bremen, polystyrene spheres are coated with a metal slurry and sintered, giving hollow metal spheres of high uniformity. The consolidation of hollow spheres gives a structure with a mixture of open and closed porosity. The ratio of the two types of porosity and overall relative density can be tailored by varying the starting relative density of the hollow spheres and the extent of densification during consolidation. Overall relative densities as low as 0.05 are feasible with a pore size in the range 100 μm to several millimeters.

### *2.2.8 Co-compaction or Casting of Two Materials, One Leachable [4]*

Two powders, neither with a volume fraction below 25 %, are mixed and compacted, forming double-connected structures of both phases. After consolidation one powder (e.g., salt) is leached out in a suitable solvent. Foams based on powder mixes of aluminum alloys with sodium chloride have successfully been made in large sections with uniform structures. The resulting cell shapes differ markedly from those of foams made by other methods. In practice the method is limited to producing materials with relative densities between 0.3 and 0.5. The cell size is determined by the powder particle size and lies in the range 10 μm–10 mm.

## References

1. Sosnik B (1948) US Patent No. 2,434,775
2. Elliot JC (1956) US Patent No. 2,751,289
3. Allen BC (1963) US Patent No. 3,087,807
4. Ashby MF, Evans A, Fleck NA, Gibson LJ, Hutchinson JW, Wadley HNG (2000) Metal foams: a design guide. Butterworth-Heinemann, Boston

# Chapter 3
# Fabrication Methods of Porous Metals with Directional Pores

**Abstract** Porous metals with directional pores were investigated from long time ago from the viewpoint to elucidate solidification or casting defects. In 1980s Shapovalov et al. indicated some applicability porous metals with directional pores fabricated through gasar process using high-pressure hydrogen. Then, Nakajima et al. carried out systematic investigations using various casting techniques such as mold casting technique, continuous zone melting technique, and continuous casting technique. The latter is the most superior to control pore size and porosity and is the most suitable for fabricating long-sized casting slabs. To overcome technical difficulty of use of high-pressure hydrogen, very simple fabrication method was invented recently by adding only gas-forming compounds into melt.

**Keywords** Continuous casting technique • Continuous zone melting technique • Gasar process • High-pressure gas method • Lotus metals • Mold casting technique • Thermal decomposition method

## 3.1 Historical Background

### *3.1.1 Ice Wormholes*

Chalmers [1] observed porous ice with directional pores called as "ice worms" and discussed the mechanism of the pore formation to the following. Air is a solute that is rejected by water ahead of the freezing process during the solidification of water. It accumulates in the advancing interface until its concentration is high enough for bubbles to nucleate. Once a bubble has formed, it grows because air diffuses into it. If the interface continues to move forward, the bubble cannot grow laterally, and so, it grows forward to form the cylindrical bubble sometimes known as an ice worm as shown in Fig. 3.1. Close inspection shows that ice worms never start at the surface of the ice, but always a little way in; this is because some freezing must

H. Nakajima, *Porous Metals with Directional Pores*,
DOI 10.1007/978-4-431-54017-5_3, 

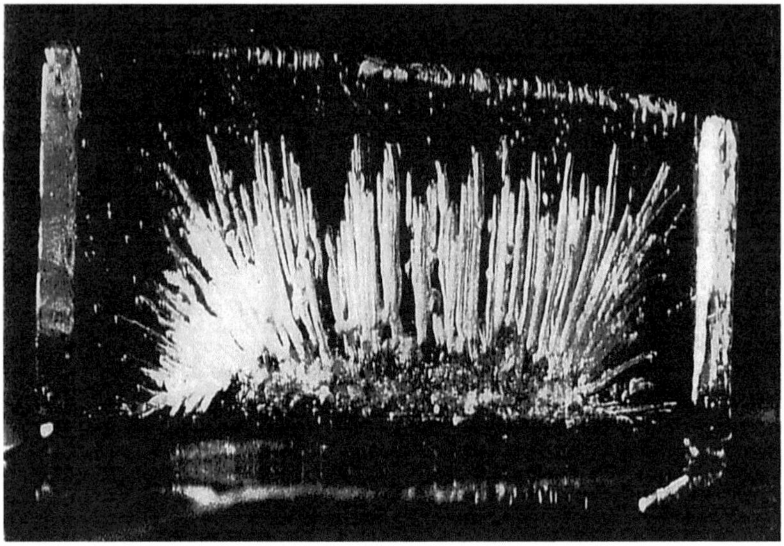

**Fig. 3.1** Ice worms form during the growth of an ice cube because the ice crystal (grew from the bottom) rejects air that was dissolved in the water (Reprinted with permission from [1] © 1959 Scientific American, Inc.)

occur before the air accumulating at the interface reaches sufficient concentration to cause the nucleation of a bubble. Here again growth occurs more easily than the nucleation that starts it.

The ice worms that grow in ice cubes frequently look like strings of pearls. This reflects the fluctuation in the freezing rate due to the intermittent operation of the refrigerator. When freezing is slow, more air diffuses into the bubbles and they grow larger; during periods of fast growth, there is less times for diffusion and the bubble decreases in cross section. Fast freezing suppresses the formation of ice worms. Because insufficient air diffuses into the bubbles to permit them to grow, the ice contains a large number of very small, round bubbles. On the other hand, very slow freezing permits the rejected air to diffuse away from the interface; because the concentration of air never reaches the nucleation point, neither bubbles nor ice worms appear. Ice grown in flowing water is also usually free of bubbles and ice worms; the continuous removal of the water prevents the buildup of a high concentration of dissolved air, hence the absence of bubbles in the ice cubes produced by some commercial freezers.

For freezing to take place continuously, the heat of fusion must somehow be carried away from the region where ice is forming. Normally, as in the growth of ice cubes or of ice on the surface of a lake, the heat is extracted by conduction through the ice that has already formed. The water–ice interface tends to remain flat, because any region of convex curvature is cooled less efficiently than the remainder, and its rate of freezing is retarded until a smooth interface is restored. However, if the ice crystal is growing in supercooled water, its growth takes an entirely

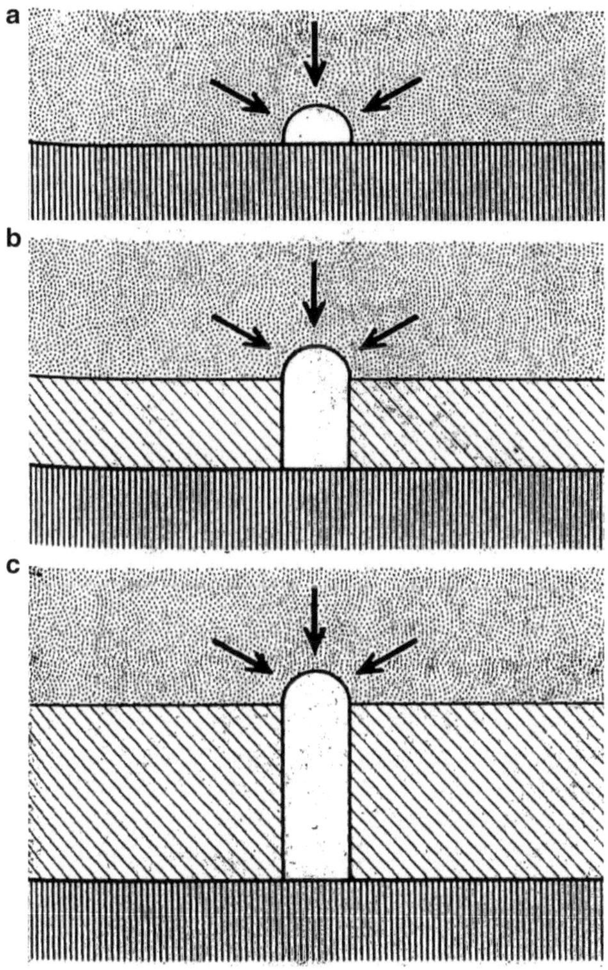


**Fig. 3.2** Growth of ice worm. Diagonal hatching shows accrual of ice; the *arrows*, dissolved air diffusing in to worm. (**a**) A pore nucleates in water in vicinity of ice. (**b**) As the interface between ice and water proceeds upward, enriched air which is rejected in the ice diffuses into the pore. (**c**) An elongated-cylindrical pore continues to grow in the direction of movement of the interface. (Reprinted with permission from [1] © 1959 Scientific American, Inc.)

different form. Because the latent heat of fusion flows outward into the supercooled liquid, cooling is more efficient at a convex region; convex regions therefore become even more convex. The latent heat produced by the growth of a convexity suppresses growth in the vicinity, and so a projection becomes an isolated spike as illustrated in Fig. 3.2. The direction in which such branching or "dendritic" growth may go in various substances depends on the characteristic molecular arrangement of the crystal. In the case of ice, the spike may grow in any one of six directions. Once a spike has formed, the supercooled environment may permit the growth of lateral secondary, ternary, and even quaternary branching spikes, always in the proper crystallographic directions. This explanation is highly suggestive to investigate the following porous metal science.

### *3.1.2 Porous Metals with Directional Pores*

Recently, a new type of porous metals whose long-cylindrical pores are aligned in one direction has been fabricated by unidirectional solidification under a pressurized hydrogen, nitrogen, or oxygen gas. Many gas pores are evolved from insoluble hydrogen (nitrogen or oxygen) in solids while hydrogen (nitrogen or oxygen) dissolves significantly in liquids. Formation of elongated gas pores during solidification has been investigated by Imabayashi et al. [2], by Svensson and Fredriksson [3], and by Knacke et al. [4]. Furthermore, Shapovalov et al. [5] fabricated longer cylindrical pores by adopting a unidirectional solidification technique at a high pressure of hydrogen. Nakajima et al. [6, 7] also produced porous iron, copper, magnesium, nickel, and those alloys in high-pressure hydrogen or nitrogen atmosphere and porous silver in high-pressure oxygen by means of the Czochralski method and a unidirectional solidification method. The processing technique is different from that of cellular and foamed metals, since it allows a control of pore size, pore direction, and overall porosity. High-pressure hydrogen or other gas is filled in the pores of the porous metals during an invariant reaction of the so-called gas-evolution crystallization reaction, where the melt is solidified to transform into a solid solution and a gas phase [8]. During solidification, the gas is rejected from the solid metal at the solid–liquid interface and forms long pores that are aligned parallel to the solidification direction. Shapovalov et al. named this method as gasar, which means a Ukrainian acronym for gas-reinforced composite metals [9]. However, it turns out by the systematic investigations by Nakajima et al. that gas pores do not affect to reinforce the porous metals. Therefore, a term from their shape is considered to be more suitable for the name of the porous metals. Nakajima group called their materials as "lotus-type porous metals" (hereafter we call lotus metals) because the morphology of the material resembles that of a lotus root. The strength of lotus metals is superior to that of conventional porous metals such as cellular and foamed metals [7, 10].

Thus, such lotus and gasar metals are expected as a new category of engineering materials.

## 3.2 High-Pressure Gas Method (PGM)

### *3.2.1 Mold Casting Technique*

As mentioned above, the gas pores are evolved by insoluble gas atoms in the solid when the melt dissolving gas is solidified because of the gas solubility difference between liquid and solid. Figure 3.3 shows the temperature dependence of hydrogen solubility in both states of various metals [11]. The solubility of hydrogen increases with increasing temperature in both solid and liquid. For the metals in which a discontinuous abrupt decrease is observed in the solubility at the melting

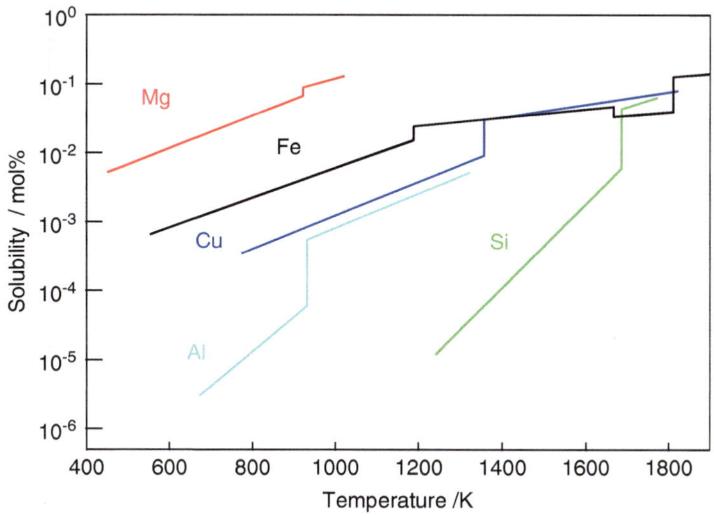


**Fig. 3.3** Temperature dependence of hydrogen solubility in solids and liquids of various metals under the hydrogen pressure of 0.1 MPa [11]

point when the temperature is lowered, significant pore evolution occurs in the solidified metals so that the lotus and gasar structure can be easily fabricated during the unidirectional solidification.

So far, casting method is usually adopted as the fabrication technique of the lotus and gasar metals. The peculiar difference of lotus and gasar metals from the cellular and foamed metals is the pore morphology; while the former has elongated cylindrical pores aligned in the solidification direction, the latter has always almost spherical pores distributed randomly in the matrix. Figure 3.4a shows schematic drawing of the mold casting technique to fabricate lotus and gasar metals; a metal inside a crucible is melted by an induction heating in a high-pressure gas atmosphere. The gas is dissolved up to the equilibrium gas concentration into the molten metal under a given gas pressure according to the Sieverts law [12]. The melt saturated with gas is poured into the mold. When some part of the mold is cooled down by a chiller or circulated water, the melt can be solidified unidirectionally from the vicinity of the cooling part. Figure 3.4b shows the overviews of the fabrication apparatus of mold casting technique installed at Nakajima Laboratory of Osaka University. The elongated pores can evolve and grow by the influence of the unidirectional solidification. The pore growth direction can be controlled by changing the location of the cooling part (chiller). Figure 3.5 illustrates three different types of the molds. When the bottom of the mold is water-cooled, the molten metal is unidirectionally solidified upwards from the bottom part so that the directional pores grow in the upward direction. When the lateral side of the mold is cooled down, the solidification takes place inwardly from the surrounding, and thus, the pore distribution becomes radial. If a particular cooling part is not set up in the

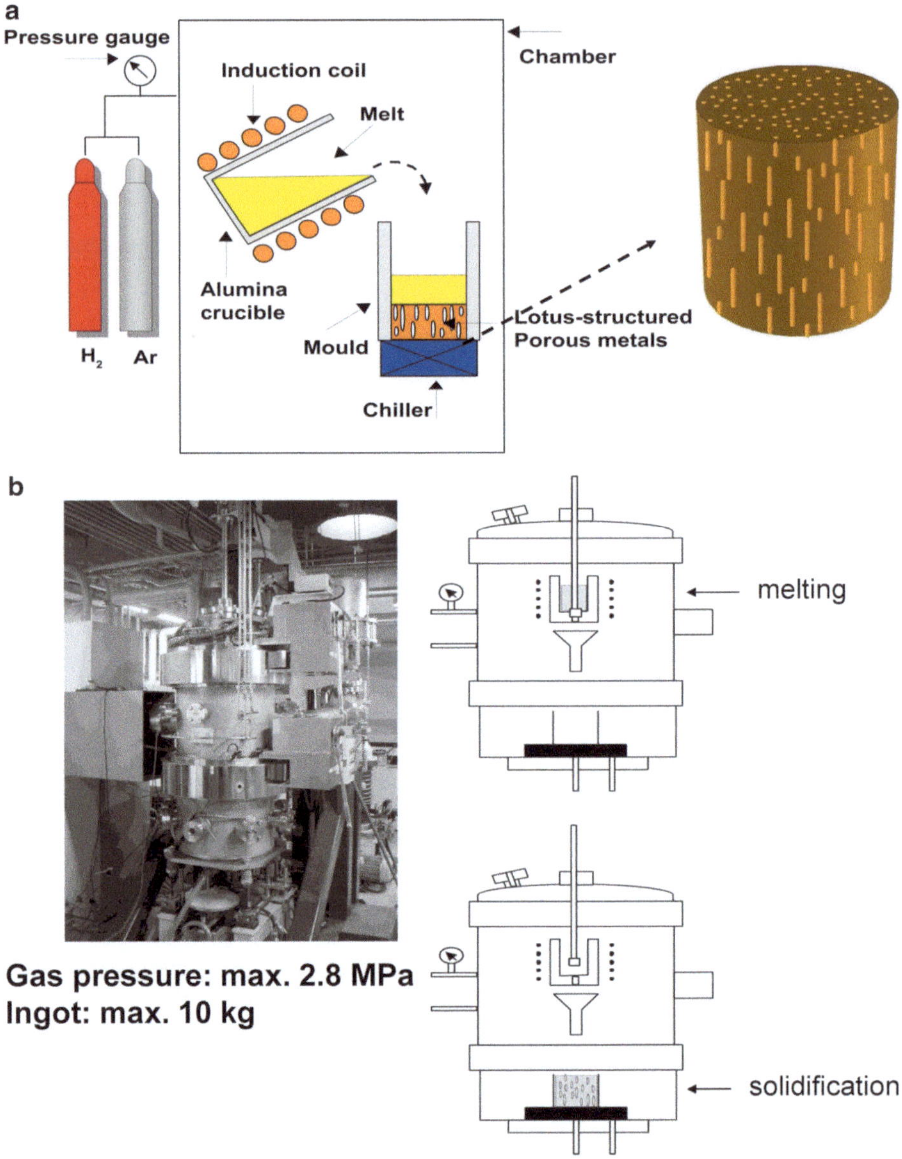


**Fig. 3.4** (**a**) Schematic drawing of principle for fabrication of lotus metals through mold casting technique (Reprinted with permission from literature of Prog Mater Sci 52(2007) 1091–1173, ©2006 Elsevier Ltd). (**b**) Overview of the apparatus for fabrication of lotus metals through mold casting technique installed at Nakajima Laboratory of Osaka University. The *right-hand upper figure* shows that the metal is melted by the crucible by heating radio-frequency induction coil. The *right-hand lower figure* shows that after the stopper rod is lifted to open the bottom hole of the crucible, the melt is dropped down through a funnel into the hearth whose bottom is cooled by a chiller. Unidirectional solidification takes place to produce the lotus metals

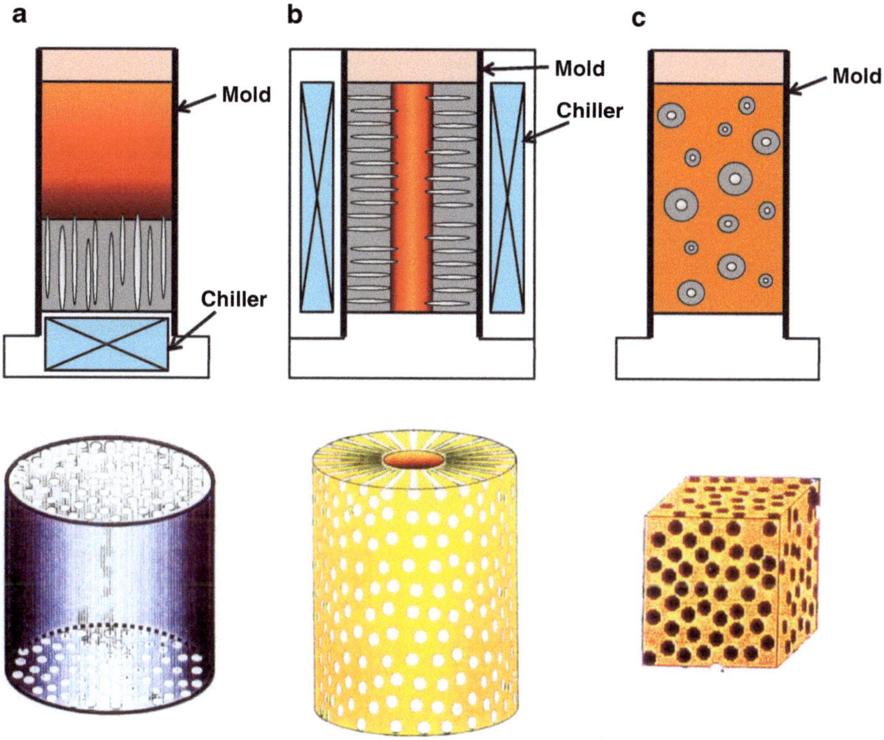


**Fig. 3.5** Three different types of molds for casting and resulting pore configurations in the porous metals: (**a**) unidirectional pores, (**b**) radial pores, and (**c**) isolated spherical pores. (**a**) and (**b**) are anisotropic pores, while (**c**) is isotropic pores (Reprinted with permission from literature of Prog Mater Sci 52(2007) 1091–1173, © 2006 Elsevier Ltd)

mold, the spherical pores start to grow everywhere and distributed randomly whose morphology resembles pumice stones.

In general, the porous, cellular, and foamed metals are characterized by pore growth direction, pore size, and porosity. As mentioned earlier, while the pore direction, pore size, and porosity of the cellular and foamed metals are not easy to control by the fabrication methods, those of the lotus and gasar metals are relatively easier by those fabrication methods because the direction of the solidification, gas pressure, gas content, and the solidification velocity can be controlled. The parameters to control the pore morphology can be listed up as

the melt temperature
the solidification rate
the temperature gradient of the interfacial region of liquid–solid phase
the dissolving gas pressure during melting and solidification
the inert gas pressure during melting and solidification

**Table 3.1** Lotus metals to be fabricated and available gases

| Available gas | Lotus metals to be fabricated |
|---|---|
| Hydrogen | Iron, carbon steel, stainless steel, aluminum, magnesium, nickel, copper, cobalt, tungsten, manganese, chromium, beryllium, and those alloys |
| Nitrogen | Iron, carbon steel, stainless steel |
| Oxygen | Silver |

The lotus metals to be fabricated and available gases are compiled in Table 3.1. Hydrogen gas is used to fabricate various lotus metals and alloys: iron, nickel, aluminum, copper, magnesium, cobalt, tungsten, manganese, chromium, beryllium, and those alloys. Fortunately, most of the base metals for commercially available practical alloys can be made porous. However, there are some problems; as the hydrogen concentration dissolving in molten aluminum is small, the porosity of lotus aluminum is less than 40 % by the present technique. On the other hand, although fabrication of porous titanium is possible, the pore shape is not cylindrical, but spherical, because of high dissolving concentration of hydrogen.

Since hydrogen is inflammable and explosive when oxygen is present, its use is not convenient from the industrial point of view. Use of other gases than hydrogen is desirable. It is well known that nitrogen is an important alloying element widely used to improve corrosion resistance and mechanical properties of steels. The temperature dependence of nitrogen solubility in solid and liquid of iron is similar to that of hydrogen, which exhibits a large nitrogen solubility difference between solid and liquid of iron at the melting temperature [13–15]. Moreover, it is known that an invariant reaction [8] of "gas-evolution crystallization reaction" takes place in the Fe-N system, in which the iron melt dissolving nitrogen is solidified to transform into a primary solid solution and nitrogen gas. Utilizing the nitrogen solubility difference between liquid and solid, lotus iron was fabricated by the abovementioned technique [16]. Using oxygen gas, porous silver can be produced, because metallic oxide of silver cannot be formed in oxygen atmosphere when the molten silver is solidified [17].

Figure 3.6 shows typical examples of optical micrographs on the cross section (above) and the longitudinal sections (below) of lotus copper.

### 3.2.2 *Continuous Zone Melting Technique*

The mold casting technique was usually used to produce lotus-type porous metals. For a few porous metals such as copper [6, 7] and magnesium [18] as shown in Fig. 3.6, long-cylindrical pores were grown in the same direction as the unidirectional solidification. The pore size in the section perpendicular to the solidification direction depends on the solidification velocity; the higher the solidification velocity, the smaller the pore size [19]. Since these metals exhibit high-thermal

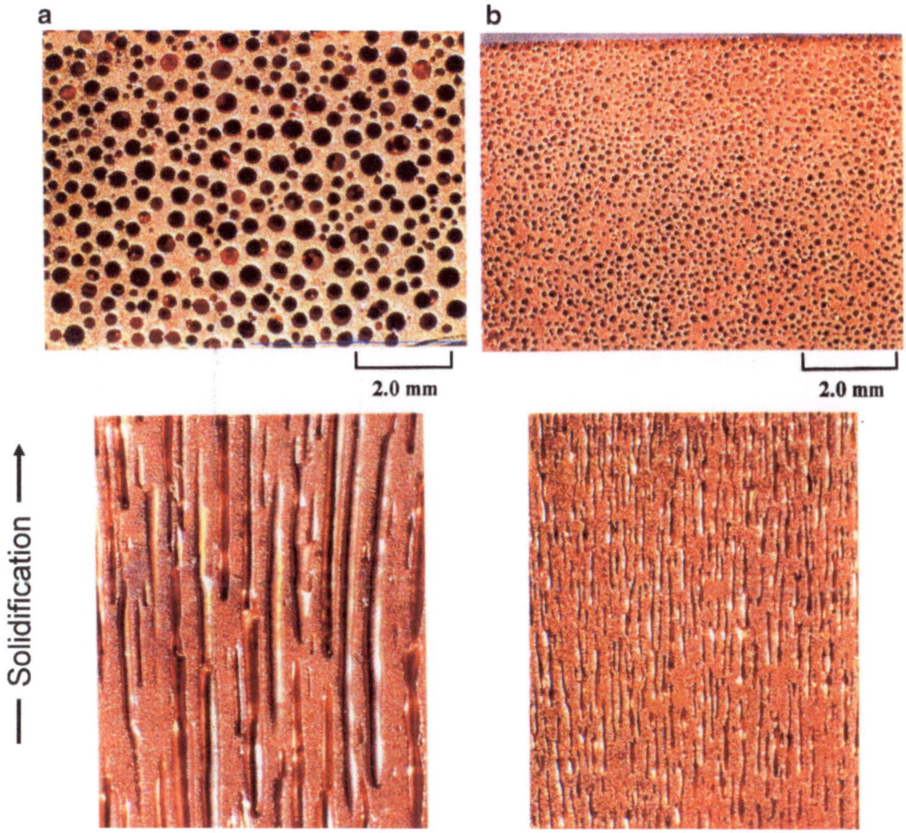


**Fig. 3.6** Typical examples of optical micrographs on the cross sections of lotus-type porous copper fabricated through mold casting technique at different hydrogen pressure. The above is the cross sections perpendicular to the solidification and the below is the cross sections parallel to the solidification. (**a**) 0.4 MPa hydrogen pressure, porosity 44.9 %, and (**b**) 0.8 MPa hydrogen pressure, porosity 36.6 % (Reprinted with permission from literature of Prog Mater Sci 52 (2007) 1091–1173, © 2006 Elsevier Ltd)

conductivity, the solidification proceeds with almost constant solidification rate through the whole of ingot. Thus, porous metals with uniform pore size and porosity can be produced by this technique as shown in Fig. 3.6. However, the same technique cannot be applied to fabricate porous metals and alloys with low-thermal conductivity such as stainless steel [20]. For the metals and alloys with low-thermal conductivity, although the heat from the melt is easily dissipated to the water-cooled plate during the solidification process, the cooling becomes slower at the upper part of the solidified ingot where it is far from the cooling part and thus the pores become coarse. As a result only porous metals and alloys with nonuniform pore size and porosity can be produced as illustrated in Fig. 3.7. In order

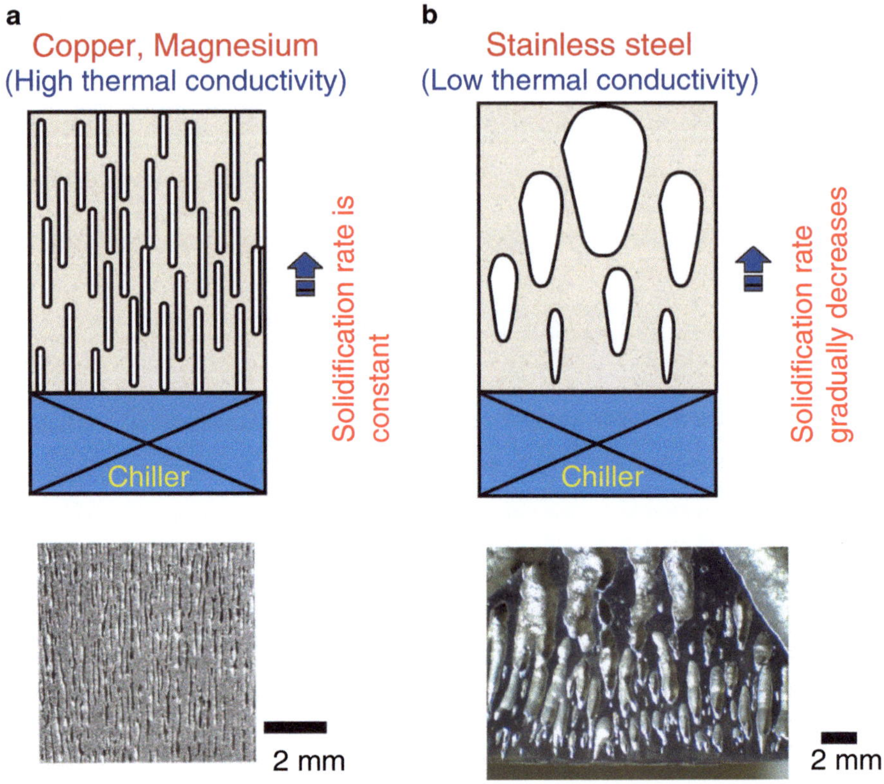


**Fig. 3.7** Comparison of the evolution of pores in porous metals fabricated by mold casting technique in gas atmosphere. The *above* is schematic drawings of pore evolution during the unidirectional solidification. The *below* is optical micrographs of the sectional views perpendicular to the solidification. The *left* is lotus copper, and the *right* is stainless steel. (**a**) The uniform pore size and porosity are observed in copper and magnesium with high thermal conductivity, and (**b**) various pore size and porosity are found in stainless steel with low-thermal conductivity. The magnitude of the thermal conductivity affects the solidification velocity of the melt (Reprinted with permission from literature of Prog Mater Sci 52(2007) 1091–1173, © 2006 Elsevier Ltd)

to overcome the shortcoming, a novel technique was invented by the present author to fabricate pore-elongated lotus metals and alloys even with low-thermal conductivity [21, 22]. Figure 3.8 illustrates the schematic setup of the continuous zone melting technique, which consists of radio-frequency induction coil, blowers, specimen rod, and movable specimen holders; the induction coil is used for the zone (restricted area) melting of the rod-shaped specimen, while the blower is helpful for further cooling of the melt metal. These components are placed into a high-pressure chamber filled with gases such as hydrogen (or nitrogen) and argon. While a part of the specimen rod is melted by induction heating, the hydrogen (or nitrogen) gas is absorbed into the melt up to the gas equilibrium solubility in the pressurized gas atmosphere according to Sieverts' law. Concurrently, the specimen

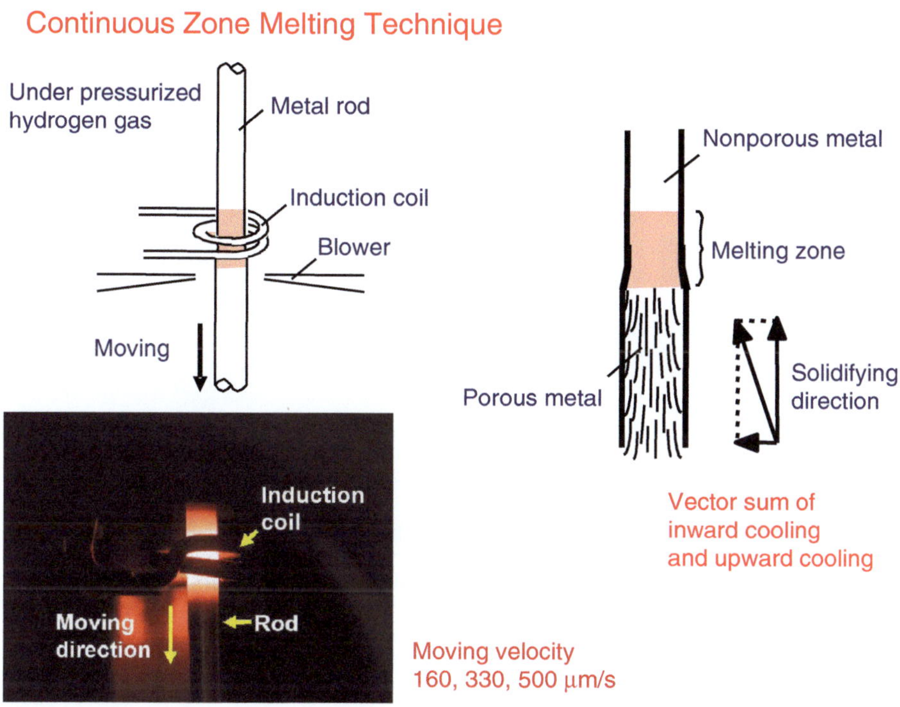


**Fig. 3.8** Schematic drawings and a photograph of the overview for the melting part of continuous zone melting technique. After passing the melting zone, the rod in the lower part is expanded to form lotus metals (Reprinted with permission from literature of Prog Mater Sci 52(2007) 1091–1173, © 2006 Elsevier Ltd)

rod is moved downward at a given transfer velocity. In the lower part of the melt zone, the solidification takes place simultaneously. Then, directional elongated pores are evolved by precipitation of insoluble gas of hydrogen (or nitrogen) in the solidified specimen rod. If the transfer velocity is kept constant, the solidification velocity becomes constant so that the pore size should be constant. The direction of the growing pores is determined by the vector sum of two kinds of solidification directions: an inner cooling vector and an upward cooling vector as shown in the same figure. It is noticed after observation of the sectional view of the specimen that the upward cooling is usually dominant in the interior part of the rod, while the inner cooling effect cannot be ignored near the surface region of the specimen rod. The stainless steel (e.g., SUS304L) exhibits a low-thermal conductivity so that porous stainless steel with homogeneous pore size and porosity is impossible by the conventional mold casting technique as shown in Fig. 3.7. However, lotus stainless steel with homogeneous pore size and porosity is successfully fabricated by so-called continuous zone melting technique. This technique has an advantage that we can control the solidification velocity by changing the

**Fig. 3.9** Overview of the fabrication apparatus of lotus metals through continuous zone melting technique installed at Nakajima Laboratory of Osaka University

lowering speed of the specimen rod regardless of the magnitude of thermal conductivity [23]. On the other hand, the solidification velocity is not controllable to a wide range by the conventional mold casting technique and is almost uniquely determined by its own inherent thermal conductivity. Figure 3.9 shows the photographs of the overview of the fabrication apparatus for continuous zone melting technique. Figure 3.10 shows a sectional view of a lotus stainless steel rod (SUS304L) and the pore size and the porosity are almost identical everywhere through the solidified specimen rod more than 300 mm in length. The length of the lotus metals and alloys to be fabricated by the continuous zone melting technique is essentially endless, but in the present chamber, the movable height is limited to less than 300 mm. This continuous zone melting technique is a promising technique in order to produce long-sized lotus rods for commercial application.

### *3.2.3 Continuous Casting Technique*

The continuous casting technique is extensively used as a mass-production method for ferrous and nonferrous metals and alloys. In this conventional continuous casting process, the solidified ingot can pass through the mold smoothly due to the solidification shrinkage from the melt. However, for fabrication process of lotus

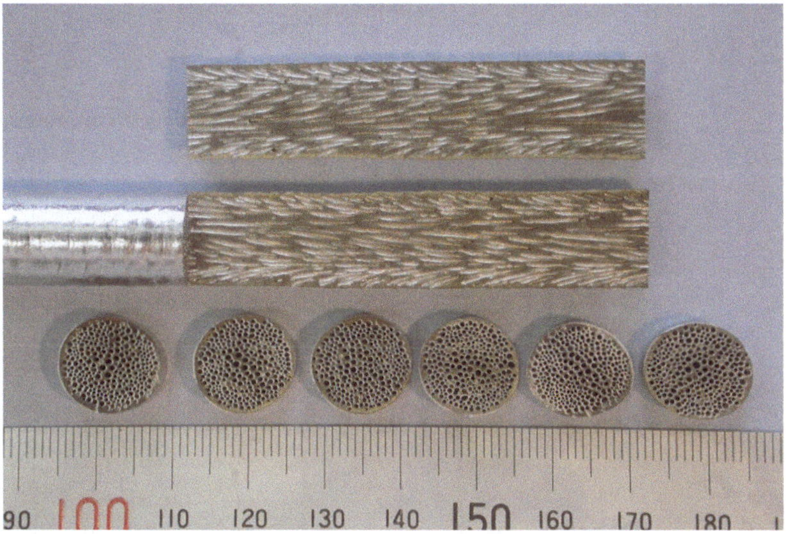


**Fig. 3.10** Sectional views of lotus stainless steel fabricated by continuous zone melting technique in the 2.0 MPa hydrogen atmosphere. The transfer velocity of the rod is 330 μm s$^{-1}$. Resulting porosity and average pore size are 40 % and 320 μm, respectively, both of which are almost uniform in the whole part of the ingot (Reprinted with permission from literature of Prog Mater Sci 52(2007) 1091–1173, © 2006 Elsevier Ltd)

metals, a large volume expansion due to large amount of pore formation occurs when the solidified ingot is passed through the mold. It was initially thought that such an expansion would present problems with the stacking of the ingot in the mold and that consequently this technique could not be applied to the fabrication of lotus metals. However, it was later realized that such a large expansion inherent from the pore evolution was released to push the volume toward the copper part of the molten metal, so that the melt can accommodate the large strain of the solidified ingot. That is why the continuous casting technique is applicable to the fabrication of lotus metals.

Through this technique, the solidification velocity can be controlled by the transfer velocity in hydrogen gas atmosphere. Since the pore morphology of lotus metal is related to the solidification velocity, it is suggested that the pore morphology of lotus metals can be easily controlled by the technique. Park et al. successfully fabricated long-sized lotus copper by controlling its pore size and porosity [24].

Slabs of lotus copper were fabricated by a vacuum-assisted and pressurized continuous casting apparatus, as illustrated in Fig. 3.11. The apparatus consists of a crucible with a rectangular hole at the bottom, a dummy bar for preventing the melt from flowing through the hole, and induction heating coil, and a mold which is surrounded by a water-cooled chill block and pinch rollers to control the transfer velocity of the dummy bar. The solidified ingot bar (slab) can be produced to 700 mm long. Pure copper was melted in the crucible by radio-frequency induction

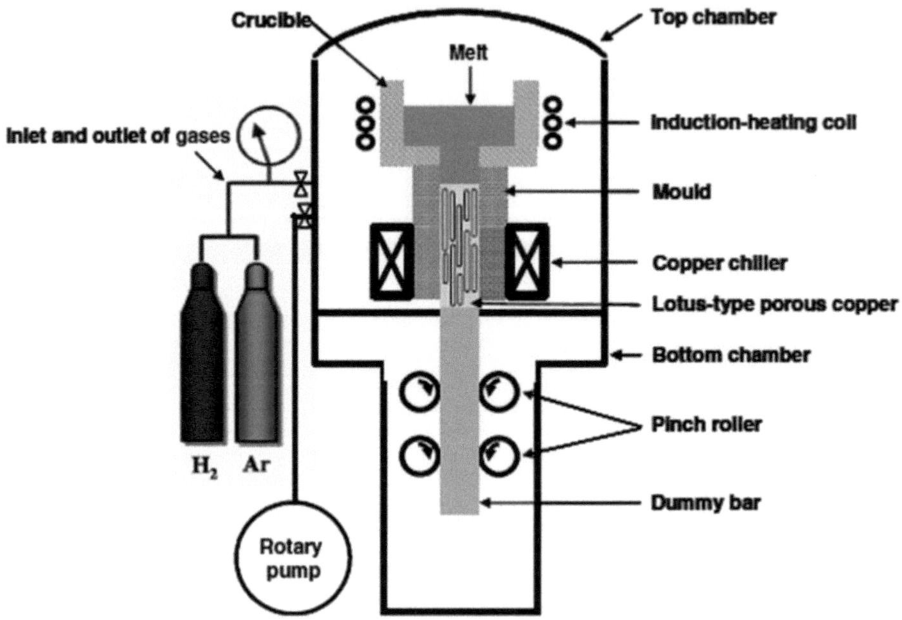


**Fig. 3.11** Schematic drawing of continuous casting apparatus (Reprinted with permission from [24]. © 2007 Acta Materialia Inc.)

heating under a hydrogen gas pressure of 2.0 MPa. The temperature of the molten copper in the crucible was monitored by a W-5Re/W-26Re thermocouple, which was available in the hydrogen atmosphere and was set to be 1,573 K. The melt was pulled down by the dummy bar of nonporous copper through the cooled mold at a given transfer velocity. By the cooling of the melt through both the mold and the connected dummy copper bar, the melt was simultaneously and continuously solidified. Then, the hydrogen in the melt was rejected at the solid–liquid interface due to the solubility gap of hydrogen between liquid and solid, and cylindrical pores aligned parallel to the solidification direction were formed. Figure 3.12 shows the photographs of the overview of fabrication apparatus for continuous casting technique installed at Nakajima Laboratory of Osaka University.

Figure 3.13a, b shows outer view and cross-sectional view of lotus copper rod fabricated in the mixture gas of hydrogen 0.25 MPa and argon 0.15 MPa under transfer velocity of 100 mm $min^{-1}$, respectively. The pore size and porosity are almost identical everywhere through the solidified specimen plate more than 700 mm. The length of the fabricated lotus metals should be essentially endless, but in this chamber, there is some size limitation; the movable height was limited to less than 1,000 mm. Figure 3.14 shows the cross-sectional views of the lotus copper fabricated at various transfer velocities under various hydrogen gas pressure; these views are parallel and perpendicular to the transfer direction. In the observed cross

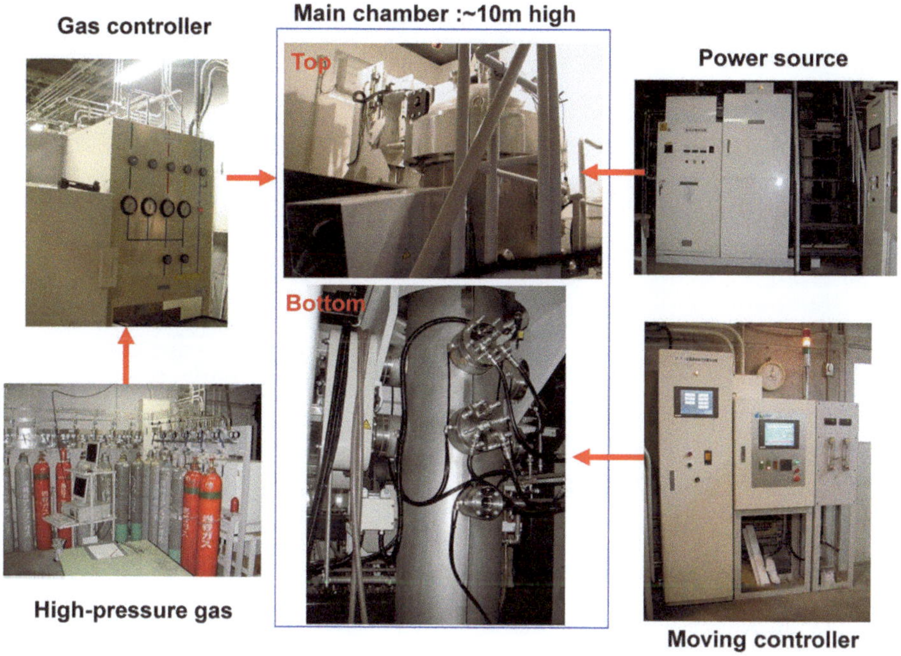


**Fig. 3.12** Photographs of the overview of fabrication apparatus for the continuous casting technique installed at Nakajima Laboratory of Osaka University. The *central upper photo* shows the upper chamber set up by heating part and solidification part, while the *central lower photo* shows lower chamber set up by mechanical pinch roller to pull down the ingot slab

section of the slabs fabricated at 1 mm min$^{-1}$ under 1.0 or 2.0 MPa hydrogen. Three large (diameter $\geq$ 5 mm) and many small pores are distributed inhomogeneously; many small pores are also found in the interior of the large pores. The shapes of the pores are irregular. Thus, it is considered that the large pores are found by the cohesion of a few small pores; the volume of the small pores increases during solidification when the amount of hydrogen diffused from the solid to the pores increases with decreasing transfer velocity. However, the slabs fabricated at a high-transfer velocity possess long-cylindrical pores distributed homogeneously. The pore size decreases and the number density of the pores increases with an increase in not only the transfer velocity but also the hydrogen gas pressure.

Figure 3.15 shows the relationship between the porosity and the transfer velocity under hydrogen gas pressures of 1.0 and 2.0 MPa. The porosity decreases with increasing hydrogen gas pressure, while the porosity is almost constant and independent of the transfer velocity. Such a tendency is consistent with the results reported by Hyun and Nakajima [19] and Ikeda et al. [23]; they respectively fabricated lotus copper and stainless steel using the mold casting and continuous zone melting techniques at various solidification velocities under various gas pressure. Figure 3.16 shows the effect of transfer velocity on the average pore

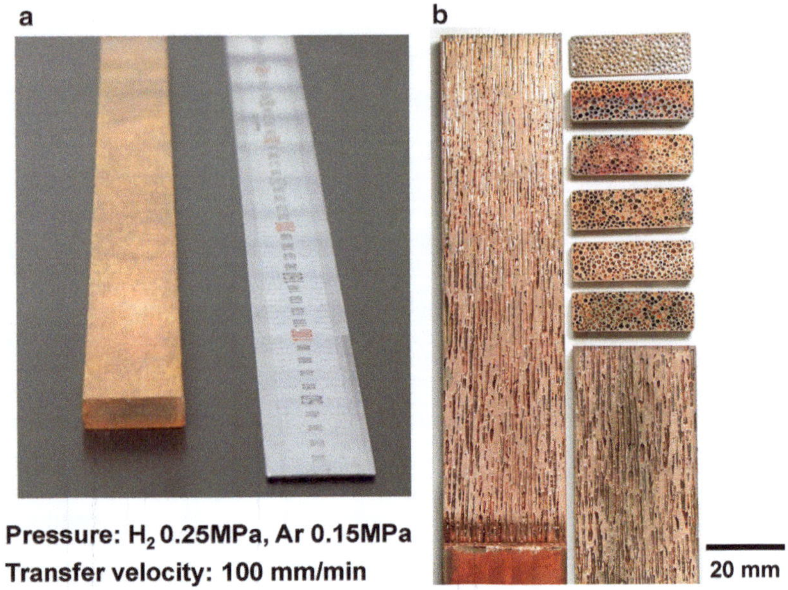


**Fig. 3.13** (**a**) A photograph of the overview of lotus copper fabricated using the continuous casting technique and (**b**) cross sections parallel and perpendicular to the transfer direction of the lotus copper fabricated in mixture gases of hydrogen 0.25 MPa and argon 0.15 MPa by continuous casting technique (Reprinted with permission from literature of Prog Mater Sci 52(2007) 1091-1173, © 2006 Elsevier Ltd)

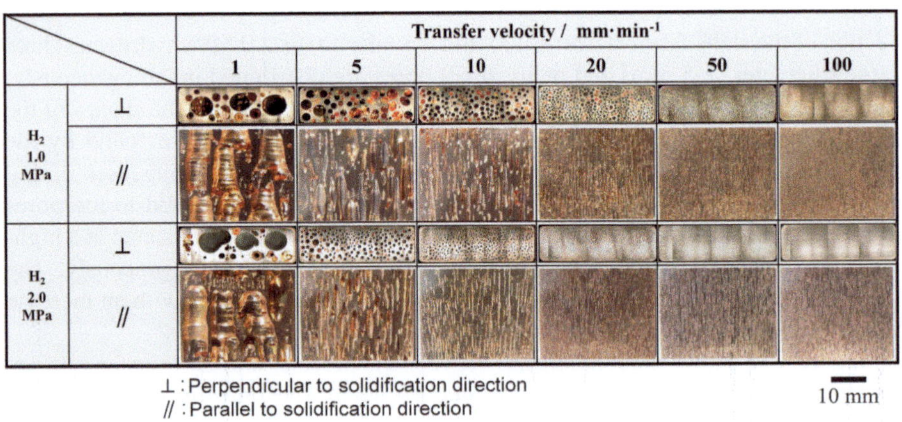


**Fig. 3.14** Cross sections perpendicular and parallel to the transfer direction of lotus copper fabricated at various transfer velocities under hydrogen gas pressure of 1.0 and 2.0 MPa (Reprinted with permission from [24]. © 2007 Acta Materialia Inc.)

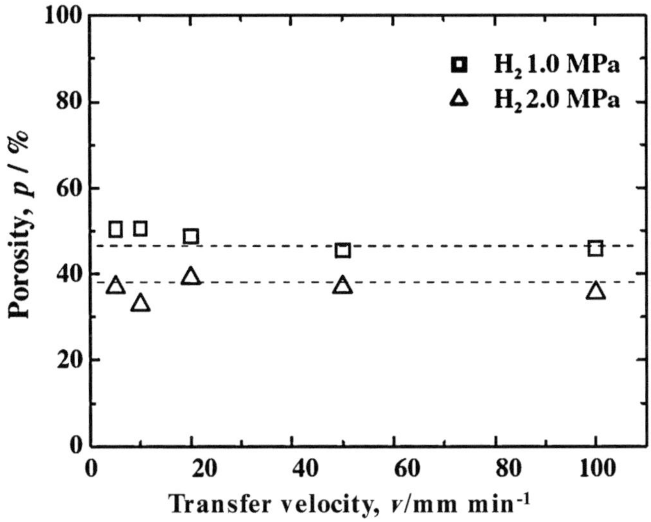


**Fig. 3.15** Porosity against the transfer velocity of lotus copper fabricated under hydrogen gas pressure of 1.0 and 2.0 MPa (Reprinted with permission from [24]. © 2007 Acta Materialia Inc.)

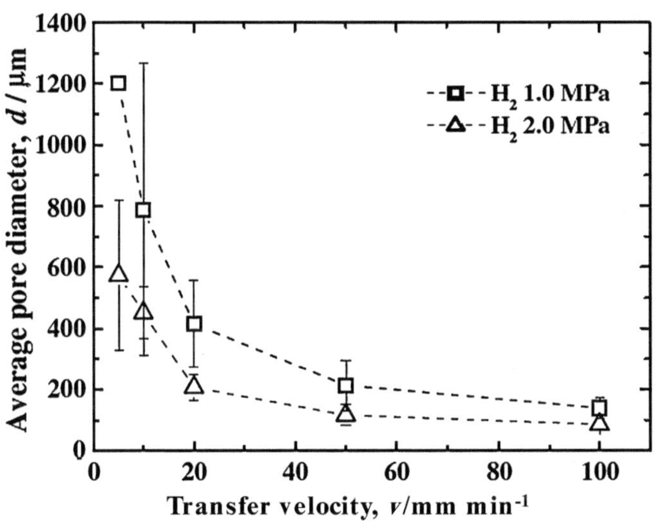


**Fig. 3.16** Average pore diameter against the transfer velocity of the lotus copper fabricated under hydrogen gas pressure of 1.0 and 2.0 MPa (Reprinted with permission from [24]. © 2007 Acta Materialia Inc.)

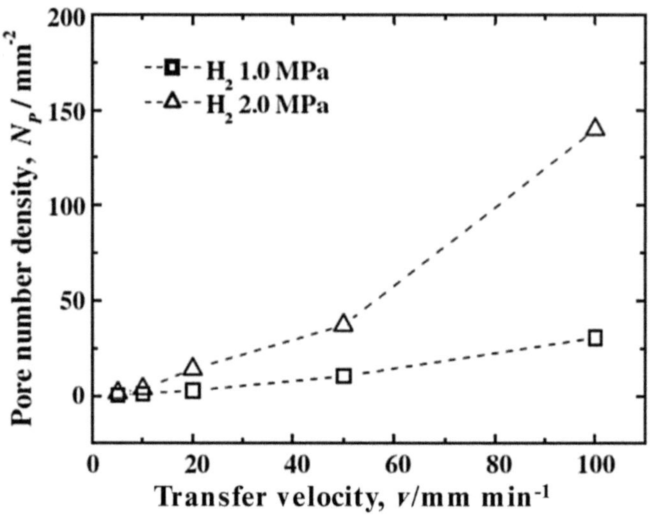


**Fig. 3.17** Pore number density calculated using the average pore diameter and the porosity in a unit area (1.0 × 1.0 mm$^2$) on a cross section perpendicular to the transfer direction at various transfer velocities under hydrogen gas pressure of 1.0 and 2.0 MPa (Reprinted with permission from [24]. © 2007 Acta Materialia Inc.)

diameter $d$ under each hydrogen gas pressure. The average pore diameter decreases with an increase in not only hydrogen gas pressure but also the transfer velocity. In addition, the density of the pore number increases with increasing transfer velocity as shown in Fig. 3.14. Figure 3.17 shows the change of the pore number density $N_p$ calculated using the average pore diameter and the porosity in a cross-sectional area of 1.0 × 1.0 mm$^2$, with the transfer velocity. The pore number density increases with increasing transfer velocity, while an increase in hydrogen gas pressure brings about an increase in the pore number density for a given transfer velocity. These indicate that the pore number density is affected not only by the transfer velocity but also by the total gas pressure. The pore length $l$ was also affected by the transfer velocity as shown in Fig. 3.18, which decreases with increasing transfer velocity.

The pore growth direction is affected by the transfer velocity. Figure 3.19 shows the angle $\theta$ between the pore growth direction and the transfer direction in various positions of the slab in a direction perpendicular to transfer direction. In the center of the slab, the angle is almost zero and independent of the transfer velocity. On the other hand, the pore growth angle increases when the position of the slab moves from center to near the surface, and the pore growth angle increases more with increasing transfer velocity. Thus, the pore growth angle is affected by the transfer velocity, and the effect increases with increasing transfer velocity. Figure 3.20 shows the magnifications and the schematics of the pore formation position near the surface. The thickness $t$ decreases with increasing transfer velocity as shown in Fig. 3.21.

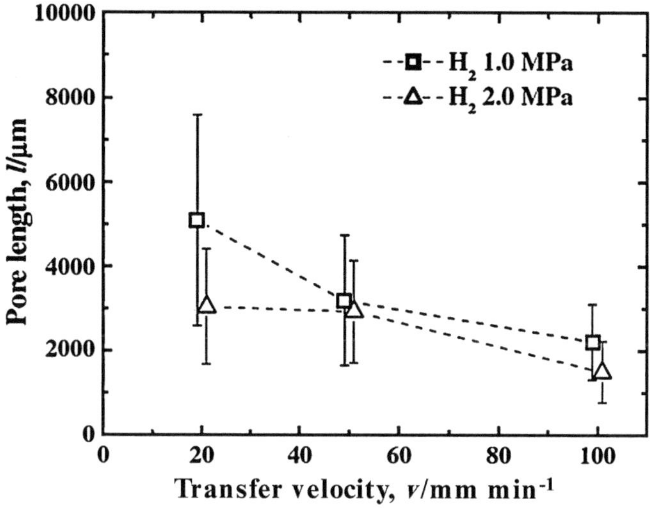


**Fig. 3.18** Pore length against the transfer velocity of lotus copper fabricated under hydrogen gas pressure of 1.0 and 2.0 MPa (Reprinted with permission from [24]. © 2007 Acta Materialia Inc.)

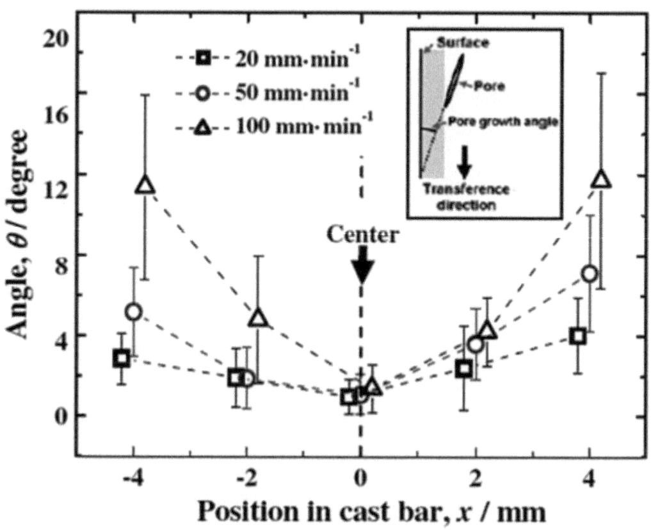


**Fig. 3.19** Angle between the pore growth direction and the transfer direction in various positions of the slab in a direction perpendicular to the transfer direction; the slab was fabricated at various transfer velocities under a hydrogen gas pressure of 1.0 MPa (Reprinted with permission from [24]. © 2007 Acta Materialia Inc.)

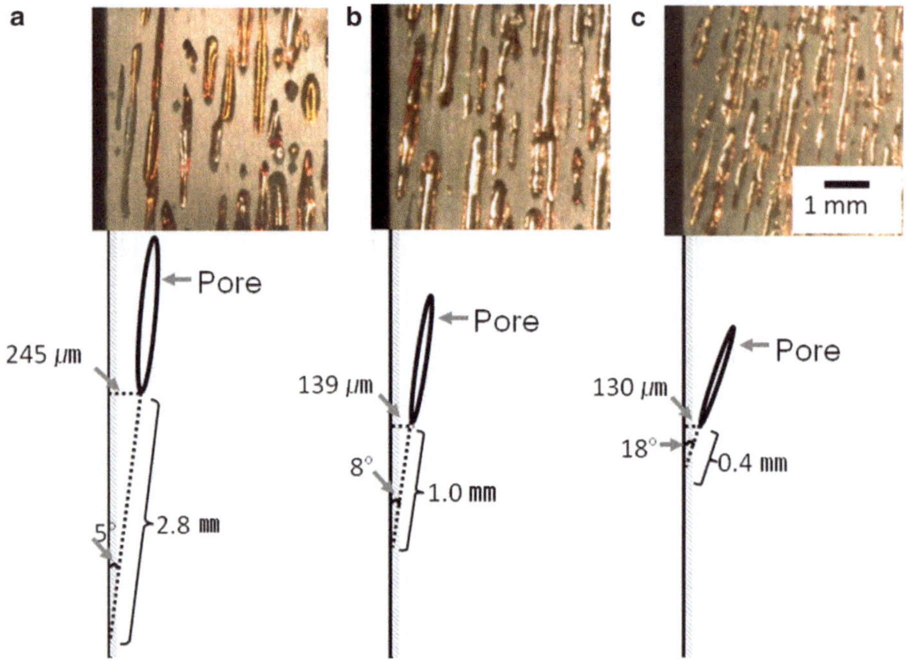


**Fig. 3.20** Photographs (*upper row*) and schematic (*lower row*) of the pore formation position near the surface fabricated by the continuous casting technique at the transfer velocity of (**a**) 20 mm min$^{-1}$, (**b**) 50 mm min$^{-1}$, and (**c**) 100 mm min$^{-1}$ under a hydrogen pressure of 1.0 MPa (Reprinted with permission from [24]. © 2007 Acta Materialia Inc.)

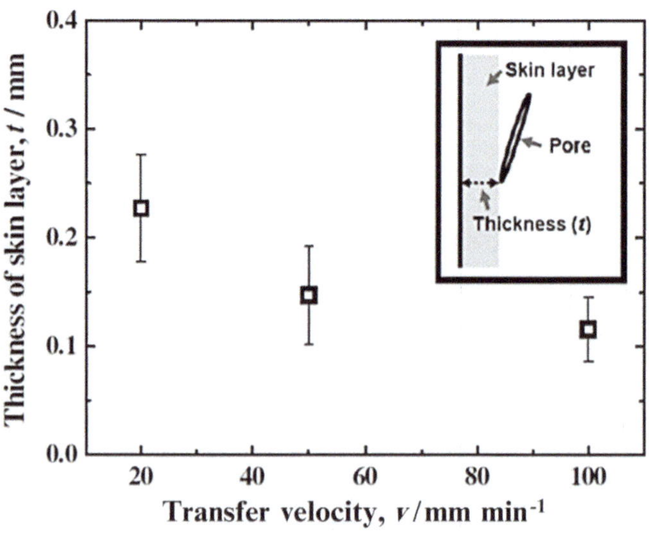


**Fig. 3.21** Thickness of the skin layer plotted against the transfer velocity of lotus copper fabricated under a hydrogen gas pressure of 1.0 MPa (Reprinted with permission from [24]. © 2007 Acta Materialia Inc.)

In order to discuss the pore diameter and the pore length, the pore nucleation mechanism must be taken into consideration. Pore nucleation in the liquid has been investigated by many researchers [25, 26], and it has been reported that inhomogeneous pore nucleation occurs because the surface Gibbs free energy is lowered by the existence of impurities and inclusions. Fisher [26] suggested the following relationship between the pore nucleation rate $I$ and the critical Gibbs free energy $(\Delta G^*_{\text{hetero}})$ for heterogeneous pore nucleation:

$$I = \frac{NkT}{h} \exp\left( -\frac{\Delta G_\text{a} + \Delta G^*_{\text{hetero}}}{kT} \right), \tag{3.1}$$

where $N$ is the number of atoms in the liquid, $k$ is Boltzmann's constant, $T$ is the temperature, $h$ is Planck's constant, and $\Delta G_\text{a}$ is the activation energy for motion of an individual atom. Here, $\Delta G_\text{a}$ is negligible compared to $\Delta G^*_{\text{hetero}}$, and one can obtain the expression

$$I = \frac{NkT}{h} \exp\left( -\frac{1}{kT} \frac{16\pi}{3} \frac{\gamma^3}{\Delta P^2} f(\theta_\text{c}) \right), \tag{3.2}$$

where $\gamma$ is the surface energy of the pore, $\Delta P$ is the difference between the ambient and the internal pressure of the pore, and $f(\theta_\text{c})$ a function of the surface energy that depends on the contact angle $\theta_\text{c}$ between the solid and the pore. The pore nucleation rate is closely related to $\Delta P$, and $\Delta P$ is proportional to the undercooling $\Delta T$ through the Clausius–Clapeyron equation [27]:

$$\Delta P \propto \Delta T. \tag{3.3}$$

In addition, it is generally reported that the solidification velocity $v$ affects the undercooling $\Delta T$ during the solidification process as follows:

$$v \propto \Delta T^n \quad (1 \leq n \leq 2), \tag{3.4}$$

where $n$ is an empirical constant. Thus, the following expression can be finally obtained from Eqs. (3.3) and (3.4):

$$v \propto \Delta P^n. \tag{3.5}$$

Thus, the pore nucleation rate in Eq. (3.2) increases with increasing transfer velocity because of the relationship between $v$ and $\Delta P$ in Eq. (3.5); the solidification velocity $v$ is assumed to be equal to the transfer velocity.

On the other hand, since the hydrogen content in the melts is almost constant at the melting point under a constant hydrogen gas pressure, the hydrogen content diffused in each pore during the solidification decreases with an increase in the pore nucleation rate. Thus, it is considered that the pore diameter decreases with a decrease in each pore volume. This is in good agreement with the results of pore length and pore aspect ratio; the changing in the pore length with the transfer velocity is similar to the change in the pore diameter, and therefore, the pore aspect ratio is not changed very much. This indicates that the pore dimensions such as diameter and length are determined by the relationship between the pore nucleation rate $I$ and hydrogen solubility in the melt.

The pore grows in the direction perpendicular to the solid–liquid interface. Thus, the pore growth direction depends on the shape of the interface during solidification. The shape is determined by the flow of the heat emitted from the liquid during solidification. Assuming that the heat flow rate is constant in a unit area and the amount of extracted heat during solidification increases with increasing transfer velocity during unit time, it is considered that the heat extracted from the liquid at lower velocities is sufficient even at a flat interface. However, if the amount of the heat increases by an increase in transfer velocity, then a large interface area may be required at higher velocities in order to emit the additional heat. Therefore, it is thought that the interface shape changes from flat at lower velocities to concave at higher velocities. Because the depth at the center in the concave shape is proportional to the transfer velocity, the depth increases with increasing velocity. Thus, the pore growth angle shows a change such as that displayed in Fig. 3.19.

It is considered that the thickness of the skin layer is related not only to the pore growth angle but also to the distance between the pore and the surface of the slab; the thickness of the skin layer changes with a relationship of sine function between the angle and the distance as shown Fig. 3.20. The distance decreases with increasing transfer velocity as shown in Figs. 3.20 and 3.22. Assuming that the hydrogen rejected from the solid was only transported away by diffusion when solidification began, it is considered that the concentration profile of the hydrogen just before the pore formation can be expressed as shown in Fig. 3.23. There will be a buildup of the hydrogen ahead of the solid, which is accelerated by increasing solidification velocity since it is more difficult to diffuse the hydrogen from the solid to the liquid by increasing solidification velocity. If a critical hydrogen concentration $C_{\mathrm{H}}^{\mathrm{pore}*}$ for the pore formation exists, then the pore forms and grows when the hydrogen concentration ahead of the solid reaches $C_{\mathrm{H}}^{\mathrm{pore}*}$ by hydrogen buildup. When the pore is formed, the solidified distance $x_c$ is expressed as follows [28, 29]:

$$x_c = \frac{D}{k_0 v}, \tag{3.6}$$

where $D$ is the hydrogen diffusivity in the liquid copper at the melting point and $k_0$ is the equilibrium distribution coefficient; $k_0$ is 0.31, which is evaluated as the proportion of the hydrogen solubility in the solid and liquid copper at 1,357 K under $P_{\mathrm{H}_2} = 0.1$ MPa by using an equation suggested by Fromm and Gebhardt [30]. Assuming that the solidification velocity is equal to the transfer velocity, the

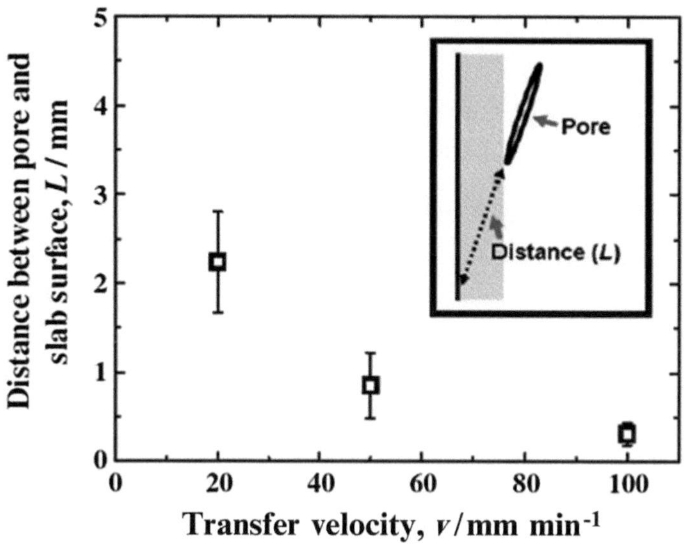


**Fig. 3.22** Distance between the pore and the surface plotted against the transfer velocity of lotus copper fabricated under a hydrogen pressure of 1.0 MPa (Reprinted with permission from [24]. © 2007 Acta Materialia Inc.)

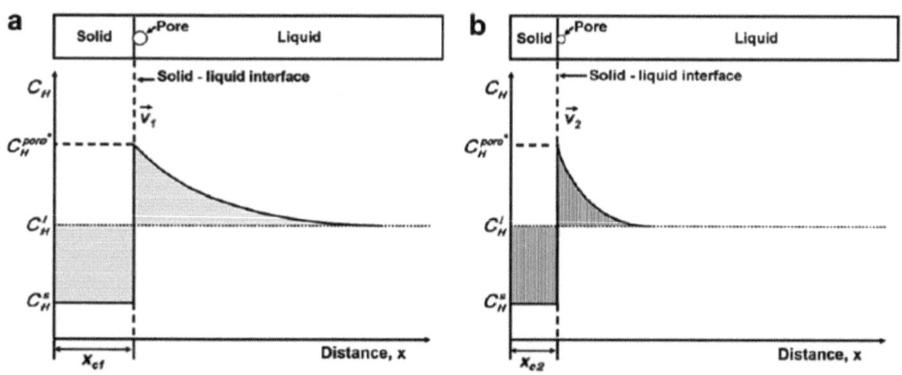


**Fig. 3.23** Schematic of the hydrogen concentration profile from the beginning of solidification to pore formation in (**a**) lower solidification velocity $v_1$ and (**b**) higher solidification velocity $v_2$ (x: solidified distance, $C_{\mathrm{H}}^{\mathrm{l}}$: hydrogen concentration in liquid, $C_{\mathrm{H}}^{\mathrm{s}}$: hydrogen concentration in solid, $C_{\mathrm{H}}^{\mathrm{pore}*}$: critical hydrogen concentration for pore nucleation, and $x_{\mathrm{c}}$: the solidified distance until the pore is formed) (Reprinted with permission from [24]. © 2007 Acta Materialia Inc.)

distance described in Fig. 3.22 can be plotted as a function of $v^{-1}$, as shown in Fig. 3.24, where the distance is in inverse proportion to the transfer velocity as shown in Eq. (3.6). The $D$ value calculated from the slope of the dashed line is $2.28 \times 10^{-7}\ \mathrm{m^2\ s^{-1}}$, which is close to the value of $0.99 \times 10^{-7}\ \mathrm{m^2\ s^{-1}}$ at 1,374 K reported by Wright and Hocking [31].

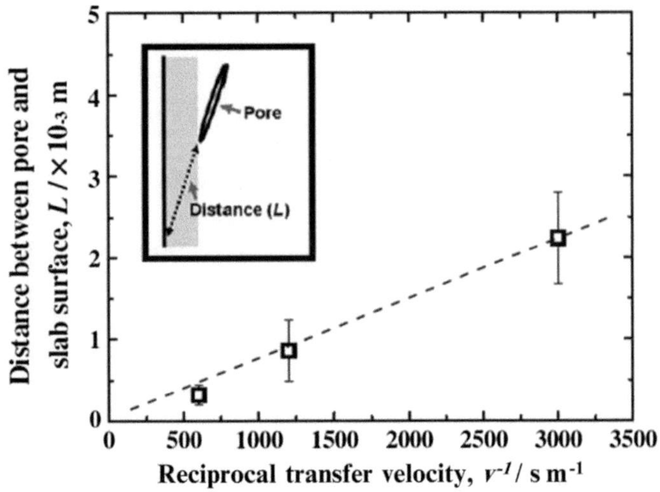


**Fig. 3.24** Distance from the surface to the pore plotted as a function of reciprocal transfer velocity $v^{-1}$ in lotus copper fabricated under a hydrogen gas pressure of 1.0 MPa (*Error bars* are the standard deviation, and the *dotted line* is the fitted line.) (Reprinted with permission from [24]. © 2007 Acta Materialia Inc.)

Consequently, the thickness of the skin layer decreases with increasing transfer velocity, which is affected by a change not only in the pore growth but also in the distance between the pore and the surface of the slab.

## 3.3 Thermal Decomposition Method (TDM)

As already mentioned above, the high-pressure gas method is a convenient method to supply the gas source in the molten metals for evolving gas pores in the solidified metals. However, the PGM possesses two disadvantages: (1) necessity of strongly built high-pressure chamber and application of high-pressure regulation and (2) necessity of safety procedure to protect against hydrogen as an inflammable and explosive gas. Mass production and commercialization in industries are hesitated from such disadvantages. Therefore, a technique that does not require high-pressure hydrogen to fabricate lotus metals is highly desirable. In order to overcome these difficulties, Nakajima group [32] proposed an alternative, but simple, method to fabricate such lotus metals by using a thermal decomposition method (TDM) of compounds containing gas elements in a non-hydrogen atmosphere under nearly atmospheric pressure.

The compounds suitable for TDM can be listed up to the following [33]:

$TiH_2$, $MgH_2$, $ZrH_2$, $Fe_4N$, TiN, $Mn_4N$, CrN, $Mo_2N$, $Ca(OH)_2$, $Cu_2O$, $B_2O_3$, $CaCO_3$, $SrCO_3$, $MgCO_3$, $BaCO_3$, and $NaHCO_3$.

The compounds should be decomposed into gas element and another type of metallic compound at the temperature (dissolving temperature) just below the melting point of the solvent metal to be solidified. If the dissolving temperature of the compounds is higher than the melting point of the metal, sufficient dissolution of the gas element into the melt cannot be ensured. The principle of TDM and the first versatile method to control the pore morphology, including pore size and porosity of lotus metals are described.

## *3.3.1 Mold Casting Technique through TDM*

### 3.3.1.1 Fabrication of Lotus Copper through TDM

Copper (99.99 % pure) was melted by radio-frequency induction heating in a graphite crucible under an argon atmosphere from 0.1 to 0.5 MPa. The melt was poured into the mold, which had a copper bottom plate cooled by water and lateral side walls made of 0.1-mm-thick cylindrical stainless steel. A few pellets of titanium hydride, which ranged between 0.075 and 0.25 g in mass for melting copper of 200 g, were set on the bottom plate of the mold. Unidirectional solidification occurred in the mold so that a lotus copper ingot was obtained, as illustrated in Fig. 3.25a. The ingot size was 28 mm in diameter and had a height of maximum 90 mm. The ingot was cut by a spark-erosion wire-cutting machine parallel and perpendicular to the solidification direction.

Another method was also used to supply the hydrogen source in the melt. As shown in Fig. 3.25b, a few pellets of titanium hydride were set in the crucible along with copper. During the copper melt, hydrogen, which decomposed from titanium hydride, was dissolved into the melt. The molten copper was poured from the crucible into the mold without pellets of titanium hydride for unidirectional solidification. The ingots were investigated using the same method described previously.

Figure 3.26 shows the optical micrographs of the cross-sectional views of lotus copper parallel and perpendicular to the solidification direction [32]. The pore growth direction is coincident to the direction of the unidirectional solidification, which is in agreement with that of the conventional high-pressure gas method (PGM). Therefore, the principal mechanism of this pore evolution is suggested to be similar to that of PGM; when the molten metal dissolving a gas is solidified, insoluble gas precipitates and evolves the gas pores in the solid metals.

Figure 3.27 shows dependence of the porosity and the average pore diameter on the mass of titanium hydride for the lotus copper fabricated by TDM in a 0.1 MPa argon atmosphere. The porosity abruptly increases by the addition of 0.10 g of titanium hydride and becomes constant to be about 55 % in the porosity on further addition. When 0.10 g of titanium hydride was added to 200 g of molten copper, the concentration of hydrogen in the melt was evaluated as 0.128 at. pct. Directly comparing the hydrogen concentration from the titanium hydride to the available data on the solubility in molten copper just above the melting temperature ($T_m$),

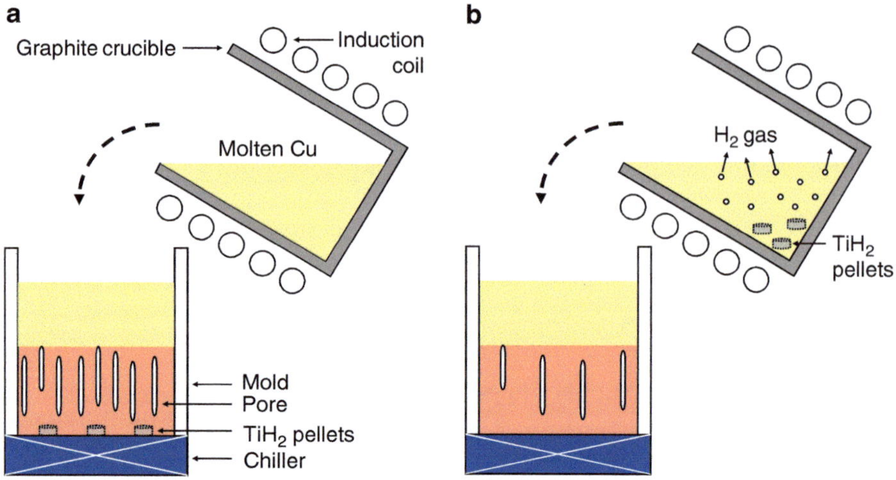


**Fig. 3.25** Schematic drawings of the principle to fabricate lotus metals by a mold casting technique: (**a**) pellets of titanium hydride are set in the mold and (**b**) pellets of titanium hydride are set in the crucible (Reprinted with permission from [32]. © 2008 The Minerals, Metals & Materials Society and ASM International)

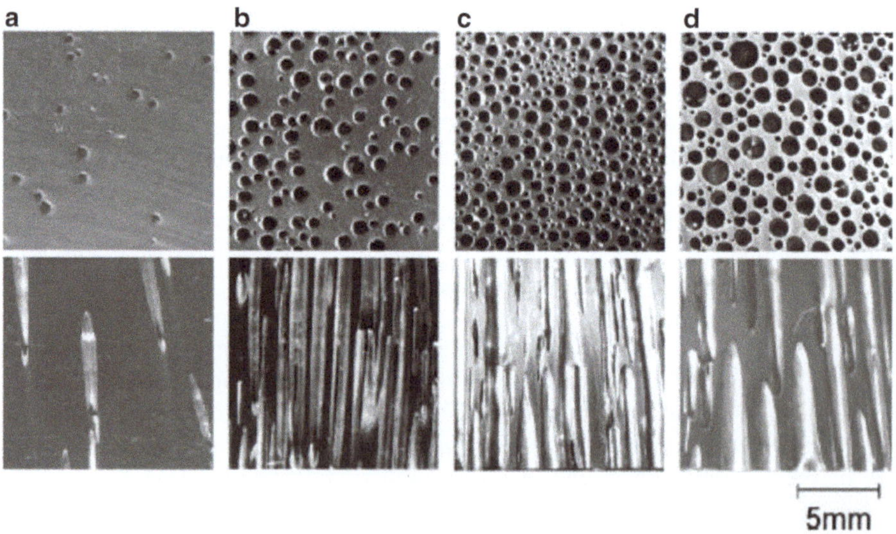


**Fig. 3.26** Optical micrographs of cross-sectional views of lotus copper (*upper views*) perpendicular and (*lower views*) parallel to the solidification direction. Mass of titanium hydride added to the mold is (**a**) 0.075 g, (**b**) 0.10 g, (**c**) 0.125 g, and (**d**) 0.25 g. Melting and subsequent solidification were carried out in 0.1 MPa argon atmosphere (Reprinted with permission from [32]. © 2008 The Minerals, Metals & Materials Society and ASM International)

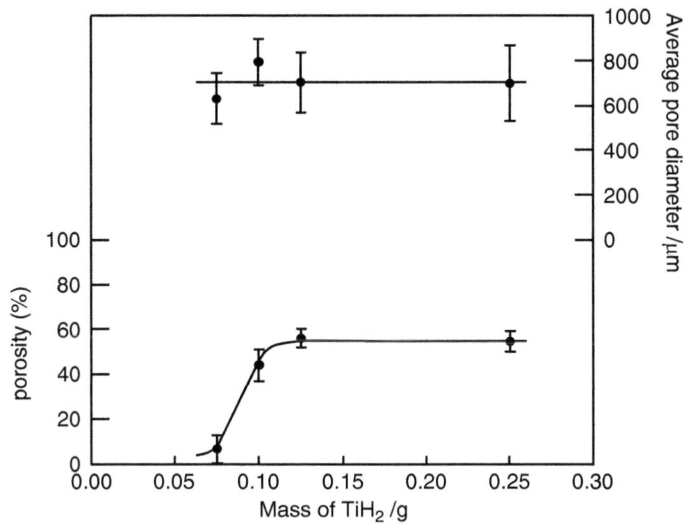


**Fig. 3.27** Dependence of the porosity and the average pore diameter on the mass of titanium hydride. Melting and subsequent solidification were carried out in 0.1 MPa argon atmosphere (Reprinted with permission from [32]. © 2008 The Minerals, Metals & Materials Society and ASM International)

1,083 K, is impossible, because the atmospheric hydrogen pressure of the TDM process differs from that of the solubility measurement (0.1 MPa). However, the concentration of dissolving hydrogen in the TDM process is comparable to the solubility of hydrogen in the liquid phase of copper near $T_m$ [30]. If more than 0.1 g of titanium hydride is added to the melt, the supersaturated hydrogen may generate gas bubbles, which are then liberated from the melt to the atmosphere. The hydrogen dissolved in the melt evolves pores at the solid–liquid interface during solidification. If less than 0.1 g of titanium hydride is added, then all hydrogen can dissolve in the melt without bubbling so that some of the hydride forms pores in the solid–liquid interface, but the porosity may be smaller than that when more than 0.1 g of the hydride is added. Such redistribution of hydrogen in the liquid phase to the solid phase and the atmosphere is exhibited in Fig. 3.28a, b, depending on the mass of titanium hydride. For comparison, Fig. 3.28c shows the PGM case. The TDM process is based on a nonequilibrium state, while the PGM process is based on an equilibrium state in the chamber where a constant high-pressure gas is maintained. On the other hand as shown in Fig. 3.27, the average pore diameter is constant and nearly independent of the mass of titanium hydride. It has been previously confirmed that the pore size is affected by the supercooling rate from the liquid to the solid, which can be controlled by the solidification velocity in the PGM [19, 23, 24]. The same reasoning can be used to interpret the present data.

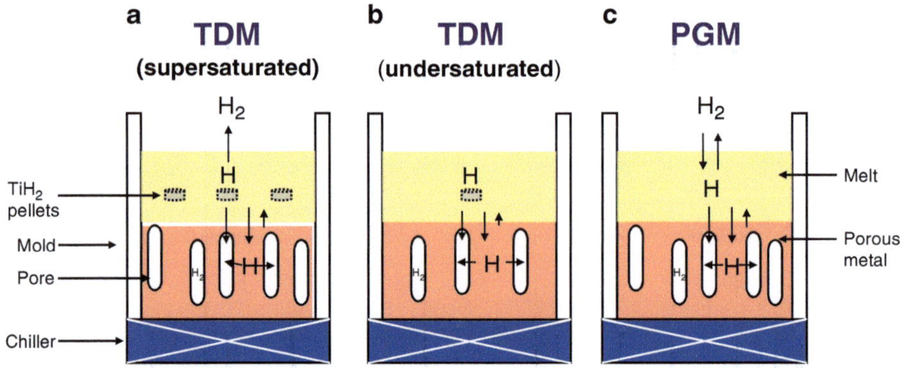


**Fig. 3.28** Schematic drawings of the redistribution of hydrogen dissolved in the melt to the solid phase and atmosphere. (**a**) Thermal decomposition method: the melt contains supersaturated hydrogen. (**b**) Thermal decomposition method: the melt contains undersaturated hydrogen. (**c**) High-pressure gas method. In all cases, hydrogen dissolved in the melt near the solid–liquid interface moves directly into the pores, but some of the hydrogen moves in the solidified metal, while insoluble hydrogen diffuses into the pores via the solid or diffuses back to the melt (Reprinted with permission from [32]. © 2008 The Minerals, Metals & Materials Society and ASM International)

Next, the atmospheric pressure effect was investigated. For this study, the mass of titanium hydride was constant at 0.25 g, and argon was selected as the atmospheric gas, which served as the external pressure. The argon pressure was varied from 0.1 to 0.5 MPa. Figure 3.29 shows sectional views of lotus copper parallel and perpendicular to the solidification direction as a function of argon pressure. The effect of external pressure is obvious, and the pore growth is suppressed at higher pressure. Figure 3.30 shows the dependence of the porosity and the average pore diameter as a function of argon pressure. Both the porosity and the average pore diameter decrease with increasing argon pressure. The pore volume $v$, which is equal to the porosity, is inversely proportional to the external argon pressure $P$, which can be described by the Boyle–Charles law, $v = nRT/P$, where $n$, $R$, and $T$ are the hydrogen molar number, the gas constant, and the temperature, respectively. Therefore, the pore diameter can be written as $d \propto P^{-1/3}$. The tendency of the pressure dependence of the porosity and the pore diameter are explained by the law. However, it seems that such a pressure effect may be more significant than that predicted by the Boyle–Charles law. This difference may be attributed to the possibility that the molar number of hydrogen is not constant under changing pressure. The decomposition rate of titanium hydride should be retarded as the external pressure of argon increases. Thus, the porosity and the pore diameter may decrease remarkably.

Finally, we examined the role of the other metallic element, Ti, through the thermal decomposition of titanium hydride. Titanium hydride is decomposed into hydrogen and titanium. The latter is a very reactive element, which easily reacts with residual oxygen in the molten copper. Consequently, titanium oxide particles are formed and dispersed, which may serve as the nucleation sites for the hydrogen

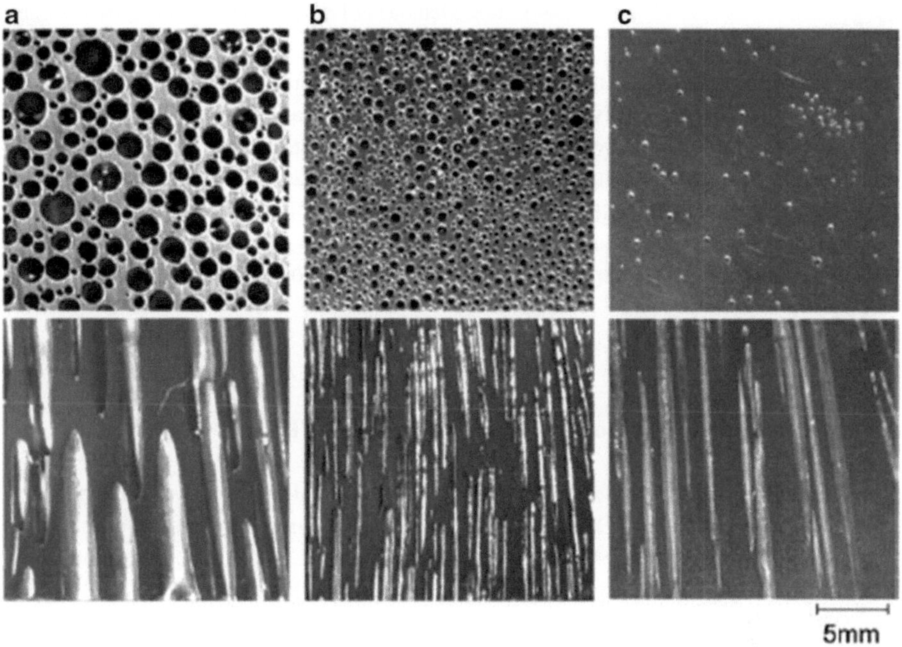


**Fig. 3.29** Cross-sectional views of lotus copper fabricated under different argon pressure. (**a**) 0.1 MPa, (**b**) 0.25 MPa, and (**c**) 0.5 MPa. *Upper* and *lower views* are the cross section perpendicular and parallel to the solidification direction, respectively. Titanium hydride (0.25 g) was added to the mold during solidification (Reprinted with permission from [32]. © 2008 The Minerals, Metals & Materials Society and ASM International)

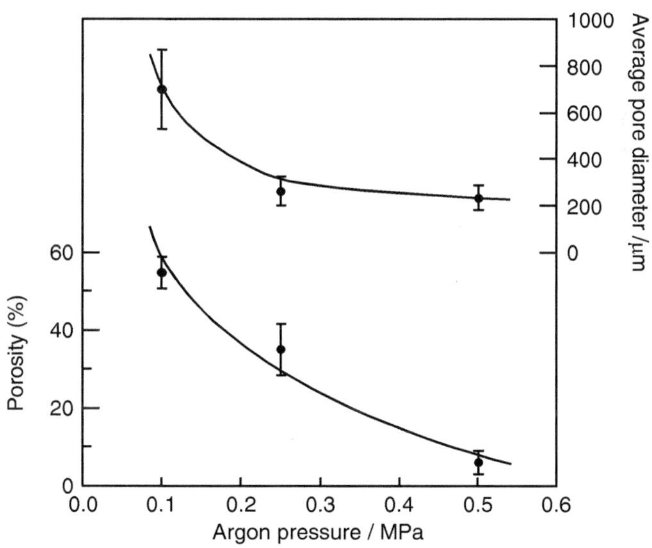


**Fig. 3.30** Dependence of the porosity and the average pore diameter on argon pressure. Titanium hydride (0.25 g) was added to the mold during solidification (Reprinted with permission from [32]. © 2008 The Minerals, Metals & Materials Society and ASM International)

pores in the solid–liquid interface during unidirectional solidification. It is well known that the pores evolve by heterogeneous nucleation in metal melts in the presence of small amounts of foreign particles [34]. Therefore, the same reasoning can be applied to the present case, and it is surmised that the pore size and porosity become homogeneous by uniformly distributed nucleation sites. Thus, TDM exhibits another advantage to produce lotus metals with more homogeneous pore size and porosity than those by PGM, which does not provide intentional nucleation sites.

#### 3.3.1.2 Fabrication of Lotus Aluminum through TDM

The investigation was undertaken to fabricate lotus aluminum with the porosity more than 10 % through TDM [35]. For this purpose, three types of the compounds containing hydrogen were used: calcium hydroxide, sodium bicarbonate, and titanium hydride. The pore morphology is usually characterized with the pore size and porosity, which were controlled by the temperature of melt, amount of compounds, and external argon pressure in this work.

Pure aluminum (99.99 % pure) of about 100 g was melted in a graphite crucible by an induction heating coil in vacuum. The bottom of the mold was copper plate cooled by a circulated water chiller, while the side was made of stainless steel. 0.2 g of the compounds such as calcium hydroxide, sodium bicarbonate, or titanium hydride was usually wrapped with aluminum foil and was set on the bottom plate of the mold. The temperature of the liquid in the graphite crucible was monitored by an infrared pyrometer. After pouring the molten aluminum at 1,023 K in the crucible into the mold, hydrogen decomposed from the compounds dissolves in the melt, which is then solidified so that insoluble hydrogen precipitates to evolve the hydrogen pores. In order to investigate the dependence of the pore size and porosity on the mass of calcium hydroxide, the amount of compounds was changed from 0.1 to 1.0 g. In order to investigate the external pressure dependence of the pore size and porosity, the argon pressure was changed from vacuum to 0.04 MPa.

When the solidification took place unidirectionally, lotus aluminum with directional cylindrical pores was fabricated. The size of the obtained ingot was about 25 mm in diameter and about 60 mm in height. The specimens were cut using a spark-erosion wire-cutting machine in both directions parallel and perpendicular to the solidification direction. Each cross section was polished with a series of emery papers and was observed using an optical microscope.

The constituent elements of pores were investigated using gas analyzers. After the specimen kept into salt-saturated distilled water was cut, the gas bubbles were collected into a vial in the distilled water. The collected gas was analyzed by a gas chromatograph (GC-14A, Shimadzu Co., Kyoto, Japan) in order to identify hydrogen, carbon monoxide, and carbon dioxide. Oxygen was analyzed by a gas analyzer (TC-300, LECO Co., St. Joseph, MI).

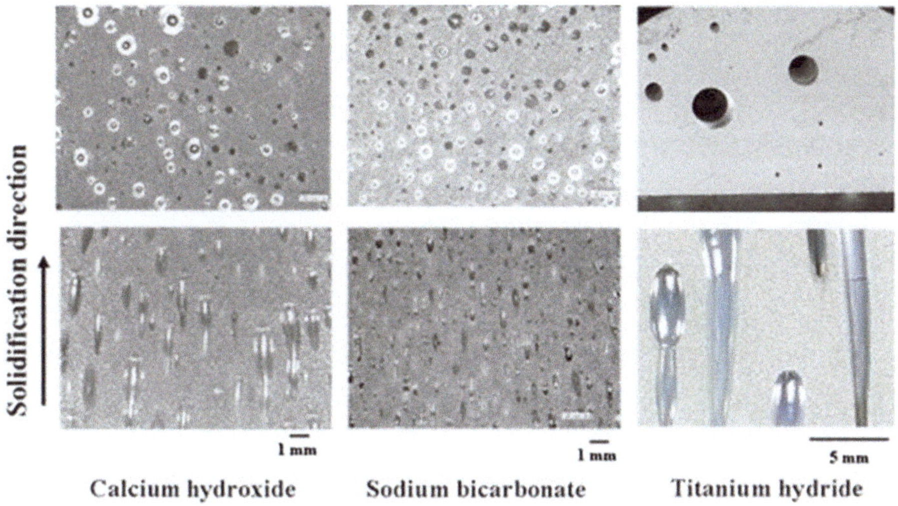


**Fig. 3.31** Optical micrographs of lotus aluminum fabricated by mold casting technique with different compounds in vacuum at 1,023 K. The *upper* and *lower* micrographs are the cross sections perpendicular and parallel to the solidification direction, respectively (Reprinted with permission from [35]. © 2009 The Minerals, Metals & Materials Society and ASM International)

Figure 3.31 shows the microstructure of lotus aluminum on cross sections parallel and perpendicular to the solidification direction. The pore morphologies perpendicular to the solidification direction are shown on the upper photos, while those parallel to the solidification direction are shown in the lower photos. In all specimens, the aligned pores parallel to the solidification direction were observed. It is considered that the cylindrical gas pores are evolved by TDM from the compounds during solidification of aluminum containing hydrogen gas.

Table 3.2 compiled the decomposition reactions of three compounds in the aluminum melt. Calcium hydroxide and sodium bicarbonate decompose into vapor and compounds and then, the vapor decomposes to hydrogen and metallic oxide. On the other hand, titanium hydride decomposes into hydrogen and titanium. These decomposed gas elements can dissolve in aluminum melt. When the melt is solidified, lotus aluminum can be produced.

Figure 3.32 shows the pore size and porosity of lotus aluminum with different compounds using the mold casting technique in vacuum. It is seen that the pores with average size 400, 268, and 1,086 μm are evolved by using calcium hydroxide, sodium bicarbonate, and titanium hydride, respectively. The porosity of lotus aluminum was similar, being as much as 20 % regardless of the kinds of the compounds. The pore size was small, as much as 400 μm, and its distribution was homogeneous in porous aluminum through TDM of calcium hydroxide and sodium bicarbonate, while the pore size was large as much as 1,000 μm and its distribution is not uniform through the TDM of titanium hydride.

**Table 3.2** Decomposition reaction and temperature of gas-forming compounds

| Reactions | Decomposition temperature (K) | Gas atoms or molecules to be dissolved |
|---|---|---|
| $Ca(OH)_2 \rightarrow CaO + H_2O$<br>$H_2O \rightarrow$ metallic oxide + 2H | 853 | H |
| $2NaHCO_3 \rightarrow Na_2CO_3 + H_2O + CO_2$<br>$H_2O \rightarrow$ metallic oxide + 2H | 473 | H, CO, $CO_2$, O |
| $TiH_2 \rightarrow Ti + 2H$<br>$Ti \rightarrow$ dissolved in liquid | 723 | H |

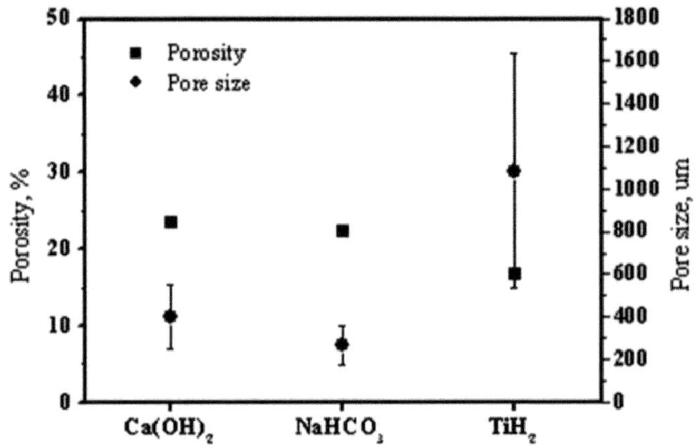


**Fig. 3.32** Pore size and porosity of lotus aluminum fabricated by mold casting technique using compounds in vacuum (Reprinted with permission from [35]. © 2009 The Minerals, Metals & Materials Society and ASM International)

Calcium hydroxide or sodium bicarbonate decomposes into compounds (CaO, $Na_2CO_3$), carbon dioxide, and vapor, the latter of which furthermore decomposes into metallic oxide and hydrogen. Small pore size with homogeneous distribution was observed through these reactions. On the other hand, titanium hydride decomposes directly into titanium and hydrogen. Large pores with inhomogeneous distribution were observed. It is not clear at present, but evolution of small pores with homogeneous distribution may be attributed to the existence of the oxide particles, which may serve as the nucleation sites of the pores. In order to investigate the content of pore gas, the gas analysis was carried out. Figure 3.33 shows the results of gas analysis of lotus aluminum using different compounds. It can be seen that hydrogen was the pore-formation source in the case of titanium hydride, calcium hydroxide, and sodium bicarbonate. Although the absolute values of concentration of each gas element cannot be compared in the analysis, it is apparent that the major gas is hydrogen in three kinds of the specimens. The contents of

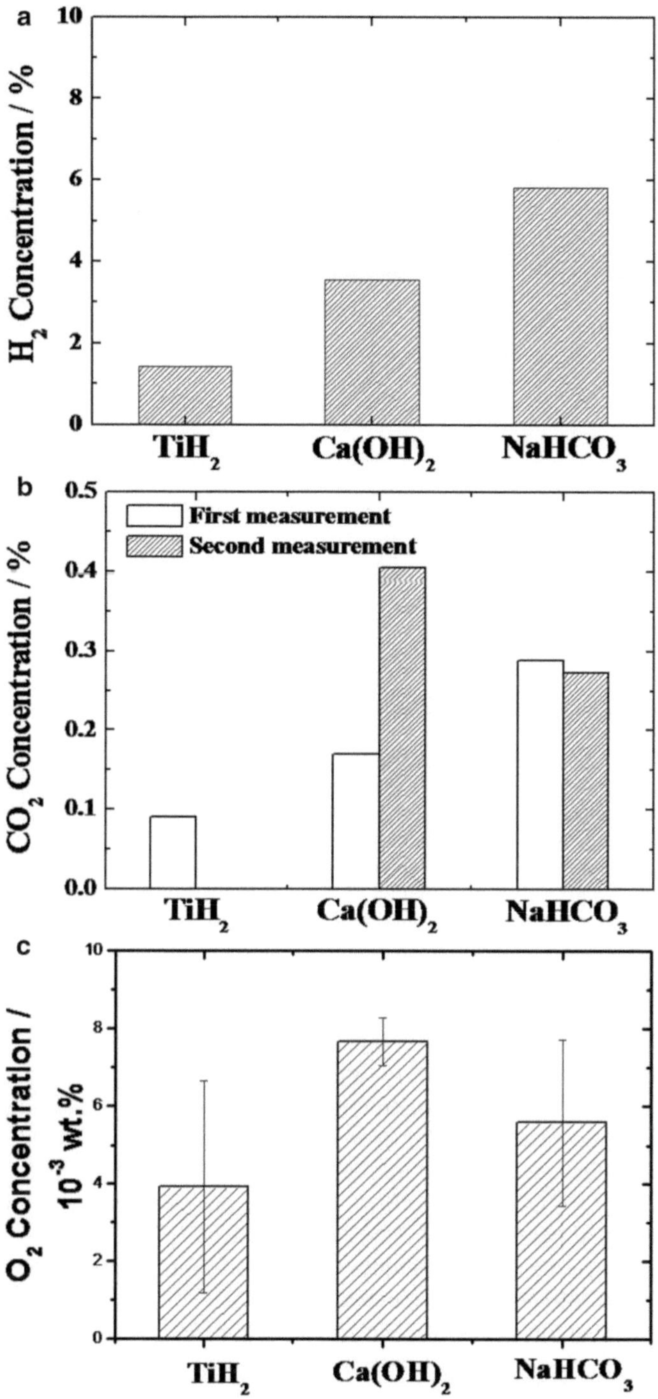


**Fig. 3.33** Results of gas analysis in the pore in lotus aluminum. (**a**) Hydrogen, (**b**) carbon dioxide, and (**c**) oxygen (Reprinted with permission from [35]. © 2009 The Minerals, Metals & Materials Society and ASM International)

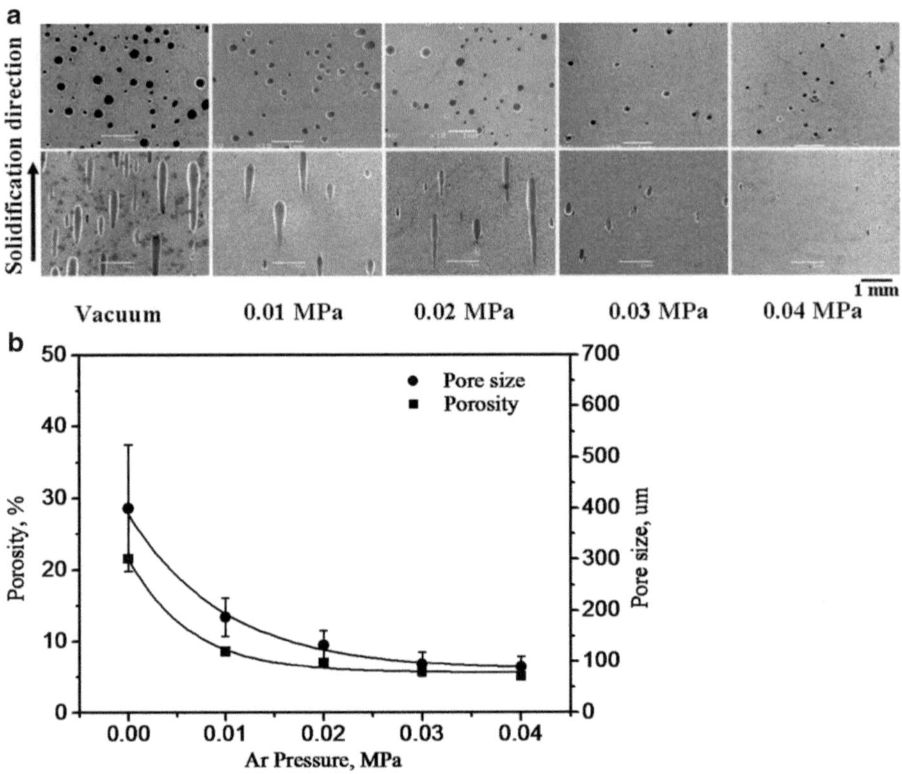


**Fig. 3.34** Lotus aluminum fabricated using calcium hydroxide in vacuum or under argon pressure (0.01–0.04 MPa). The amount of calcium hydroxide was kept to be 0.2 g. (**a**) Pore morphology. The *upper* and *lower* micrographs are the cross sections perpendicular and parallel to the solidification direction, respectively. (**b**) Variation of the porosity and pore size as a function of argon pressure (Reprinted with permission from [35]. © 2009 The Minerals, Metals & Materials Society and ASM International)

oxygen and carbon dioxide are very small; some amount of them may be contained in the collection handling procedure of the pores.

Figure 3.34a shows the pore structures on cross sections of lotus aluminum perpendicular (upper) and parallel (lower) to the solidification direction. The lotus aluminum was fabricated using calcium hydroxide in vacuum and under argon atmosphere. The aligned pores formed under the pressure less than 0.03 MPa Ar; however, spherical pores formed over 0.04 MPa argon. This may indicate that the applied pressure in the chamber reduces the driving force for pore nucleation and growth. Figure 3.34b shows the dependence of the porosity and the average pore size as a function of argon pressure. The effect of external pressure is obvious, and the pore growth is suppressed under higher pressure. Both the porosity and the average pore size decrease with increasing argon pressure. The pore volume $v$, which is equal to the porosity, is inversely proportional to the external argon pressure $P$, which can be described by the Boyle law, $v = nRT/P$, where $n$, $R$, and $T$ are the hydrogen molar number, the gas constant, and the temperature, respectively. Therefore, the pore diameter can be written as $d \propto P^{-1/3}$. The tendency of the pressure dependence of the porosity and pore size are explained by the law.

### 3.3.2 Continuous Zone Melting Technique through TDM

Recently, Makaya and Fredriksson produced porous Fe-based materials by dissolving CrN compound into a metallic melt of Fe-based alloy in an argon atmosphere [36]. The decomposition of the compound leads to dissolution of a gas into the melt. They solidified the melt in the crucible to produce porous metals with an isotropic pore structure. However, no investigations were carried out to produce porous metals with an anisotropic cylindrical pore structure until Nakajima and Ide performed theirs. Nakajima and Ide studied fabrication of lotus copper using titanium hydride in an argon atmosphere instead of a pressurized hydrogen atmosphere, which is called as TDM [32]. In order to fabricate the lotus metals, unidirectional solidification was usually carried out using the mold casting technique, as mentioned already. However, this technique cannot be applied to the metals and alloys with low- thermal conductivity. The solidification rate changes from the bottom to the top: the cooling speed near the bottom is faster, while it becomes slower near the top where the pore is coarsened because the heat cannot be dissipated. Thus, the pore size significantly changes with the height; the pore size increases with increasing height of the mold. In order to solve this problem, Ikeda et al. developed a continuous zone melting technique [23]. The partial melting zone is moved to retain the constant solidification velocity so that uniform distribution of porosity and pore size are obtained. This technique was adopted to fabricate lotus iron by Wada et al. [37].

Iron rods of 10.0 mm in diameter and 80 mm in length were used as specimens. In order to supply a constant source of chromium nitride, a hole of 2.0 mm in diameter in a central part of the cross section was made by an electric drill as shown in Fig. 3.35. The compound was filled into the hole. The chromium nitride was selected as a gas dissolving compound, since the temperature of gas releasing chromium nitride is the closest to the melting temperature of iron. The $Cr_{1.18}N$ and $Cr_2N$ powders were used. The former was a mixture of 81.4 mass pct CrN and 18.6 mass pct $Cr_2N$. 1.0 mass pct of $Cr_{1.18}N$ powders was filled into the permeable straight hole of the iron rods. The continuous zone melting was carried out by controlling the transfer velocity from 80 to 580 μm $s^{-1}$.

Figure 3.36 shows the cross-sectional views of lotus iron perpendicular and parallel to the solidification direction as a function of the transfer velocity. No pores were observed in the ingot fabricated with the velocity of 80 μm $s^{-1}$. Figure 3.37 exhibits the transfer velocity dependence of the porosity and average pore diameter. The porosity increases with increasing transfer velocity, while the variation of the

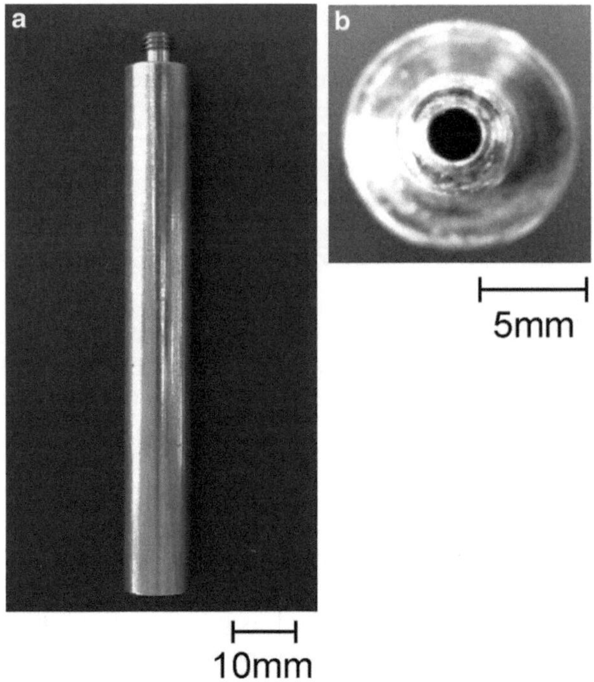


**Fig. 3.35** Outer view of iron rod: (**a**) longitudinal direction and (**b**) cross section perpendicular to the longitudinal direction (Reprinted with permission from [37]. © 2009 The Minerals, Metals & Materials Society and ASM International)

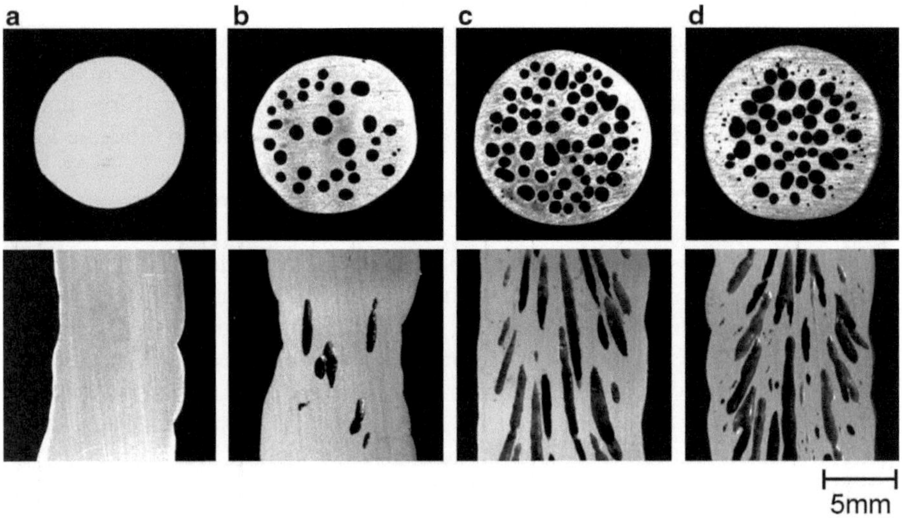


**Fig. 3.36** Cross-sectional views of lotus iron perpendicular (*upper*) and parallel (*lower*) to the solidification direction. Transfer velocity: (**a**) 80 μm $s^{-1}$, (**b**) 250 μm $s^{-1}$, (**c**) 410 μm $s^{-1}$, and (**d**) 580 μm $s^{-1}$. Atmosphere is 0.5 MPa helium (Reprinted with permission from [37]. © 2009 The Minerals, Metals & Materials Society and ASM International)

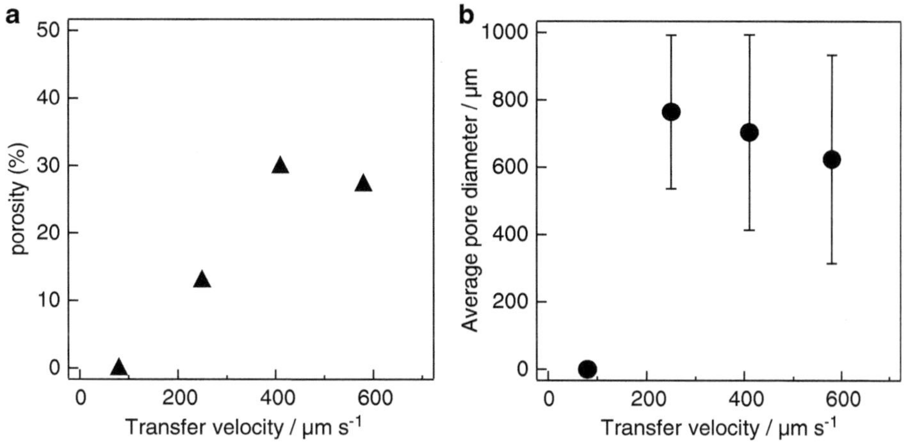


**Fig. 3.37** Transfer velocity dependence of (**a**) the porosity and (**b**) average pore diameter of lotus iron. Atmosphere is 0.5 MPa helium (Reprinted with permission from [37]. © 2009 The Minerals, Metals & Materials Society and ASM International)

pore diameter with the velocity is relatively limited compared with previous results [23, 24]. This suggests that the mechanism of pore formation obtained by this study by TDM is different from that obtained by PGM.

In the mold casting, the gas compound reacts and simultaneously dissolves a gas into the melt during the casting without preheating of the gas compound. However, in the continuous zone melting, when the rod is solidified unidirectionally, it is preheated by heat conduction before melting so that it has a temperature distribution. In general, the behavior of thermal decomposition of compounds is strongly affected by the heating rate [38]. The possibility of forming pores by continuous zone melting will be discussed, taking into consideration the transfer velocity dependence of the temperature distribution in the iron rod and the thermal decomposition of nitride. The relation between the temperature $T$ of preheated rod and the distance $x$ from the solid–liquid interface under the transfer velocity $v$ can be expressed by applying an equation of redistribution of solute in material solidified in the zone melting at constant velocity to a heat conduction [39].

$$\frac{d}{dx}\left(\lambda\frac{dT}{dx}\right)+\rho cv\frac{dT}{dx}=0, \tag{3.7}$$

where $\rho$, $c$, and $\lambda$ are the density of iron, the specific heat of iron, and the thermal conductivity of iron, respectively. Figure 3.38 shows the temperature and the heating rate as a function of the distance from the interface between liquid and solid at different transfer velocities. Both the temperature gradient and heating rate increase with increasing transfer velocity. Therefore, it is considered that the position in the rod where the gas is released from chromium nitride varies with the transfer velocity.

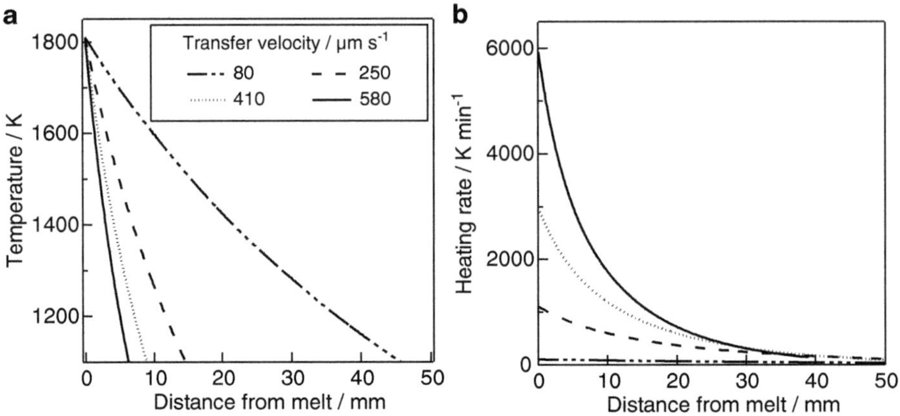


**Fig. 3.38** Changes in (**a**) temperature and (**b**) heating rate as a function of the distance from the interface between liquid and solid at different transfer velocities (Reprinted with permission from [37]. © 2009 The Minerals, Metals & Materials Society and ASM International)

Derivative thermogravimetry (DTG) of chromium nitrides of $Cr_{1.18}N$ and $Cr_2N$ powders was carried out using a thermal analyzer in an argon flow of 0.1 MPa in the temperature range from room temperature to 1,773 K. The heating rate was changed from 10, 20, 30, and 40 K $min^{-1}$. Figure 3.39 shows DTG curves with the heating rate of 10 K $min^{-1}$ for $Cr_{1.18}N$ and $Cr_2N$ powders. Two peaks for gas release from $Cr_{1.18}N$ were observed at 1,240 K($T_{p1}$) and 1,570 K($T_{p2}$), while one peak from $Cr_2N$ was found at 1,680 K. Thus, the 1,570 K peak observed in higher temperature is attributed to the gas release from $Cr_2N$. Since $Cr_{1.18}N$ is composed of CrN and $Cr_2N$, the 1,240 K peak of lower temperature is due to the gas release from CrN. According to the DTG measurement of $Cr_{1.18}N$ at different heating rate, the starting temperature and peak temperature of the gas release are shifted to higher temperature with increasing heating rate.

According to Kissinger [40], the heating rate $\beta$ is related to the peak temperature $T_p$ for gas release by the following equation:

$$\frac{d \ln\left(\frac{\beta}{T_p^2}\right)}{d\left(\frac{1}{T_p}\right)} = -\frac{E_a}{R}, \tag{3.8}$$

where $E_a$ is an activation energy of reaction and R is the gas constant. The Kissinger plots of the peak temperature were obtained from the measured DTG curves. It is predicted from the plots that the heating rates with which the peak temperature for the gas release is equal to the melting temperature (1809 K) of iron are $3.58 \times 10^4$ K $min^{-1}$ (H.$R_1$) for CrN and $5.70 \times 10^2$ K $min^{-1}$ (H.$R_2$) for $Cr_2N$. In order to have decomposition of the nitrides into the melt, the transfer velocity has to be such that the heating rate in the rod exceeds the predicted value, while there is an optimal heating rate. Figure 3.40 shows the relation between the temperature of the rod and

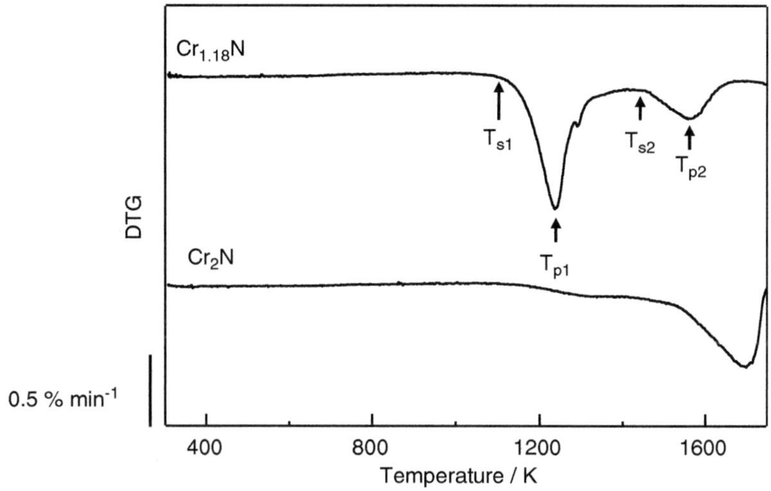


**Fig. 3.39** DTG curves measured by the DTG with the heating rate of 10 K min$^{-1}$ for $Cr_{1.18}$ N and $Cr_2N$ powders. The $T_s$ and $T_p$ are the starting temperature and peak temperature of the gas release, respectively (Reprinted with permission from [37]. © 2009 The Minerals, Metals & Materials Society and ASM International)

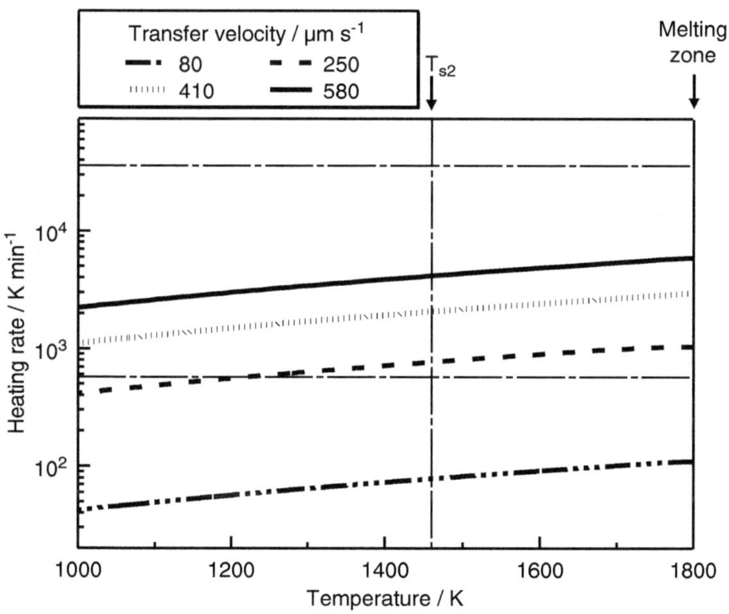


**Fig. 3.40** Relation between heating rate and temperature of the rod at different transfer velocities (Reprinted with permission from [37]. © 2009 The Minerals, Metals & Materials Society and ASM International)

heating rate, which is obtained from the results of Fig. 3.38a, b. Since the predicted $H.R_1$ is higher than the heating rates at all transfer velocity, gas from CrN contained in $Cr_{1.18}N$ is released before melting and does not contribute to pore evolution. On the other hand, $H.R_2$ in higher temperature than the starting temperature (about 1,450 K slightly depending on the heating rates) for gas release is slower than the heating rates in the transfer velocity more than 250 μm $s^{-1}$. Therefore, it is surmised that no pore formation at lower transfer velocity of 80 μm $s^{-1}$ is attributed to insufficient nitrogen release into the molten iron, because most of nitrogen gas released from the nitride escapes to the atmosphere; the rod is heated at high temperature by heat conduction from the melt part of a longer time. It is considered that gas from $Cr_2N$ contained in $Cr_{1.18}N$ cannot be released until part of the rod melts and is used effectively to evolve the pores. Thus, for transfer velocity such that the decomposition peak temperature lies just above the melting point, decomposition of the nitrides in the melt is optimal.

Different from PGM, a gas equilibrium between the atmosphere (nitrogen partial pressure ~ 0) and the melt is not maintained in TDM [32]. During melting, nitrogen escapes from the melt to the atmosphere. Since the holding time of the melting condition decreases with increasing transfer velocity, the amount of escaped nitrogen may decrease; thus, the porosity increases with increasing transfer velocity, as shown in Fig. 3.37a.

### 3.3.3 Continuous Casting Technique through TDM

As mentioned above, TDM is desirable to avoid inherent risk of employing high-pressure hydrogen gas, and the continuous casting technique is suitable to produce large-sized lotus metal slabs with controllable pore size and porosity. Combining both techniques is the most suitable to fabricate lotus metals. Nakajima and Ide [41] fabricated long-sized lotus copper slabs through TDM by continuous casting technique. Figure 3.41a illustrates the continuous casting apparatus for TDM; for comparison, the apparatus of continuous casting technique through PGM is shown in Fig. 3.41b. The apparatus consists of three parts. (1) The first part is the melting part; the copper were melted in the graphite crucible by the induction heating. The upper cylindrical graphite bar was pushed down. (2) The overflowing copper melt dropped down into a tundish, in which pellets of titanium hydride were supplied through a nozzle at a constant time interval in order to maintain the dissolving hydrogen constant. (3) The third part is the mold cooled by water and pinch rollers which pulled the dummy bar and the solidified slab ingot horizontally. Figure 3.42 shows one example of lotus copper fabricated with addition of a pellet, 0.2 g, per minute at the melt temperature of 1,573 K in argon atmosphere of 0.1 MPa with different transfer velocity. The concentration of titanium hydride in the melt added in the tundish was evaluated as 0.001mass%. Almost uniform pore distribution in size and position was observed in the cross section parallel to the

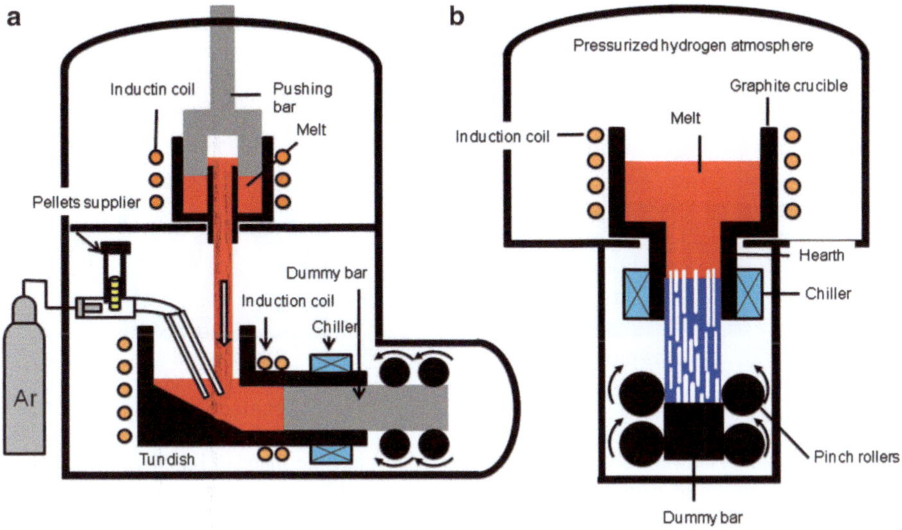


**Fig. 3.41** Apparatus of continuous casting technique to fabricate lotus metals through (**a**) TDM and (**b**) PGM (Reprinted with permission from [41]. © 2012 Deutsche Gesellschaft fur Materialskunde e.V.)

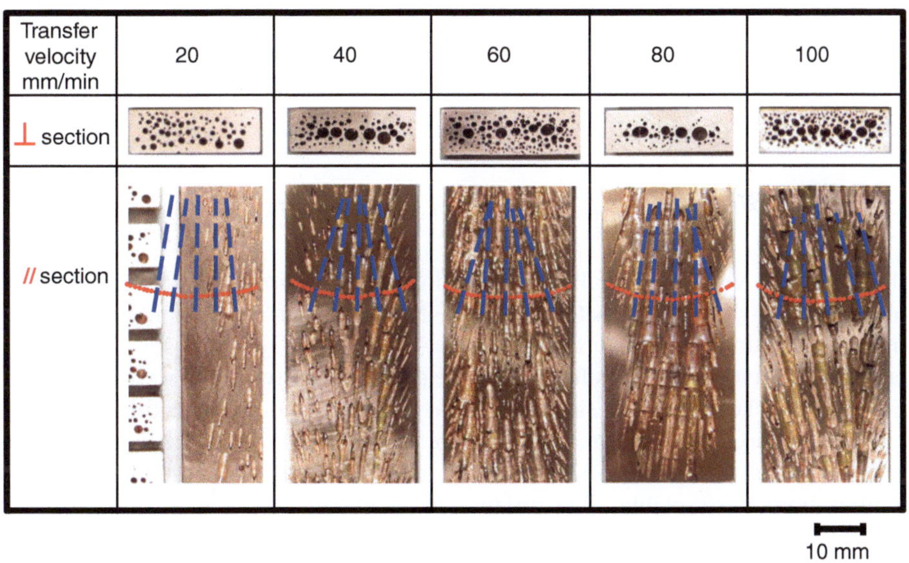


**Fig. 3.42** Lotus copper fabricated by continuous casting technique through TDM. *Upper* and *lower* are the cross sections perpendicular and parallel to the solidification direction, respectively

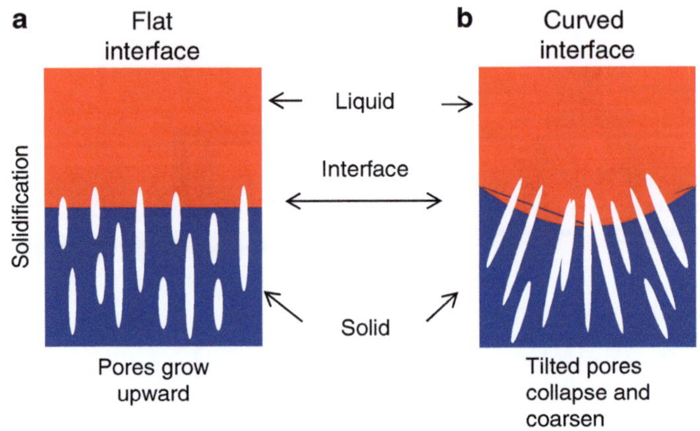


**Fig. 3.43** Relation between alignment of evolving pores and the shape of liquid–solid interface. (**a**) The alignment of the evolving pores is unidirectional. (**b**) The alignment is tilted. The tilted pores collapse to coarse pores in the central part

solidification direction, but not uniform in the section perpendicular. This is because tilted pores collapsed to cause coarse pores in the central part, as shown in Fig. 3.43b. The alignment of the evolving pores should be unidirectional as shown in Fig. 3.43a. If a flat liquid–solid interface can be obtained by controlling the chiller temperature in the hearth, this problem will be solved.

## 3.4 Moisture Decomposition Method

### *3.4.1 Fabrication of Lotus Nickel by Moisture*

Lotus nickel was successfully fabricated by unidirectional solidification using molybdenum-sheet mold coated by $Al_2O_3$–$Na_2SiO_3$ in an argon atmosphere. During solidification, the moisture absorbed in $Al_2O_3$–$Na_2SiO_3$ at atmospheric temperature is dissolved into the melt and then the gas pores are formed at the solid–liquid interface. Nickel in a magnesia crucible was melted by induction heating at a given pressure of argon gas in the high-pressure chamber. As shown in Fig. 3.44, the molten nickel was poured into the mold whose bottom was water-cooled by a chiller from the crucible. The mixture paste of alumina, sodium silicate, and water was coated on the inside lateral surface on the molybdenum cylindrical sheet of the mold in order to suppress a reaction between molybdenum sheet and the molten nickel as insulated ceramics and also to maintain the moisture. Figure 3.45 shows the optical micrographs of lotus nickel in parallel section to the solidification direction. The moisture contents in the mold weigh (a) 0.0596 g, (b) 0.0876 g, (c) 0.1070 g, and (d) 0.1201 g. In the ingot with the smallest amount of

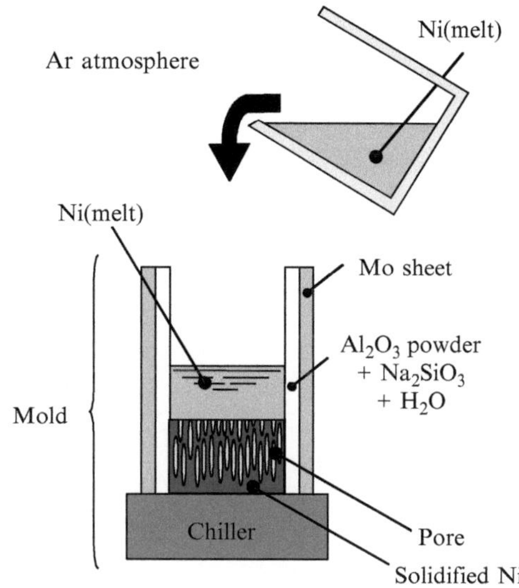


**Fig. 3.44** Schematic drawing of the process of mold casting technique for fabrication of lotus nickel with moisture in argon atmosphere (Reprinted with permission from [42]. © 2004 Japan Institute of Metals)

moisture, almost all pores are spherical and long-cylindrical pores were not observed. The number of elongated pores increases with increasing moisture. Such longer pore evolution may be attributed to increase in hydrogen content which serves as a source for pore growth.

Figure 3.46 shows the magnified optical micrographs of cross-sectional planes (a) parallel and (b) perpendicular to the solidification direction at 4.5 mm in height from the bottom of the ingot (c) in Fig. 3.45, whose porosity is 44.7 % and the average pore size is 105 μm, ranging from 5 to 200 μm in diameter. On the other hand, for example, the porosity and pore size are 49.8 % and 499 μm, respectively, in lotus nickel fabricated using the dried mold without moisture under the pressure of 0.6 MPa argon and 0.2 MPa hydrogen. Thus, the porosity in lotus nickel fabricated in the mold with moisture in argon is as high as that in lotus nickel without moisture in hydrogen, and moreover, the pores can be more minute than that using hydrogen.

The pore evolution model by moisture was proposed by Suematsu et al. [42]. Figure 3.47 illustrates schematic of pore evolution process before solidification, after pouring the melt and during solidification. The moisture is absorbed in the ceramic coat on the lateral side of the mold, and then, when the molten nickel is poured into the mold, the moisture is dissolved into hydrogen and oxygen as a result of chemical reaction between the molten metal and moisture:

$$n\ \mathrm{H_2O(gas)} + m\ \mathrm{M} \leftrightarrow \mathrm{M}_m\mathrm{O}_n + 2n\ \mathrm{H}, \qquad (3.9)$$

where M is a metallic element to be oxidized easily. Then, the hydrogen is absorbed in the molten nickel and insoluble hydrogen when nickel is solidified is

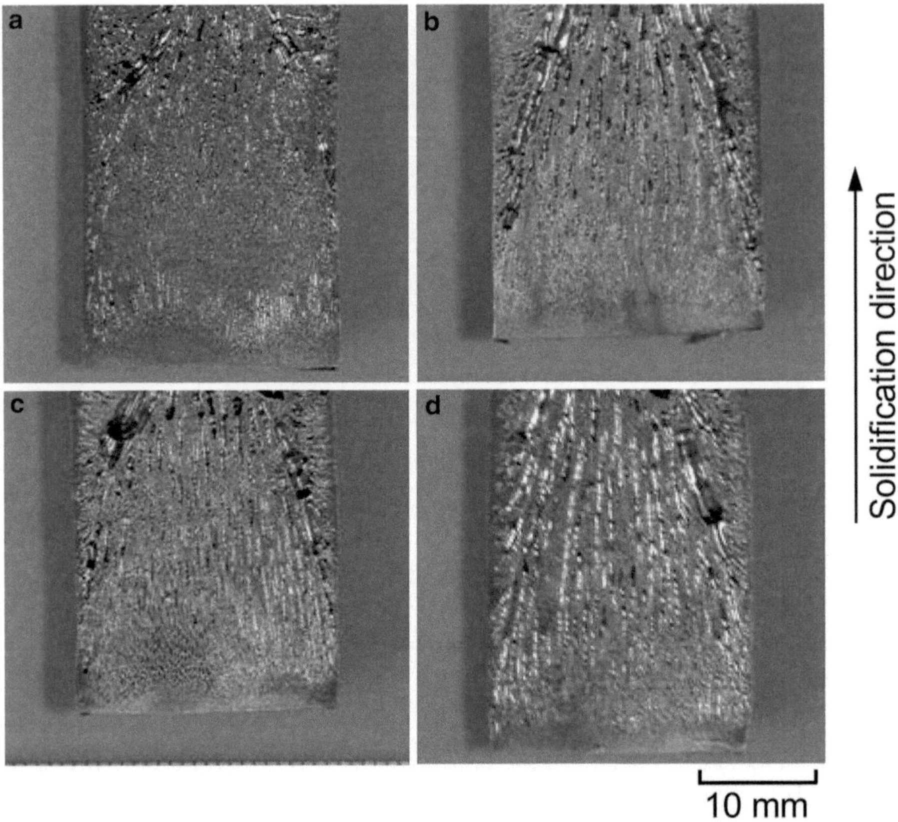


**Fig. 3.45** Optical micrographs of lotus nickel in parallel section to the solidification direction (argon pressure, 0.3 MPa); moisture contents in mold are (**a**) 0.0596 g, (**b**) 0.0876 g, (**c**) 0.1070 g, and (**d**) 0.1201 g (Reprinted with permission from [42]. © 2004 Japan Institute of Metals)

able to form pores, whose principle is the same as that in fabrication of lotus metals. On the other hand, oxygen forms oxides with, for example, nickel which may become to be heterogeneous nucleation sites. It is well known that the pore nucleation takes place by heterogeneous nucleation mechanism. Thus, dissolution of oxygen in the melt increases the number of the pore nucleation sites. Even if the same amount of hydrogen dissolves in the molten nickel, a number of more minute elongated pores can be evolved using the mold with sufficient moisture in comparison with that using the dry mold in hydrogen atmosphere. Therefore, lotus-type porous nickel with minute elongated gas pores can be successfully produced by using the mold with moisture, which is far easy and simple technique to fabricate lotus metals.

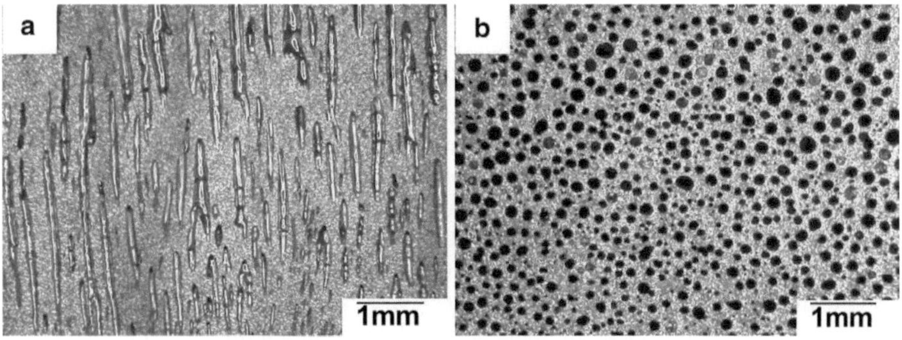


**Fig. 3.46** Optical micrographs of cross sections of lotus nickel with 44.7 % porosity (argon pressure, 0.3 MPa). The cross-sectional plane (**a**) parallel and (**b**) perpendicular to solidification direction at 4.5 mm from the bottom plane in the ingot (Reprinted with permission from [42]. © 2004 Japan Institute of Metals)

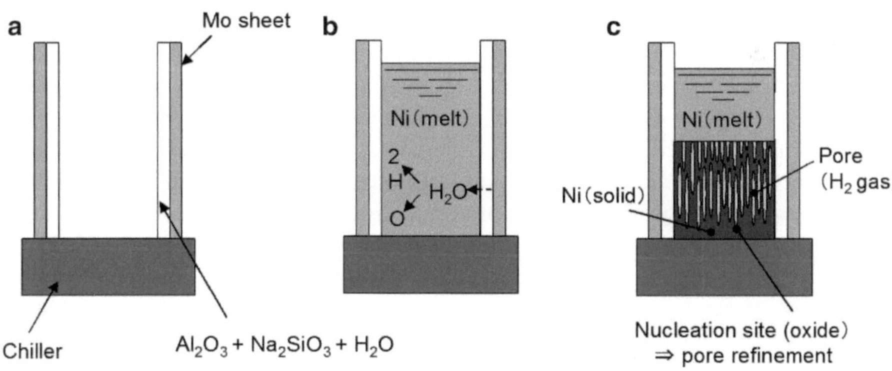


**Fig. 3.47** Schematic drawings of pore-forming process. (**a**) Mold assembly before solidification, (**b**) molten nickel in the mold just after pouring of the melt, and (**c**) process of unidirectional solidification (Reprinted with permission from [42]. © 2004 Japan Institute of Metals)

## *3.4.2 Fabrication of Lotus Cobalt and Silicon by Moisture*

Previous study as described in Sect. 3.4.1 suggests that formation of pores strongly depends on the dissociation constant of water at the casting temperature [42]. Because the free energy changes of Eq. (3.9) are known to be positive values, the dissociation constant $P_{H_2}/P_{H_2O}$ is very small [43, 44]. It is worthwhile to investigate in detail whether the feasibility of pore formation on other materials is also closely related with the magnitude of the dissociation constant of water in the moist atmosphere. The feasibility to fabricate lotus silicon, cobalt, and copper was investigated by unidirectional solidification using moisture contained in the mold [45].

The metal in the crucible was melted by an induction coil and the melted metal was poured into the mold. The bottom of the mold was cooled by a water chiller, while the lateral wall was made of molybdenum whose surface was coated with a mixture of alumina, water glass (54.5 % $H_2O$, 31.2 % $SiO_2$, and 14.3 % $Na_2O$), and water with a weight ratio of 1:1.5:1. The mold was dried in an oven at 423 K for 7.2 ks. Then, the mold was kept in a closed chamber with constant humidity (90 %), which was measured by a hydrometer attached to the chamber for a given time, and 0.15 g of water, which corresponds to 0.083 mol of $H_2O$, was reabsorbed into the coating material. This value is sufficient to produce pores. About 2 mol of silicon, cobalt, and copper ingots with 99.9 % purity were used for casting. The size of the solidified ingots was 28 mm in diameter and 50 mm in length.

Figure 3.48 shows the cross-sectional views of solidified ingots of Si, Co, and Cu perpendicular and parallel to the solidification direction under a given pressure of the argon atmosphere. The pores elongated along the solidification direction were observed for Si and Co, while no pores were observed for Cu. Moreover, lotus Si and Co were also fabricated in the hydrogen atmosphere in the mold without moisture. The pore size in the lotus Si and Co fabricated in hydrogen gas is much larger than that by moisture dissociation, as shown in Fig. 3.49. When the moisture contained in the oxides mixture coated on the molybdenum mold was completely removed by drying, only nonporous cobalt ingot was obtained so that the pore evolution is attributed to hydrogen decomposed from the moisture in the mold. According to the previous investigation [10], lotus copper was easily produced by unidirectional solidification in pressurized hydrogen atmosphere. However, lotus copper was not produced by the moisture in the mold. In order to elucidate the reason why the moisture can produce lotus cobalt and silicon, not lotus copper, the ratio of pressure of hydrogen decomposed to that of the moisture was evaluated to the following reactions:

$$\mathrm{Co} + \mathrm{H_2O} = \mathrm{CoO} + \mathrm{H_2}, \tag{3.10}$$

$$\mathrm{Si} + 2\mathrm{H_2O} = \mathrm{SiO_2} + 2\mathrm{H_2}. \tag{3.11}$$

They assume that the moisture on the mold decomposes to hydrogen and oxygen atoms at the initial stage of casting. According to the Ellingham diagram, the ratio of hydrogen to moisture, $P_{H_2}/P_{H_2O}$, can be derived from the Gibbs free energy changes [43, 44]. Figure 3.50 shows the formation free energy changes of water and nickel, silicon, cobalt, and copper oxides. Porous nickel was fabricated by the mold casting method ratio of $P_{H_2}/P_{H_2O}$, at the casting temperature was approximately $1.18 \times 10^{-2}$, as calculated using the following equation:

$$\frac{P_{H_2}}{P_{H_2O}} = \exp\left(-\frac{\Delta G}{2RT}\right), \tag{3.12}$$

where $\Delta G$, $R$, and $T$ denote the Gibbs free energy change, the gas constant, and the casting temperature, respectively. The ratio of $P_{H_2}/P_{H_2O}$, for the formation of

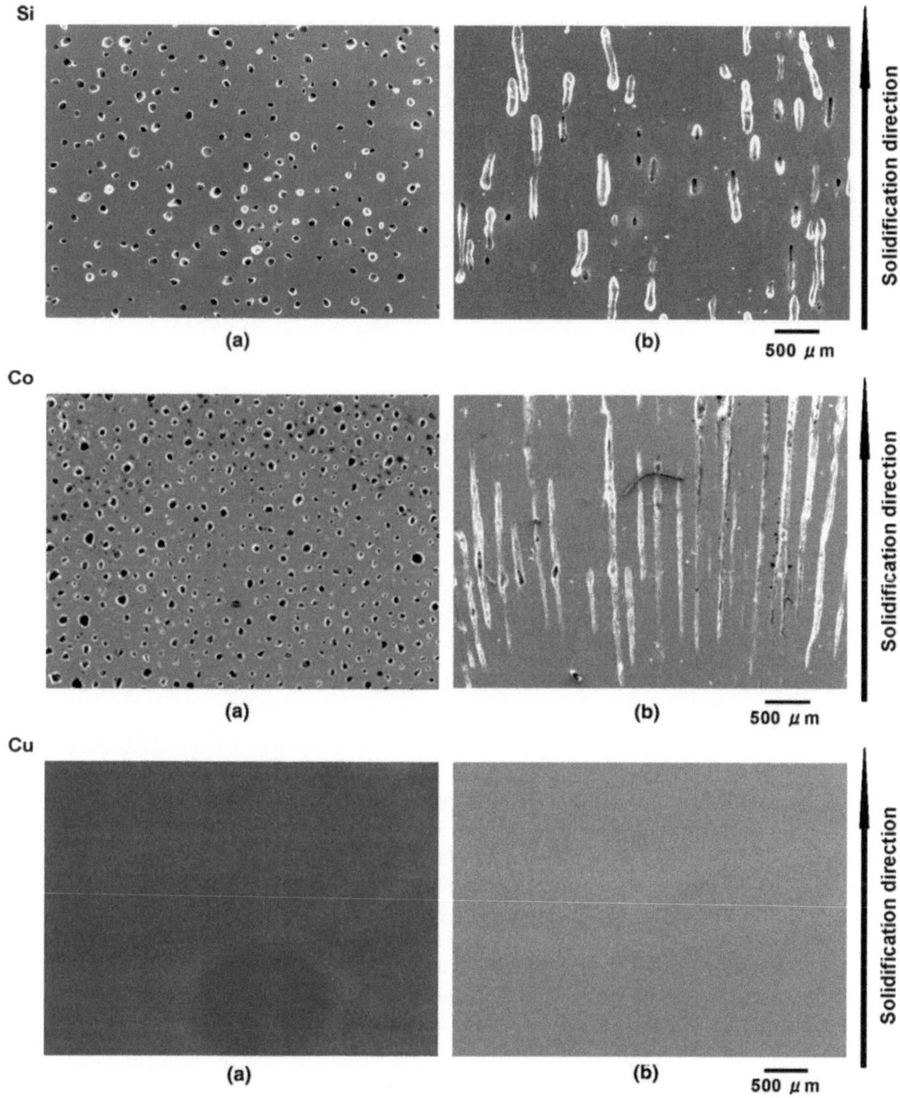


**Fig. 3.48** (**a**) Transverse cross-sectional and (**b**) longitudinal cross-sectional views of lotus silicon, cobalt, and copper fabricated under Ar pressure. Argon pressure for each fabrication is 0.4 MPa for Si, 0.8 MPa for Co, and 0.4 MPa for Cu (Reprinted with permission from [45]. © 2008 The Minerals, Metals & Materials Society and ASM International)

silicon, cobalt, and copper oxides, is estimated as $2.15 \times 10^{4}$ at 1,773 K, $6.93 \times 10^{-2}$ at 1,873 K, and $4.91 \times 10^{-4}$ at 1,673 K, respectively [43, 44], as shown in Table 3.3. Assuming that the initial stage of casting is in the equilibrium state, the amount of moisture decomposed into hydrogen and oxygen atoms and the amount of hydrogen gas dissolved into the molten metal can be evaluated by the

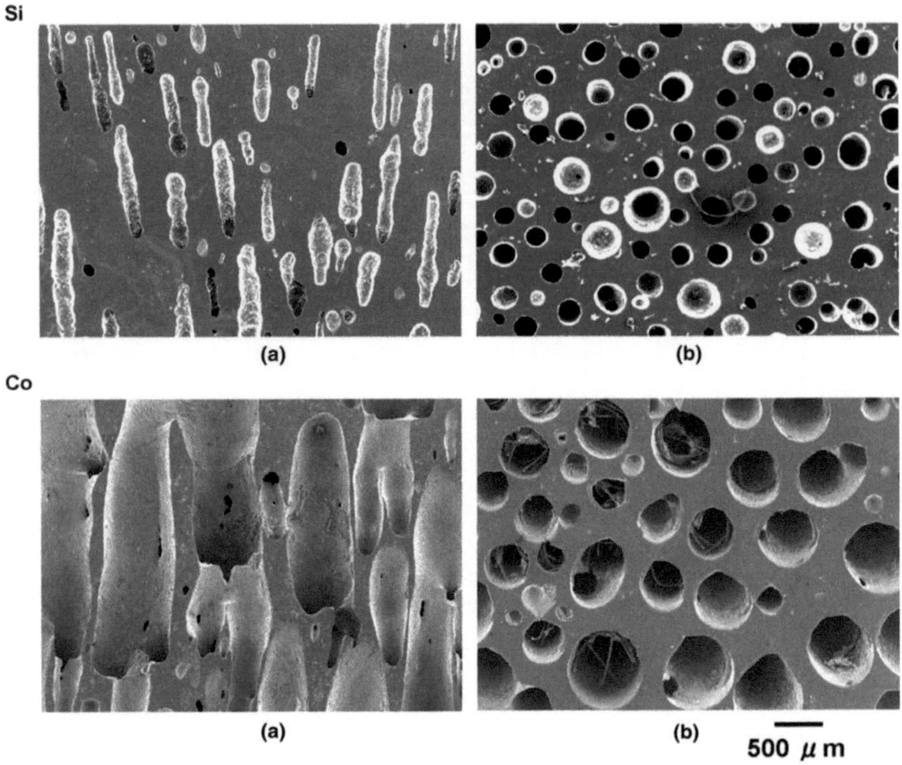


**Fig. 3.49** (**a**) Longitudinal cross-sectional and (**b**) transverse cross-sectional views of lotus silicon fabricated under hydrogen pressure of 0.4 MPa and cobalt fabricated under hydrogen pressure of 0.15 MPa and argon pressure of 0.65 MPa (Reprinted with permission from [45]. © 2008 The Minerals, Metals & Materials Society and ASM International)

ratio of $P_{H_2}/P_{H_2O}$ and the Sieverts law. The formation mechanism of the moisture decomposition into the molten metal is essentially the same as that of the method using pressurized hydrogen gas. These results indicate that the pores can be evolved, even if the value of $P_{H_2}/P_{H_2O}$ is very small. Since no hydrogen gas was introduced in the chamber, the hydrogen partial pressure in the atmosphere is considered as 0 MPa. Hence, the hydrogen gas used in the formation of pores is formed by decomposition of the moisture on the mold during solidification.

The pore diameter for the lotus silicon and cobalt fabricated in the moisture method is much smaller than that of those fabricated under pressurized hydrogen atmosphere. It is well known that heterogeneous nucleation takes place in metal melts in the presence of small amounts of foreign elements [34]. Then, it is reasonable to consider that oxygen decomposed from the moisture can result with a metallic element to form oxide clusters, which are nuclei for pore evolution as illustrated in Fig. 3.51. Such dispersive oxide clusters may promote uniform small

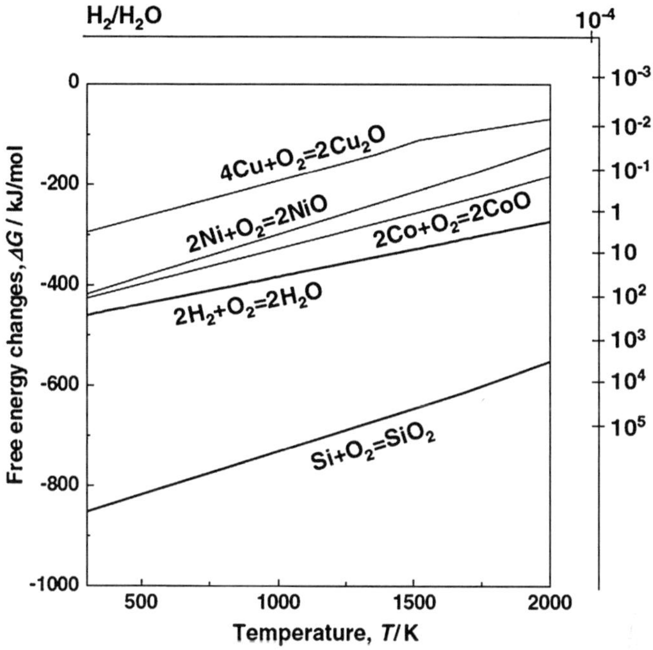


**Fig. 3.50** Temperature dependence of the change of the free energy for formation of water, silica, cobalt oxide, nickel oxide, and copper oxide [43, 44] (Reprinted with permission from [45]. © 2008 The Minerals, Metals & Materials Society and ASM International)

**Table 3.3** Free energy changes of dissociation of water for Si, Co, Cu, and Ni

| R reaction | $\Delta G$ (kJ mol$^{-1}$) | Casting temperature (K) | $P_{H_2}/P_{H_2O}$ |
|---|---|---|---|
| $Si + 2H_2O = SiO_2 + 2H_2$ | −298.5 | 1,773 | $2.15 \times 10^4$ |
| $Co + H_2O = CoO + H_2$ | 41.56 | 1,873 | $6.93 \times 10^{-2}$ |
| $2Cu + H_2O = Cu_2O + H_2$ | 53.41 | 1,673 | $4.91 \times 10^{-4}$ |
| $Ni + H_2O = NiO + H_2$ | 69.13 | – | $1.18 \times 10^{-2}$ |

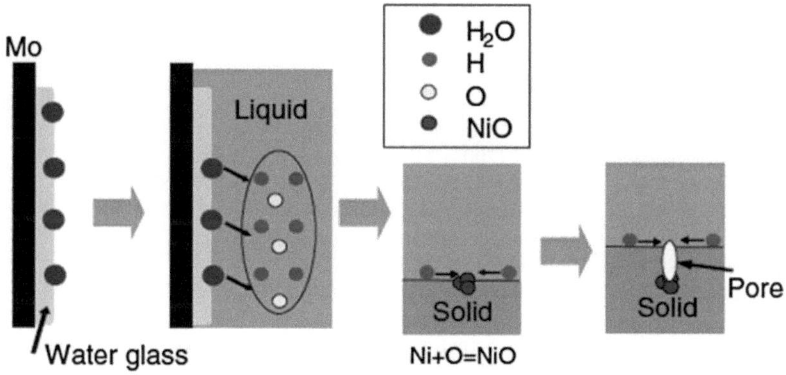


**Fig. 3.51** Schematic growing of the movement of oxide during fabrication and pore generation (Reprinted with permission from [45]. © 2008 The Minerals, Metals & Materials Society and ASM International)

pore distribution. Although quantitative comparison of the moisture decomposition method with the pressurized hydrogen method is not possible, there is a tendency that the moisture method can produce smaller-sized pores and a large number of density of pores. This tendency can be explained in terms of the oxide clusters for possible nucleation sites.

## References


1. Chalmers B (1959) Sci Am 200:114–122
2. Imabayashi M, Ichimura M, Kanno Y (1983) Trans JIM 24:93–100
3. Svensson I, Fredriksson HS (1980). In: Proceedings of international conference organized by the applied metallurgy and metals tech group of TMS, University of Warwick, pp. 376–380
4. Knacke O, Probst H, Wernekinck J (1979) Z Metallkde 70:1–6
5. Bioko LV, Shapovalov VL, Chernykh EA (1991) Metallurgiya 346:78–81
6. Hyun SK, Shiota Y, Murakami K, Nakajima H (1999) In: Koiwa M, Otsuka K, Miyazaki T (eds) Proceedings of international conference on solid-solid phase transformations '99 (JIMIC-3), Japan Inst Metals, Kyoto, pp. 341–344
7. Nakajima H, Hyun SK, Ohashi K, Ota K, Murakami K (2001) Colloids Surf A:Physicochem Eng Aspects 197:209–214
8. Nakajima H (2001) Mater Trans 42:1827–1829
9. Shapovalov VI (1994) MRS Bull XIX:24–28
10. Hyun SK, Murakami K, Nakajima H (2001) Mater Sci Eng A 299:241–248
11. Smith DP (1947) Hydrogen in metals. The University of Chicago Press, Chicago, p 34
12. Reed-Hill RE (1964) Physical metallurgy principles. D.Van Nostrand Company Inc, Princeton, p 393
13. Massalski TB (1986) Binary alloy phase diagram. American Society for Metals, Metals Park, p 1079
14. Levinsky Y (1997) Pressure dependent phase diagrams of binary alloys. ASM International, Materials Park, p.693
15. Satir-Kolorz AH, Feichtinger HK (1991) Z Metallkde 82:689–697
16. Hyun SK, Nakajima H (2002) Mater Trans 43:526–531
17. Nakahata T, Nakajima H (2005) Mater Trans 46:587–592
18. Ikeda T, Nakajima H (2002) Japan Foundry Eng Soc 74:812–816
19. Hyun SK, Nakajima H (2003) Mater Lett 57:3149–3154
20. Ikeda T, Tsukamoto M, Nakajima H (2002) Mater Trans 43:2678–2684
21. Nakajima H, Ikeda T, Hyun SK (2003) In: Banhart J, Fleck A (eds) Cellular metals: manufacture, properties, applications. MIT, Berlin, pp 191–202
22. Nakajima H, Ikeda T, Hyun CK (2004) Adv Eng Mater 6:377–384
23. Ikeda T, Aoki T, Nakajima H (2005) Metall Mater Trans A 36A:77–86
24. Park JS, Hyun SK, Suzuki S, Nakajima H (2007) Acta Mater 55:5646–5654
25. Campbell J (1991) Castings. Butterworth-Heinemann, Oxford
26. Fisher JC (1948) J Appl Phys 19:1062–1067
27. Flemings MC (1974) Solidification processing. McGraw-Hill, New York
28. Chalmers B (1964) Principles of solidification. Wiley, New York
29. Porter DA, Easterling KE (1992) Phase transformations in metals and alloys. Chapman & Hall, London
30. Fromm E, Gebhardt E (1976) Gases and carbon in metals. Springer, Berlin
31. Wright JH, Hocking MG (1972) Metall Trans 3:1749–1753
32. Nakajima H, Ide T (2008) Metall Mater Trans A 39A:390–394

33. Nakajima H, Ide T (2007) Method for manufacturing porous body. PCT/JP2007/062769 (patent pending)
34. Fredriksson H, Akerlind U (2006) Materials processing during casting. Wiley, Chichester, pp 141–142
35. Kim SY, Park JS, Nakajima H (2009) Metall Mater Trans A 40A:937–942
36. Makaya A, Fredriksson H (2005) Mater Sci Eng A 413A–414A:533–537
37. Wada T, Ide T, Nakajima H (2009) Metall Mater Trans A 40A:3204–3209
38. Murray P, White J (1955) Trans Br Ceram Soc 54:204–237
39. Tiller WA, Jackson KA, Rutter JW, Chalmers B (1953) Acta Metall 1:428–437
40. Kissinger HE (1957) Anal Chem 29:1702–1706
41. Nakajima H, Ide T (2012) In: Proceedings of cellular materials (Cellmat2012), Deutsche Gesellschaft fur Materialskunde e.V, pp. 1–4
42. Suematsu T, Hyun SK, Nakajima H (2004) J Japan Inst Metals 68:257–261
43. Kubaschewski O, Alcock CB (1979) Metallurgical thermochemistry, 5th edn. Pergamon, Oxford
44. Elliott JF, Gleiser M, Ramakrishna V (1963) Thermochemistry for steelmaking, vol 1. Addition-Wesley, New York, pp 161–215
45. Onishi H, Ueno S, Hyun SK, Nakajima H (2009) Metall Mater Trans A 40A:438–443

# Chapter 4
# Nucleation and Growth Mechanism of Pores in Metals

**Abstract** In this chapter, the nucleation and growth mechanism of pores in metals are represented in order to explain why the directional pores are grown during the unidirectional solidification. The presented models are consistent with the model experiment of ice-freezing from water containing carbon dioxide. It is shown that the mechanism of pore formation in lotus materials is much different from that of foaming process.

**Keywords** Carbon dioxide • Ice • Pore growth • Pore nucleation • Sieverts' law

## 4.1 Solubility of Gases in Metals: Sieverts' Law

The difference between the solubility of gases in a metal melt and in the solid phase after solidification causes gas precipitation in the metal at casting. The solubility of a gas in a metal melt depends on its partial pressure on the surrounding atmosphere. Most common gases are diatomic at room temperature. At high temperature (temperature of the metal melt), a dissociation of the gas $G_2$ occurs at the metal surface in the presence of metal atoms M:

$$M + G_2 \leftrightarrow M + 2\underline{G}. \tag{4.1}$$

A chemical equilibrium between the dissolved gas atoms $\underline{G}$ and the surrounding gas is established. Guldberg–Waage's law controls the equilibrium:

$$\frac{[\underline{G}]^2}{p_{G_2}} = \text{constant}, \tag{4.2}$$

where the value of the constant depends on the temperature. The gas atoms then diffuse into the melt and/or the solid phase and stay there in the state of a solution. We denote the concentration in the solution with $[\underline{G}]$, where the underlined letter

H. Nakajima, *Porous Metals with Directional Pores*,
DOI 10.1007/978-4-431-54017-5_4, 

means that the gas is dissolved in the metal as G atoms and not $G_2$ molecules. At equilibrium, the concentration of the dissolved gas atoms in the melt or in the solid phase is proportional to the square root of the partial pressure of the gas in the surrounding atmosphere, which can be expressed as Sieverts' law [1]:

$$[\underline{G}] = \text{constant} \times \sqrt{p_{G_2}}. \tag{4.3}$$

## 4.2 Evolution Process of Directional Gas Pores

Figure 4.1 shows two different modes of gas pore evolution processes: (a) gas nucleation and growth of pores in the case of water boiling or supersaturation in soda in a glass and (b) pore nucleation and growth in unidirectional solidification [2]. In the former case, for example, when the soda is poured into a glass cup, heterogeneous nucleation occurs to nucleate the gas bubbles on the glass wall.

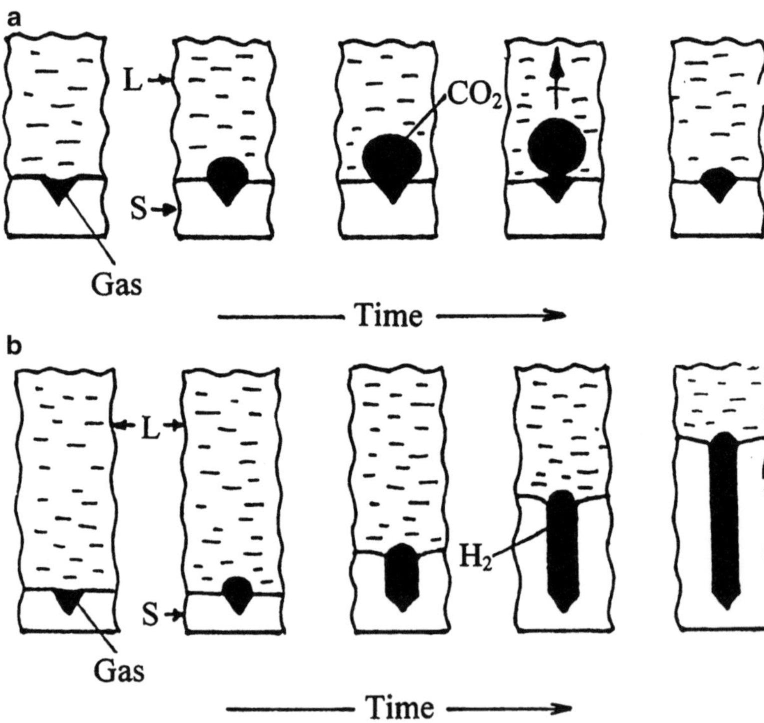


**Fig. 4.1** Two different modes of gas pore evolution processes. (**a**) Gas nucleation and growth of pores during water boiling and supersaturation in soda in a glass and (**b**) pore nucleation and growth in unidirectional solidification in gas atmosphere (Reprinted with permission from [2]. © 1998 Materials Research Society)

If the pore continues to absorb the carbon dioxide to coarsen, the interfacial surface area, that is, the interfacial energy, between liquid and gas phases above the critical size of the bubble becomes too large to maintain the stable gas pores' condition, and finally, the pores are detached from the glass wall and floated upward in order to minimize the interfacial energy. Such phenomena are repeated in the soda everywhere. We can see many bubbles evolution in the soda. On the other hand, in the case of (b) during unidirectional solidification, the gas pores nucleate in the interface between solidified metal and liquid metal by precipitation of insoluble gas in the solid. If the pore growth rate is just identical to the solidification velocity, the interfacial area (the interfacial energy) between liquid and gas phases can be kept constant so that the coarsening so as to increase the pore diameter does not occur. As a result, cylindrical pores can grow upward in the direction of the unidirectional solidification. However, in practice, split and adhesion of the isolated pores are often observed which are attributed to imperfection of flatness of the interface between solid and liquid, the existence of impurity inclusion, nonuniformity of the temperature gradient, convection of liquid, etc. In order to suppress such irregular configuration of pores, precise control of temperature gradient during solidification, suppression of liquid convection, and impurity control is necessary.

### 4.2.1 Pore Nucleation

According to the nucleation theory by Fisher [3], when the pores of hydrogen are formed in the liquid phase during solidification by homogeneous nucleation, the gas pressure required for the pore nucleation is evaluated to be a few GPa. This magnitude of the pressure is not realistic [4]. Therefore, the pores are considered to nucleate heterogeneously. In most cases, a few ppm of oxygen will be dissolved in the metal melt, because the metal is melt in a vacuum of $10^0$–$10^{-1}$ Pa. As the solidification proceeds, oxygen may concentrate at the solid–liquid interface to form an oxide layer. When the hydrogen amount in the metal melt is large in the case of high hydrogen pressure, the oxide layer at the solid–liquid interface is resolved by hydrogen. However, in the case of low hydrogen pressure, the oxide layer may remain at the solid–liquid interface. This oxide layer becomes a nucleation site and causes formation of large size pores.

### 4.2.2 Pore Growth

The pore growth model is considered in the copper–hydrogen system on the basis of the idea that the gas pores are formed by the hydrogen released at the solidification front owing to the solubility difference between liquid and solid copper and by the hydrogen diffused out from the solid copper due to the decrease in hydrogen

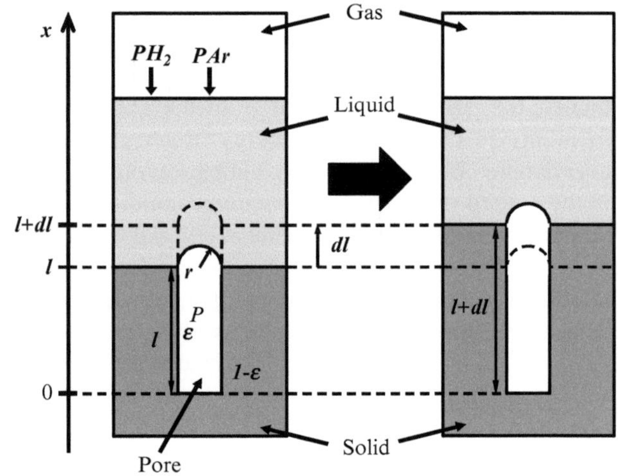


**Fig. 4.2** Schematic drawing for unidirectional solidification process of liquid under the pressure of hydrogen $P_{H2}$ and argon $P_{Ar}$. The long pore grows upward into which the gas atoms dissolving in the solid and liquid diffuse (Reprinted with permission from [5]. © 2001 Elsevier Science B.V.)

solubility of solid copper with deceasing temperature [5]. The following assumptions are made to develop the model:

1. The solidification front advances at a constant velocity of the interface between liquid and solid.
2. The solubilities of argon in liquid and solid copper are negligible.
3. The hydrogen concentration in solid copper is constant at a given distance from the solidification front and is equal to that at the inner surface of the pore.
4. The temperature gradient in the ingot is constant.

Figure 4.2 shows a schematic drawing for the unidirectional solidification process of the liquid copper. While the liquid copper in the crucible is saturated with hydrogen during melting, the hydrogen partially escapes from the liquid to the atmosphere during pouring and/or solidification. Let $a$ be the ratio of the amount of the escaped hydrogen to that of the hydrogen remaining in the liquid copper. As the length of the columnar pore increases from $l$ to $l + dl$, the mass balance of hydrogen is written as follows:

(initial amount of hydrogen contained in the liquid copper in the volume element)
+ (amount of hydrogen contained in the liquid copper flowing into the volume element to compensate for solidification shrinkage)
− (amount of hydrogen contained in the liquid copper out of the volume element due to the formation of pores)
+ (amount of hydrogen in the pore of a length of $l$)
+ (amount of hydrogen dissolved in the solid copper around the pore length of $l$)
= (amount of hydrogen in the pore of a length of $l + dl$)
+ (amount of hydrogen dissolved in the solid copper around the pore of length of $l + dl$)

$$\rho_L(l \cdot dl)\frac{C_m}{a+1} + \rho_L\{\beta(1-\varepsilon)(l \cdot dl)\}\frac{C_m}{a+1} - \rho_L\{\varepsilon(l \cdot dl)\}\frac{C_m}{a+1} + W_l + S_l = W_{l+dl} + S_{l+dl}, \quad (4.4)$$

where $\rho_L$, liquid density, $8.00 \times 10^3$ kg m$^{-3}$; $\rho_s$, solid density, $8.40 \times 10^3$ kg m$^{-3}$; $\beta$, solidification shrinkage of copper, 0.0476; $C_m$, hydrogen concentration of copper melt in equilibrium with the mixture gas of argon and hydrogen at the holding temperature; $T_m$, holding temperature of copper melt in the mixture gas of argon and hydrogen (1,523 K); $W_l$, mass of hydrogen in the pore of length $l$; $W_{l+dl}$, mass of hydrogen in the pore of length $l + d$; $S_l$, mass of hydrogen dissolved in solid copper around the pore of length $l$; and $S_{l+dl}$, mass of hydrogen dissolved in solid copper around the pore length $l + dl$.

First, the amount of hydrogen in the pore was calculated. Imagine a vertical copper cylinder that has a cylindrical bore along its center and has a length of $l$ as shown in Fig. 4.2. Cross-sectional area is unity for the copper cylinder, whereas it is $\varepsilon$ for the cylindrical bore. This means that the porosity of the cylinder is equal to $\varepsilon$. The bore is filled with hydrogen at a pressure of $P$. The temperature at the top end of the cylinder corresponds to the nonvariant temperature $T_n$ at which the reaction of liquid copper→solid copper + gaseous hydrogen occurs. When the vertical temperature profile in the cylinder is described as $T = G \cdot x + (T_m - G \cdot l)$, the mass of hydrogen $dw$ in a small volume element of the bore with a length of $dx$ is given by

$$dw = \frac{M \times P(\varepsilon dx)}{R\{Gx + (T_n - Gl)\}}, \quad (4.5)$$

where $M$, molecular weight of hydrogen, 2.016 kg mol$^{-1}$; $R$, gas constant; and $G$, temperature gradient in the bore. The total mass of hydrogen $W$ in the bore with a length of $l$ is obtained by integrating Eq. (4.5) from $x = 0$ to $x = l$; thus,

$$W = \int_0^l dw = \frac{MP\varepsilon}{RG} \ln\left(\frac{T_n}{T_n - Gl}\right). \quad (4.6)$$

The increase in the mass of hydrogen $dW$ in the bore resulting from its growth by $dl$ is given by

$$dW = \frac{dW}{dl}dl = \frac{MP\varepsilon}{R(T_n - Gl)}dl. \quad (4.7)$$

Next, the amount of hydrogen in the solid phase of copper around the pores was calculated. Since the solubilities of argon in the solid and liquid phases of copper are negligibly small, the principal gas species in the pore can be assumed to be hydrogen. The hydrogen concentration in the solid phase of copper $C_s$ is written as

$$C_s(T) = \eta(T)\sqrt{P}. \quad (4.8)$$

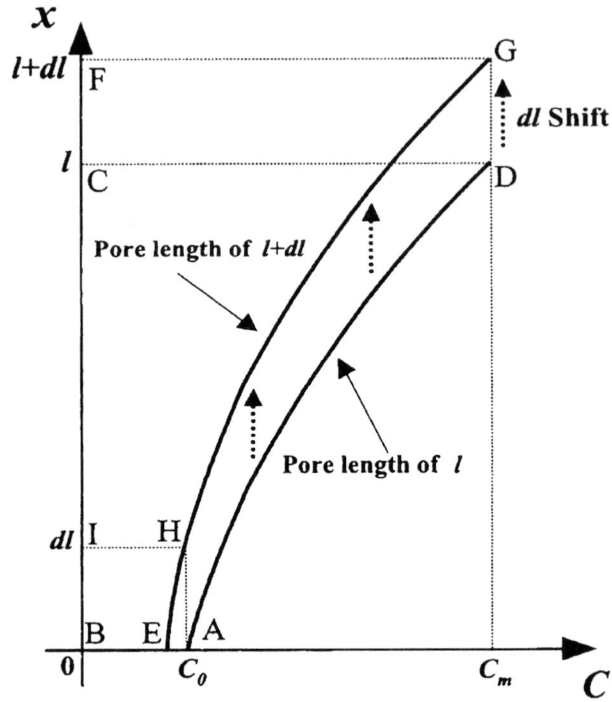


**Fig. 4.3** Distributions of hydrogen concentration in solid copper with pore length of $l$ and $l + dl$ (Reprinted with permission from [5]. © 2001 Elsevier Science B.V.)

$\eta(T) = 4.34 \times 10^{-7} \exp(-5.888 \times 10^3/T)$ wt. fraction $\mathrm{Pa}^{-1/2}$. The amounts of hydrogen $S_l$ and $S_{l+dl}$, in the solid phase of copper around the pores of length of $l$ and $l + dl$ can be given as

$$S_l = \rho_s(1 - \varepsilon) \int_0^l C_s(T)dx \tag{4.9}$$

and

$$S_{l+dl} = \rho_s(1 - \varepsilon) \int_0^{l+dl} C_s(T)dx, \tag{4.10}$$

respectively. Figure 4.3 shows the variation of the hydrogen concentration in the solid phase with the distance in the solidification direction for pore lengths $l$ and $l + dl$. Since the area ABCD equals the area HIFG, the difference in the mass of hydrogen $S_{l+sl}$–$S_l$ corresponds to the area EBIH. For a small length of $dl$, the area EBIH is approximately equal to the area of rectangle ABIH. Therefore,

$$S_{l+dl} - S_l = \rho_s dl \cdot C_0. \tag{4.11}$$

Substituting Eqs. (4.7) and (4.11) in Eq. (4.4), one finally obtains

$$P = \left[ \frac{-\rho_s(1-\varepsilon)\eta(T_m) + \sqrt{\rho_s^2(1-\varepsilon)^2 \eta_{(T_n - Gl)^2} - [4 \times M\varepsilon/R(T_n - Gl)]\left[\rho_L(1+\beta)(1-\varepsilon)\xi_{(T_m)}\sqrt{P_{H_2}}\right]/(a+1)}}{2 \times M\varepsilon/R(T_n - Gl)} \right]^2. \tag{4.12}$$

Assuming that the hydrostatic pressure of the copper melt is negligible, the pressure $P$ in the pore is given by the sum of the pressure in the chamber and the capillary pressure arising from the surface tension of the copper melt; thus,

$$P = P_{H_2} + P_{Ar} + P_r, \tag{4.13}$$

where $P_{Ar}$ is argon pressure in the chamber, and $P_r = 2\sigma/r$, where $r$ is the radius of the pore "cap" at the solidification front (assumed to be equal to the pore radius), and $\sigma$ is the surface tension of copper melt.

The relationship between the pressure of hydrogen and argon is expressed as

$$P_{Ar} = \left[ \frac{-\rho_s(1-\varepsilon)\eta_{(T_m)} + \sqrt{\rho_s^2(1-\varepsilon)^2 \eta_{(T_n - Gl)^2} - [4 \times M\varepsilon/R(T_n - Gl)]\left[\rho_L(1+\beta)(1-\varepsilon)\xi_{(T_m)}\sqrt{P_{H_2}}\right]/(a+1)}}{2 \times M\varepsilon/R(T_n - Gl)} \right]^2 . - \left(P_{H_2} - \frac{2\sigma}{r}\right) \tag{4.14}$$

The solid lines and the dotted lines in Fig. 4.4 show the results calculated from equation for the combination of the hydrogen pressure and the argon pressure to give various porosity $\varepsilon$. In the calculation, the value $a$ is assumed to be 0.52. This means that 34.3 % $(= 0.52/(1 + 0.52) \times 100)$ of the hydrogen dissolved in the copper melt at the melting temperature of 1,523 K escapes from the melt to the atmosphere during pouring and/or solidification and, therefore, that the copper melt to form the porous copper has the composition at which the nonvariant reaction occurs. Hence, in the calculation, only the hydrogen dissolved in the melt of the nonvariant composition is assumed to be incorporated in the ingot. Under such condition, the escape factor $a$ is not a function of the hydrogen pressure, but is dependent only on the holding temperature. When the holding temperature is given, the amount of hydrogen incorporated in the ingot is determined by the applied hydrogen pressure. Since the pore is filled with hydrogen whose pressure is equal to the sum of the total pressure in the chamber and the capillary pressure, the porosity decreases with increasing argon pressure at a given hydrogen pressure. This prediction is in agreement with the experimental results indicated in Fig. 4.4. The amount of hydrogen dissolved in the melt is proportional to the square root of the applied hydrogen pressure, whereas the pressure in the pore is proportional to the total pressure in the chamber. At a given argon pressure, therefore, the porosity

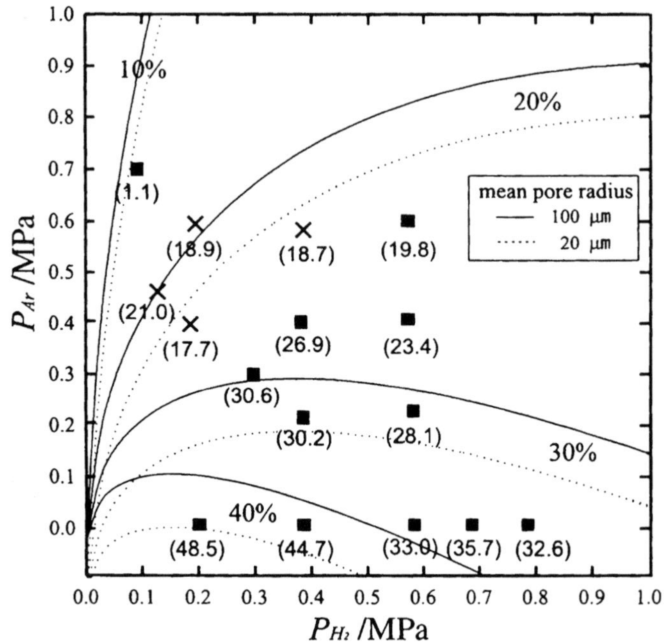


**Fig. 4.4** Porosity map of lotus copper. The numerals in the figure are the measured porosity. The *lines* are the calculated results for the combination of the hydrogen pressure and argon pressure to give various porosity values ε. In the calculation, the holding temperature of the copper melt and the escape factor are 1,523 K and 0.52, respectively (Reprinted with permission from [5]. © 2001 Elsevier Science B.V.)

first increases with increasingly applied hydrogen pressure and then decreases at higher hydrogen pressure due to densification of hydrogen in the pore as shown in Fig. 4.4.

Figure 4.5 shows the calculation for the dependence of the porosity on the holding temperature $T_m$. In the calculation, the escape factor $a$, the porosity $\varepsilon$, and the mean pore radius $r$ are assumed to be 0.52, 20 %, and 50 μm, respectively. The hydrogen solubility in the copper melt increases with increasing holding temperature, resulting in increase in the porosity. It is also known from Fig. 4.5 that the porosity decreases as the melting temperature of the copper decreases for a given set of partial pressure of hydrogen and argon and for the factor of $a = 0.52$. For the purpose of avoiding this decrease in the porosity, it is necessary to reduce the partial pressure of argon. When $T_m$ is kept constant at 1,523 K, but the escape factor $a$ is varied, the relation between the argon pressure and the hydrogen pressure to give specific porosity values is calculated from Eq. (4.14). The result is shown in Fig. 4.6. As the escape factor $a$ decreases, the porosity increases because of increasing hydrogen amount that generates pores. The value $a$ is associated with various parameters and also depends on $T_m$. The factor $a$ smaller than 0.52

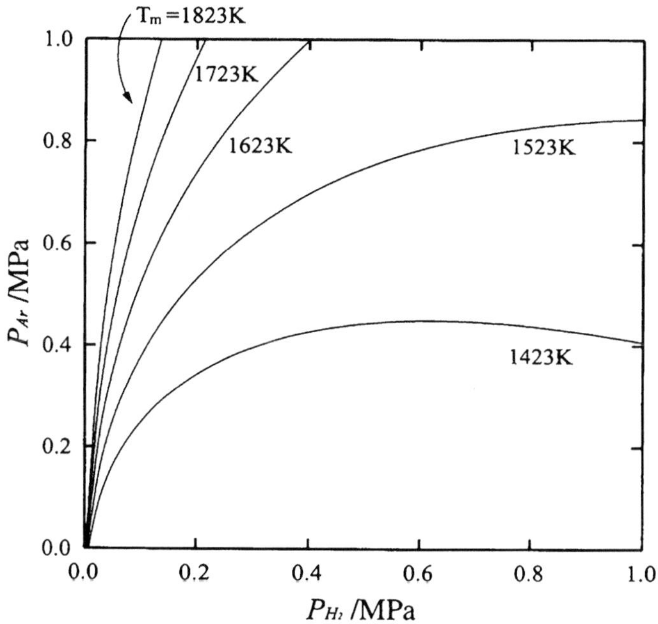


**Fig. 4.5** Calculated contribution of argon partial pressure and hydrogen partial pressure to give a porosity of 20 % for various holding temperatures of the copper melt. Escape factor and mean pore radius are 0.52 and 50 μm, respectively (Reprinted with permission from [5]. © 2001 Elsevier Science B.V.)

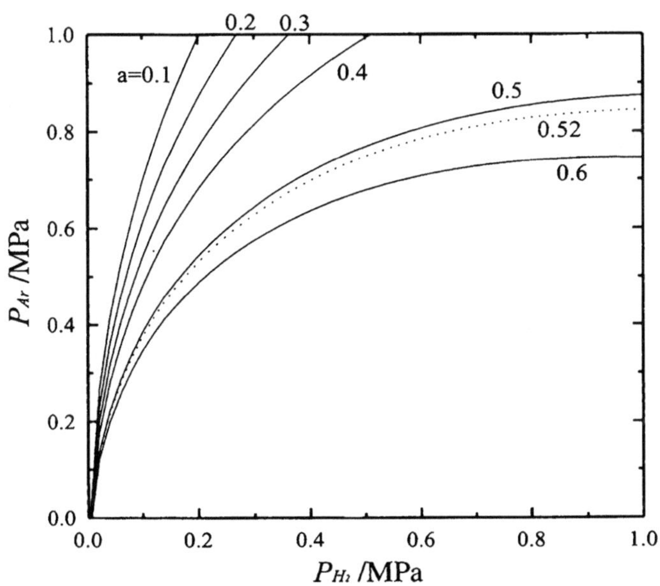


**Fig. 4.6** Calculated combination of argon partial pressure and hydrogen partial pressure to give a porosity of 20 % for various escape factors. Holding temperature of the melt and mean pore radius are 1,523 K and 50 μm, respectively (Reprinted with permission from [5]. © 2001 Elsevier Science B.V.)

corresponds to the situation where the gas phase appeared in the liquid phase just ahead of the solidification front incorporated in the solid phase.

For derivation of Eq. (4.14), the solidification is assumed to be under a steady state, but the pore nucleation is not considered. As the undercooling of the melt in the first stage of solidification will vary depending on the holding temperature $T_m$, the cooling rate, etc., the nucleation rate may change. Further investigations are necessary to elucidate the effect of the undercooling on the porosity.

Drenchev et al. [6] and Liu et al. [7] investigated a model of porosity formation and spatial distribution of pores in lotus metals.

## 4.3 Model Experiment on Unidirectional Solidification of Water Containing Carbon Dioxide

The formation of the gas pores (or blowholes) in castings and ingots has been extensively studied [4, 8–10], since the gas pores have been regarded as defects that degrade the mechanical properties of the metal products. For the purpose of studying the formation of gas pores, model experiments were widely carried out using water–gas system, and the nucleation and the growth of gas bubbles were directly observed in the course of solidification. Chalmers described the dependence of bubble formation on the freezing rate in water–air solution [11]. When the freezing is slow, more air diffuses into the bubbles from the surrounding water and they grow larger, while in the case where the growth rate is high, there is less time for diffusion and the cross section of the bubble decreases. Very slow freezing permits the air rejected at the ice/water interface to diffuse away from the interface and neither bubbles nor long columnar pores appear. It was reported [12, 13] that in the solidification of the water–air system and the water–carbon dioxide system in nonsteady conditions, the shape of the gas pores, most of which are short pores rather than long columnar pores, depends on the growth rate of ice growing from the chill. In a unidirectional solidification of water dissolving air at controlled growth rates, Geguzin and Dzuba [14] showed that the air bubbles nucleated at the advancing solidification front grow into long columnar pores when the displacement rate of the pore cap is equal to the growth rate of the solid. They observed the periodic formation of elliptical pores along the solidification direction, which they attributed to the alternative accumulation and "drop down" of the liquid concentration near the solidification front.

For this purpose of simulating the formation of gas pores in lotus metals, it is preferable to unidirectionally solidify water–gas solutions at a constant rate and to use a gas having large solubility in water since the size and the pore density can be changed over a wide range by changing the concentration of the gas. The formation of gas pores in water–carbon dioxide system solidifying at constant rates is directly observed to clarify the factors governing the morphology and the distribution of the gas pores by Murakami and Nakajima [15, 16].

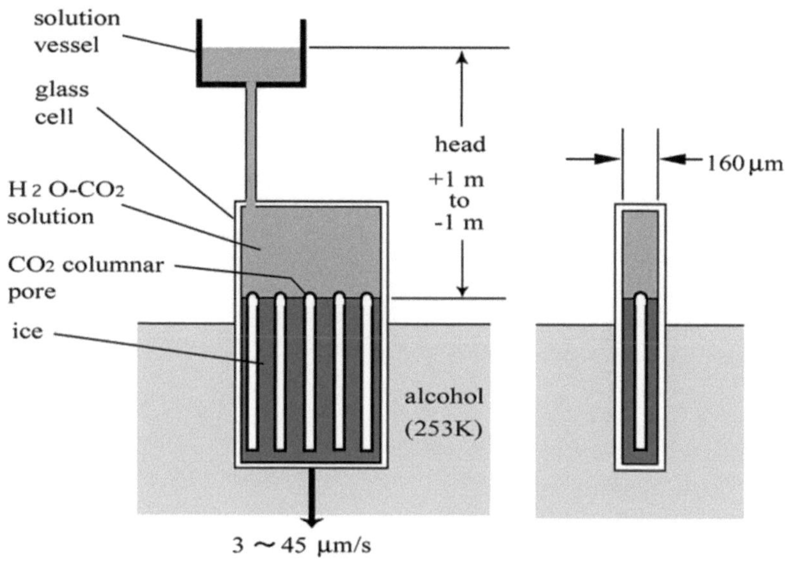


**Fig. 4.7** Schematic drawing of setup of model experiment using water–carbon dioxide solution saturated with carbon dioxide. The glass cell filled with water–carbon dioxide solution is lowered immersing into alcohol at 253 K. The gas pores are formed at the solidified ice (Reprinted with permission from [15]. © 2002 Japan Institute of Metals)

Water–carbon dioxide solution saturated with carbon dioxide was prepared by dissolving carbon dioxide in deaerated distilled water at 293 K and at atmospheric pressure [17]. The solution in the glass cell was solidified unidirectionally upwards by moving the glass cell down into an alcohol bath held at 253 K as shown in Fig. 4.7. The moving velocity of the glass cell was in the range from 3 to 45 μm $s^{-1}$. As the solidification proceeded, the pores of carbon dioxide grew in the solidification direction with ice. The experimental setup was placed in a plastic desiccator filled with flowing nitrogen to keep the low-temperature surface of the glass cell from being misted by humidity of the air. The nucleation process and growth process of the pores were observed directly by using a digital microscope through the glass window of the desiccator. The growth rate of ice was the same as that of the moving rate of the glass cell as long as the solidification front was observed under the microscope.

The morphology of the carbon dioxide pores depends on the growth rate of ice as shown in Fig. 4.8 in the case of the solution with the degree of saturation of 1/2. At the growth rate of 3 μm $s^{-1}$, spherical pores that nucleated at the solidification front grow into long columnar pores elongated in the solidification direction as shown in Fig. 4.8a. The solidification front is "planar" except that it is concave toward the solidification direction and that grain boundary grooves exist on the solidification front. As the growth rate increases, the length and the diameter of the columnar pores decrease and, at the growth rate of 12 and 18 μm $s^{-1}$, only very

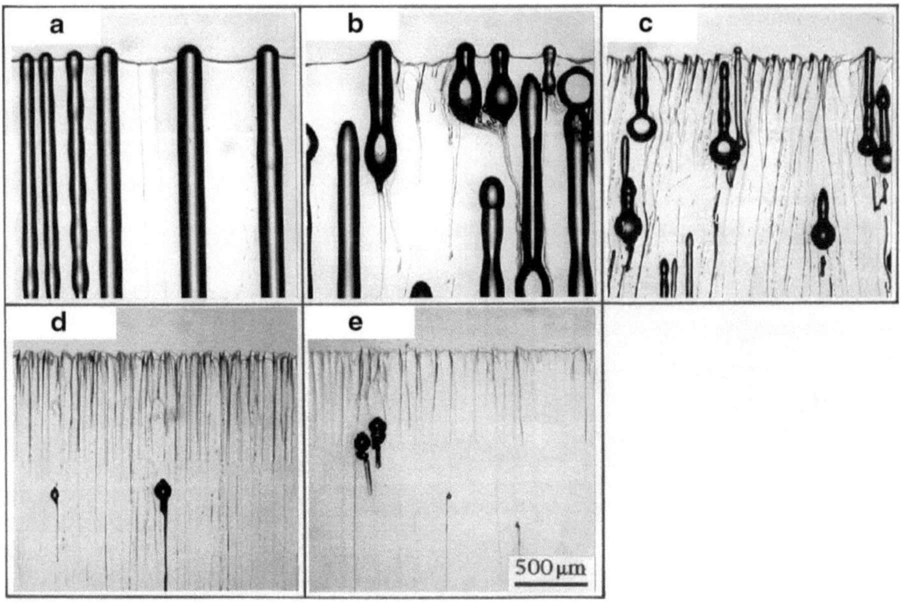


**Fig. 4.8** Effect of the growth rate on the pore morphology in ice. The degree of saturation of the solution is 1/2. The transfer velocity is (**a**) 3 μm s$^{-1}$, (**b**) 6 μm s$^{-1}$, (**c**) 9 μm s$^{-1}$, (**d**) 12 μm s$^{-1}$, and (**e**) 18 μm s$^{-1}$ (Reprinted with permission from [15]. © 2002 Japan Institute of Metals)

short columnar pores or chestnut-shaped pores are formed as shown in Fig. 4.8d, e. The morphology of the solidification front changes from planar one to cellular one with increasing growth rate due to the constitutional supercooling that occurs in the liquid adjacent to the solidification front [18].

## 4.4 Influence of Ultrasonic Agitation on Pore Morphology during Unidirectional Solidification of Water Containing Carbon Dioxide

Tane and Nakajima studied the influence of the agitation of liquid on pore formation and growth during unidirectional solidification [19]. In situ observation during solidification is useful to study the pore formation and growth. For metals, however, such observation is difficult, because pores are formed inside metals. Thus, model experiments were widely carried out using water–gas system. They also adopted water–carbon dioxide model system. The piezoelectric transducer, attached to the glass cell with an adhesive, excites ultrasonic vibration of 44 kHz in water–carbon dioxide solution; the magnitude (amplitude) of the ultrasonic

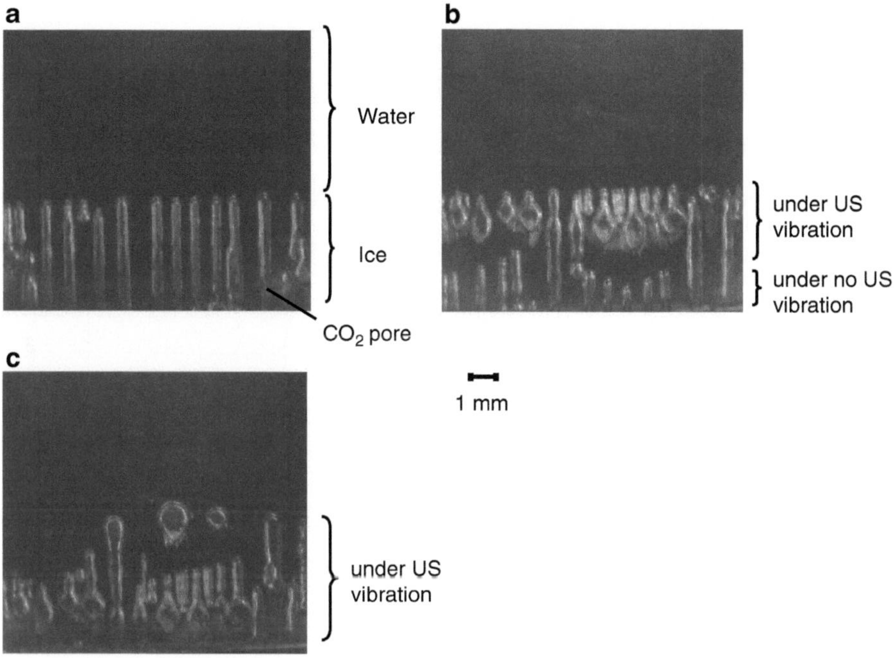


**Fig. 4.9** Morphologies of pores formed during unidirectional solidification of water–carbon dioxide solution when the moving velocity of glass cell is 3.0 μm $s^{-1}$: (**a**) No ultrasonic vibration is applied (the electric voltage applied to a piezoelectric transducer is 0 $V_{p-p}$), and (**b**) and (**c**) ultrasonic (US) vibration is applied to the solution (the voltage applied to the transducer is 20 $V_{p-p}$). The application of ultrasonic vibration to the solution was started just after the photography of (**a**). After 960 and 1,500 s, photographs (**a**) and (**b**) were taken, respectively (Reprinted with permission from [19]. © 2006 Japan Institute of Metals)

vibration is proportional to the electric voltage of alternate current signal applied to the transducer (the energy of vibration is proportional to the square of the amplitude of vibration). It is estimated that longitudinal vibration mainly is transmitted to water because transverse vibration cannot transmit in water. The solution in the glass cell solidifies unidirectionally upwards by moving the glass cell down at 3.0 μm $s^{-1}$ into an alcohol bath cooled at 253 K with a chiller. Pore formation and growth under the agitation by ultrasonic vibration were observed with a digital microscope.

Figure 4.9a shows a morphology of pores formed during unidirectional solidification when no ultrasonic vibration is applied. Cylindrical pores grow in the moving direction of the glass cell. Figure 4.9b, c shows morphologies of pore formed under ultrasonic vibration by piezoelectric transducer applied by electric voltage of 20 $V_{p-p}$. The lower pores in Fig. 4.9b are formed when no ultrasonic vibration is applied, while the upper pores are formed under ultrasonic vibration. The application of ultrasonic vibration results in the halt of pore growth. After this

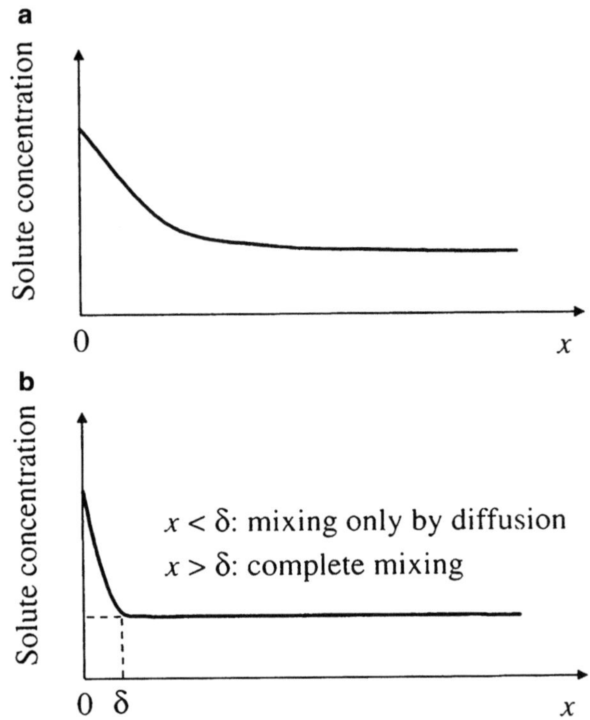


**Fig. 4.10** Schematic illustration showing the mean solute concentration in liquid near the solid–liquid interface during unidirectional solidification of a binary solution with planar solidification front: (**a**) agitation of liquid is not present and (**b**) liquid is agitated by disturbance (Reprinted with permission from [19]. © 2006 Japan Institute of Metals)

halt, new pores nucleate and grow. All the pores in Fig. 4.9c are formed under ultrasonic vibration. Flask-shaped pores are formed under the agitation by ultrasonic vibration.

The supply quantity of carbon dioxide depends on the concentration profile of carbon dioxide in the water near the solid–liquid interface. When the convection of water exists during solidification of binary solution, the concentration profile of solute in the front of the interface can be modeled by using the stagnant layer model [9, 18], which was applied to discuss the influence of agitation. When the amount of carbon dioxide absorbed by pores is balanced with the amount of carbon dioxide supplied with advance in solidification, the steady-state growth of pores is achieved and cylindrical pores are formed. On the other hand, insufficient supply of carbon dioxide prevents the steady-state growth of pores and results in the halt of pore growth. Thus, the sufficient supply of carbon dioxide is important for the steady-state growth of cylindrical pores. Figure 4.10 shows schematic illustrations exhibiting mean solute concentration in the liquid near the solid–liquid interface with and without agitation. The mean concentration with agitation drastically decreases with increasing distance in growth direction $x$. For $x > \delta$, liquid is agitated and the solute concentration is constant. For $x < \delta$, liquid is not agitated owing to the viscosity of liquid, and in this layer, the mass transport is due to only diffusion. The δ decreases with increase in the magnitude of agitation.

When ultrasonic vibration is applied to water–carbon dioxide solution, its concentration near solid–liquid interface decreases. Thus, the supply of carbon dioxide becomes insufficient for the steady-state growth of cylindrical pores and the growth of pores stops as shown in Fig. 4.9b. During the halt of pore growth, the solute concentration increases with solidification because carbon dioxide rejected with advance in solidification is not used for the growth of pores. Then, the concentration of carbon dioxide reaches a critical value and pores nucleate. Just after the nucleation, the pores grow by absorbing the carbon dioxide accumulated during the halt so that the diameter of pores at the start of growth is large. However, the carbon dioxide supplied with advance in solidification is not sufficient to maintain the initial diameter of pores owing to the decrease of carbon dioxide near the solid–liquid interface by ultrasonic agitation. Therefore, the diameter of pores decreases with advance in solidification and the flask-shaped pores are formed as shown in Fig. 4.9b.

## 4.5 Evolution of Spherical Pores during Foaming Process

In lotus metals fabricated gasar process, the directional pores are evolved by nucleation and growth of insoluble gas during solidification. Here, it is worthwhile to show growth mechanism of foamed metals during foaming process when the molten metal dissolving gas is solidified. This mechanism is due to the swelling of the isotropic spherical bubbles, which is clearly different from that of pores in lotus metals.

The evolution of the cell structure of foams produced by gas bubble expansion in a molten metal is numerically simulated by the Lattice Boltzmann Method by Korner et al. [20]. The decomposition of the blowing agent is modeled by a constant homogeneous volume source of hydrogen. The hydrogen is dissolved in the liquid aluminum and diffuses to stochastically distributed bubble nuclei already present at the beginning of foaming. The gas flow into bubbles and the surrounding atmosphere is determined by the diffusion equation and Sieverts law. Movement of the melt is given by the Navier–Stokes equations. Figure 4.11 shows the temporal evolution and subsequent collapse of the cellular structure. From the very beginning, foam expansion is accompanied by bubble coalescence if the density of the initial bubble nuclei is high enough. Coalescence is depicted in more detail in Fig. 4.12 and compared with microcomputer tomography (μCT) pictures of real foams. The number of bubbles decreases via bubble coalescence during foam expansion. If the initial density of nuclei is high, then substantial coalescence takes place during expansion from the very beginning. The maximum expansion factor is determined by two mechanism of gas loss. The first one is gas loss by diffusion to the foam surface. The contribution of this effect depends on the foam surface to volume ratio, the ambient pressure and atmosphere, and the foaming velocity. For samples smaller than 5 mm and foaming times larger than some minutes, this kind of gas loss can be substantial and suppresses expansion.

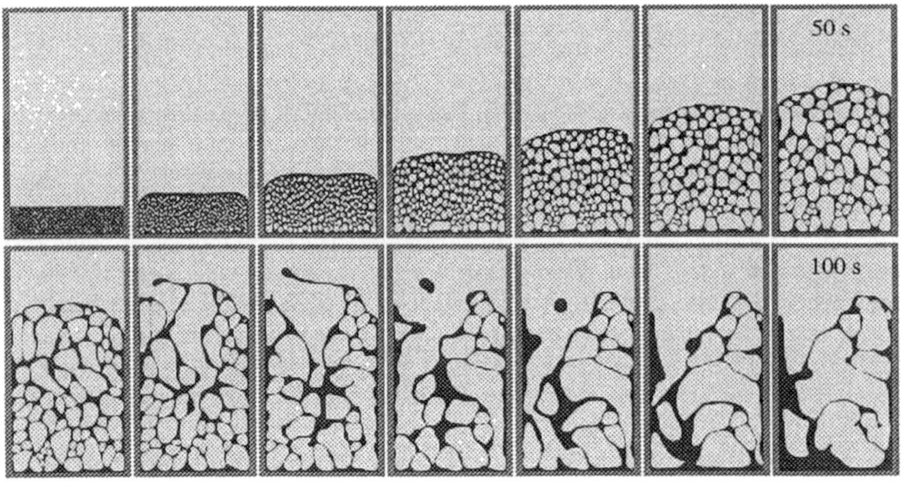


**Fig. 4.11** Temporal evolution of the foam structure. Growth of bubble nuclei, bubble coalescence, gas loss to the surrounding by bursting bubbles on the foam surface, and eventually, foam collapse (Reprinted with permission from [20]. © 2001 MIT-Verlag)

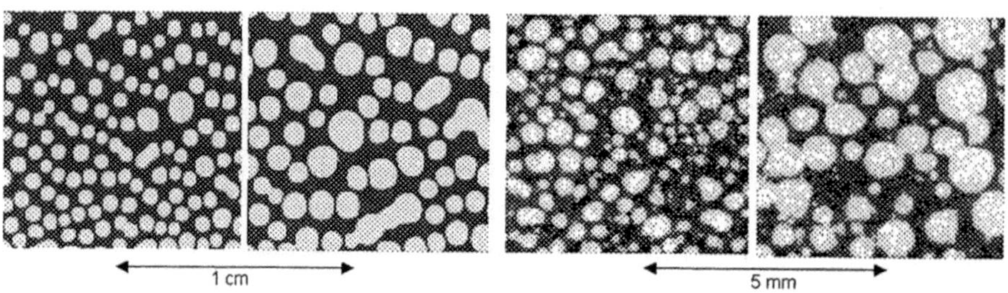


**Fig. 4.12** Coalescence. From the very beginning, foam growth is intimately correlated with bubble coalescence. (*Left*) Simulated structures for small expansion ratios (600 initial nuclei, parameter as in Fig.4.11). (*Right*) μCT pictures of a Mepura wrought alloy for two different expansion states. The samples were heated up under an external pressure of 100 bar and were subsequently foamed by decreasing the pressure (Reprinted with permission from [20]. © 2001 MIT-Verlag)

The second origin of gas loss is bursting cell walls located at the foam surface (see Fig. 4.11). In this case, the cell disappears and its gas constant is completely lost. As long as the cell size is small, this effect is insignificant. It starts to become important when the cell size reaches the order of magnitude of the foam dimension. In this case, the amount of gas lost to the environment is higher than the expansion volume itself. As a result, the foam collapses.

The in situ observation of the foaming process of aluminum was carried out by Banhart group by means of real-time radioscopy using 150 kV microfocus X-ray

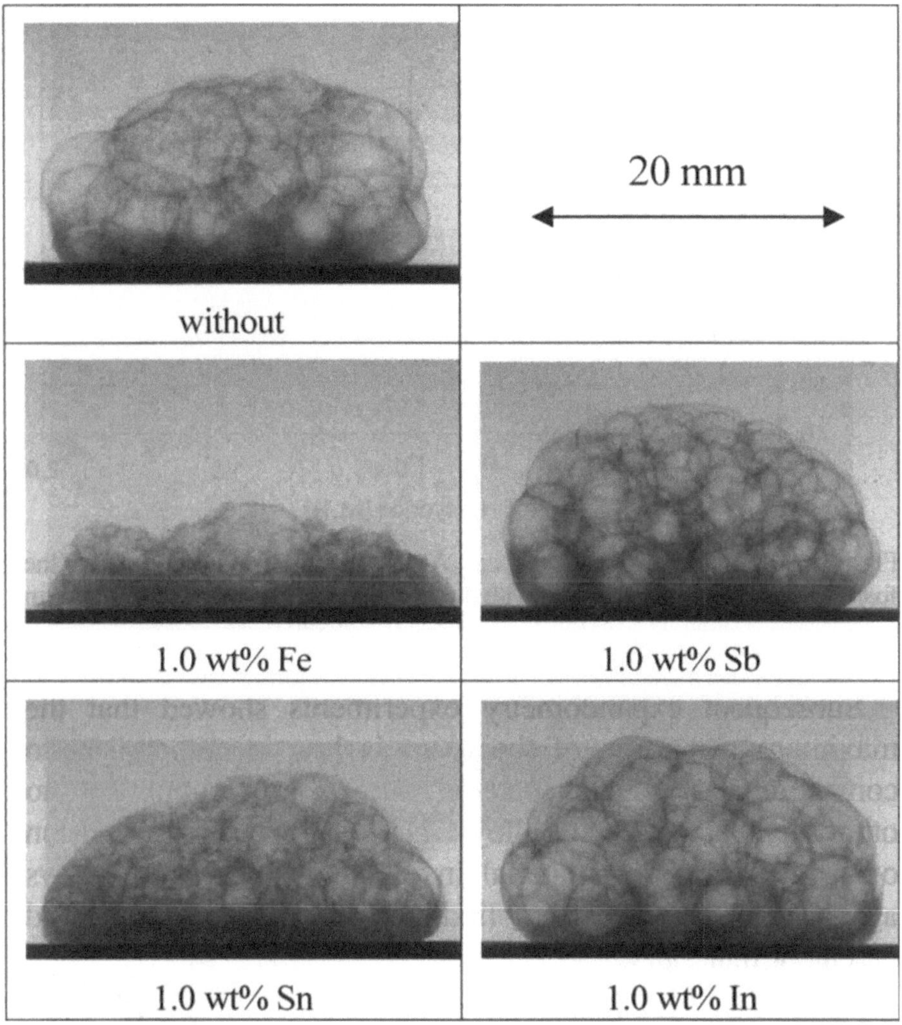


**Fig. 4.13** X-ray radiograms of AlSi11 foams with different alloying elements (Fe, Sn, Sb, and In) at 200 s after start of heating, foaming without mold (Reprinted with permission from [21]. © 2005 Japan Institute of Metals)

source with 5 μm spot size [21]. Although the micrographs are not a temporal evolution of the cellular structure, Fig. 4.13 shows examples of X-ray radiograms of AlSi11 foams with different additional alloying elements. 200 s after the heating, it foamed without mold. Obviously, there is drastic collapse of the sample with Fe, followed by the one with Sn. On the other hand, the samples with Sb and In are more homogenous than the sample without. After 200 s, no significant differences in drainage between the samples can be found.

## References


1. Fredriksson H, Akerlind U (2006) Materials processing during casting. Wiley, Chichester, pp 258–259
2. Shapovalov VI (1998). In: Scharttz DS, Shih DS, Evans AG, Wadley HNG (ed) Porous and cellular materials for structural applications, Materials Research Society, Warrendale, pp 281–290
3. Fisher JC (1948) J Appl Phys 19:1062–1067
4. Campbell J (1993) Casting. Butterworth-Heinemann, Oxford, p 163
5. Yamamura S, Shiota H, Murakami K, Nakajima H (2001) Mater Sci Eng A 318:137–143
6. Drenchev L, Sobczak J, Sobczak N, Sha W, Malinov S (2007) Acta Mater 55:6459–6471
7. Wan J, Li Y, Liu Y (2007) J Mater Sci 42:6446–6452
8. Lampman S, Moosbrugger C, DeGuire E (eds) (1998) Metals handbook, 9th edn, vol 15: Casting. ASM International, Metals Park, p 82
9. Flemings MC (1974) Solidification processing. McGraw-Hill, New York, p 208
10. Fredriksson H, Svensson I (1976) Metall Trans 7B:599–606
11. Chalmers B (1959) Sci Amer 200:114–122
12. Wei PS, Kuo YK, Chiu SH, Ho CY (2000) Int J Heat Mass Transfer 43:263–280
13. Vasconcellos KF, Beech J (1975) J Cryst Growth 28:85–92
14. Geguzin TE, Dzuba AS (1981) J Cryst Growth 52:337–344
15. Murakami K, Nakajima H (2002) Mater Trans 43:2582–2588
16. Murakami K, Nakai Y, Nakajima H (2002) Int J Cast Metals Res 15:459–463
17. Handbook of chemistry (1984). In: Chem Soc Japan (ed) Pure chemistry II, 3rd edn, Maruzen, Tokyo, p 158
18. Chalmers B (1964) Principles of solidification. Wiley, New York, p 150
19. Tane M, Nakajima H (2006) Mater Trans 47:2183–2187
20. Korner C, Thies M, Arnold M, Singer RF (2001) In: Banhart J, Ashby MF, Fleck NA (eds) Cellular metals and metal foaming technology. MIT, Bremen, pp 93–98
21. Weise J, Marchetto O, Haesche M, Garcia-Moreno F, Banhart J (2005) In: Nakajima H, Kanetake N (eds) Porous metals and metal foaming technology. Japan Inst Metals, Sendai, pp 123–128

# Chapter 5
# Control of Pore Size and Porosity in Lotus-Type Porous Metals

**Abstract** The porous metals with directional elongated pores are characterized by three important parameters: the pore direction, pore size, and porosity. The pore direction can be adjusted by changing the direction of unidirectional solidification as shown before. Here we describe the way how to control the pore size and the porosity by adjusting the solidification velocity and ambient gas pressure and by addition of oxide particles.

**Keywords** Boyle's law • Gas pressure • Moisture • Pore size • Solidification velocity

## 5.1 Control of Pore Size by Solidification Velocity

The lotus copper was fabricated with the mold casting technique by the vacuum-assisted and pressurized casting apparatus consisting of a graphite crucible with a hole on the bottom of the crucible, a stopper stick for preventing the melt flow through the hole, an induction heating coil, and a mold with water-cooled copper plate. High-purity copper was melted in the crucible by radio-frequency heating under high-pressure mixture gas of hydrogen and argon. The molten copper was poured into the mold whose bottom plate was cooled down with water circulated through a chiller. The lateral side of the mold was made of alumina-coated stainless steel tube, which was suitable for insulating in order to be solidified in one direction from the bottom to the top. During the solidification, hydrogen in the melt was rejected at the solid–liquid interface due to the solubility gap of hydrogen between liquid and solid and forms cylindrical pores that were aligned parallel to the solidification direction. The samples were solidified at different solidification velocities through the condition whether a ceramic sheet was inserted between a carbon plate and a copper plate of chiller as shown in Fig. 5.1. R-type thermocouples were inserted in the mold and coupled with a computer-controlled data acquisition system for recording the cooling curves.

H. Nakajima, *Porous Metals with Directional Pores*,
DOI 10.1007/978-4-431-54017-5_5, 

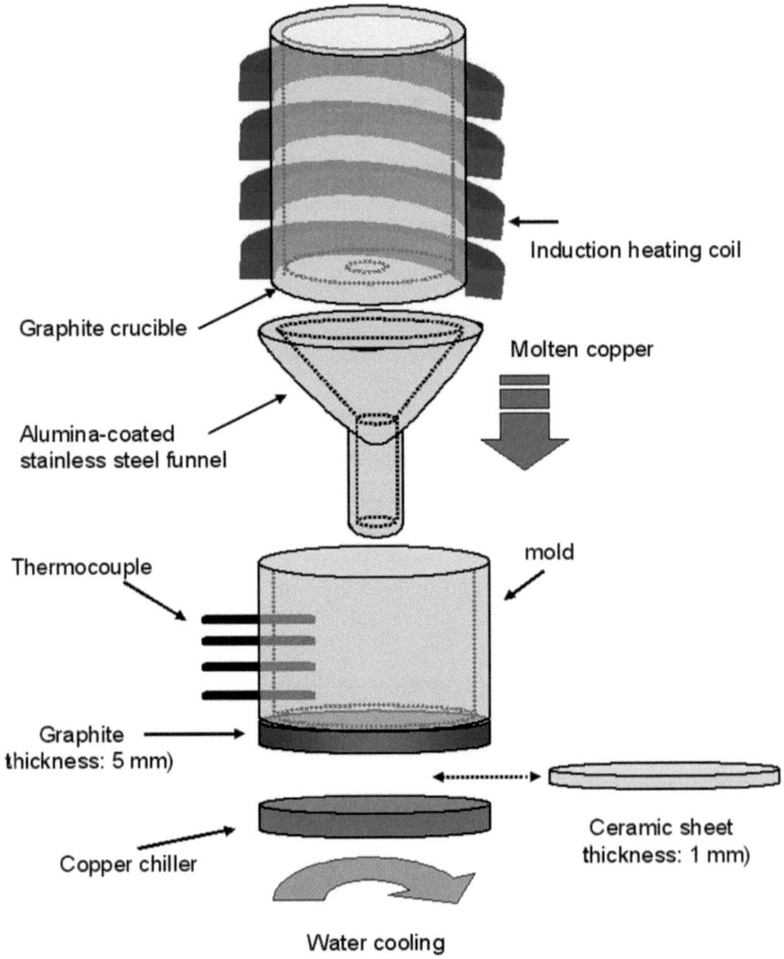


**Fig. 5.1** Schematic drawing for measurement of solidification velocity in mold casting technique. The molten metal is dropped through a funnel from the crucible into the bottom mold. Several thermocouples are set up in the vertical direction on the lateral side of the mold. In order to control the solidification velocity, ceramic sheets are inserted between the copper chiller and the mold (Reprinted with permission from [1]. © 2003 Elsevier Ltd)

It is considered that each solidification velocity is constant since the beginning time of solidification linearly increases with increasing distance from bottom of mold, in which the thermal conductivity of copper can be sufficiently high. The samples were solidified with solidification velocities of 1.185 and 0.697 mm $s^{-1}$ through different conditions of the ceramic sheet or not. The optical micrographs of the cross section of the lotus copper are shown in Fig. 5.2. It is apparent that the pore morphology is significantly affected by the solidification velocity. The average

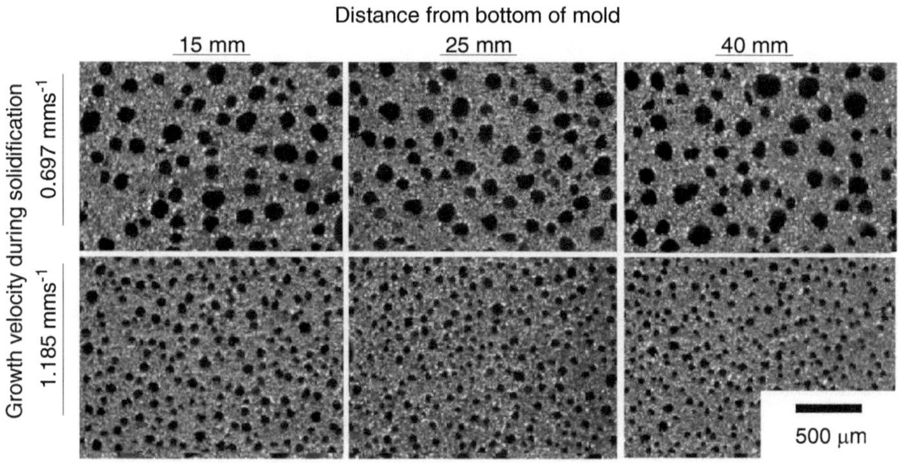


**Fig. 5.2** Microstructure of lotus copper in perpendicular section to the solidification direction. Lotus copper ingots were fabricated with two different solidification velocities by the mold casting technique. The micrographs were taken in the sections at the position from different distances from the bottom of the mold (Reprinted with permission from [1]. © 2003 Elsevier Ltd)

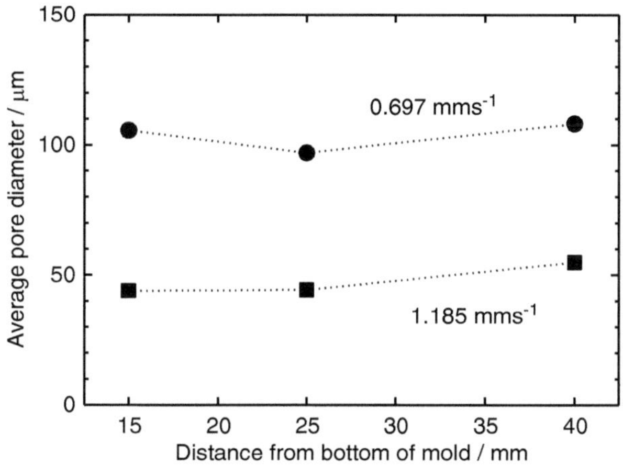


**Fig. 5.3** Average pore diameter plotted against distance from the bottom of the mold in lotus copper fabricated by mold casting technique with two different solidification velocities (Reprinted with permission from [1]. © 2003 Elsevier Ltd)

pore diameter is plotted against the distance from the bottom of the mold as shown in Fig. 5.3. The pore diameter in the lotus copper fabricated with the solidification velocity of 0.697 mm $s^{-1}$ is twice larger than that with 1.185 mm $s^{-1}$. It can be understood that the amount of hydrogen diffusing from liquid to the pores increases with decreasing solidification velocity and then the pores formed with the velocity

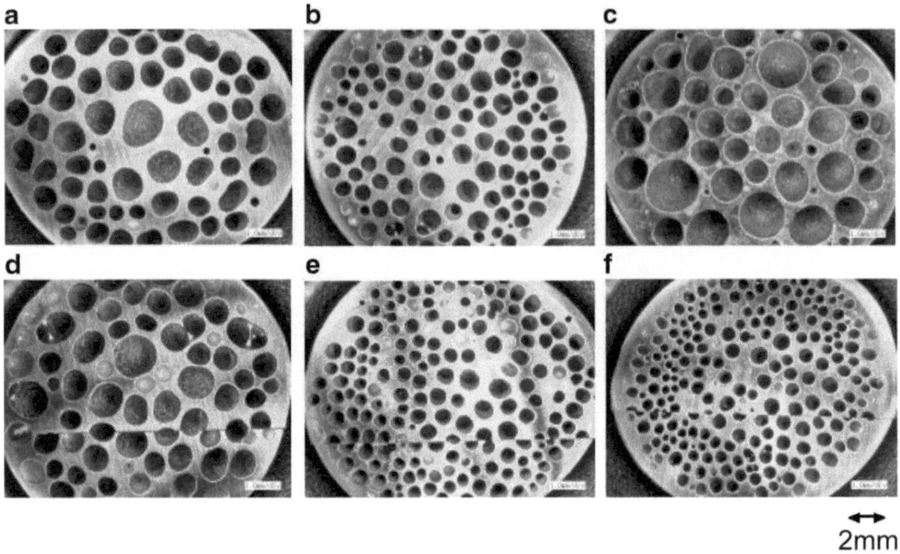


**Fig. 5.4** Cross sections perpendicular to the solidification direction of lotus stainless steel fabricated under hydrogen gas of 1.0 MPa by continuous zone melting technique. The upper figures (**a**), (**b**), and (**c**) are of the rods fabricated without gas-blow cooling and the lower figures (**d**), (**e**), and (**f**) are of those fabricated with gas-blow cooling. The transfer velocities are 160 μm $s^{-1}$ for (**a**) and (**d**), 330 μm $s^{-1}$ for (**b**) and (**e**), and 500 μm $s^{-1}$ for (**c**) and (**f**) (Reprinted with permission from [2]. © 2005 The Minerals, Metals & Materials Society and ASM International)

of 0.697 mm $s^{-1}$ become larger than those with 1.185 mm $s^{-1}$. As the solidification velocity increases, the pore size decreases but the number density of pores increases. Increasing the solidification velocity causes an increase in the hydrogen supersaturation of the solid–liquid interface and then the driving force to nucleate the pores increases. If the entire amount of hydrogen dissolved in the melt is diffused into the pores, the pore size can decrease with increasing number density of pores. Consequently it can be understood that the solidification velocity can affect the average pore diameter of lotus copper [1].

In Fig. 5.4, the cross-sectional planes perpendicular to the solidification direction in lotus stainless steel rods fabricated by the continuous zone melting technique under the hydrogen atmosphere of 1.0 MPa without or with the gas-blow cooling gas are shown. The average pore diameter and porosity determined by image analyses are shown in Table 5.1. The average pore diameter decreases with increasing transfer velocity, while the porosities are not much different. It is found that the cooling rate is almost proportional to the transfer velocity in fabricating porous metal rods with gas-blow cooling. The cause of the smaller pore diameters in the rods fabricated with the larger velocities is considered to be due to the higher cooling rates [2]. Such tendency is consistent with the experimental results in the fabrication

**Table 5.1** Porosity and average pore diameter of lotus stainless steel fabricated under hydrogen atmosphere of 1.0 MPa by continuous zone melting technique with/without gas-blow cooling

| Transfer velocity (μm $s^{-1}$) | Gas-blow cooling | Porosity (%) | Average pore diameter (μm) |
|---|---|---|---|
| 160 | × | 52 | 890 |
| 330 | × | 49 | 710 |
| 500 | × | 64 | 1,070 |
| 160 | ○ | 59 | 1,020 |
| 330 | ○ | 48 | 550 |
| 550 | ○ | 46 | 480 |

of porous copper by the casting technique [1]. Furthermore, since the temperature gradient increases by using gas-blower, it is considered that the solidification velocity increases by the gas-blower and thus the pore size becomes smaller.

## 5.2 Control of Pore Size and Porosity by Ambient Gas Pressure

The average pore diameter and porosity of the lotus stainless steel rods that were fabricated under hydrogen and argon or helium of various pressures with the transfer velocity of 330 μm $s^{-1}$ are summarized in Fig. 5.5a, b. The average pore diameter and porosity decrease with increasing pressure under the atmosphere of only hydrogen. The concentration of hydrogen dissolving in liquid metal is proportional to the square root of hydrogen pressure according to Sieverts' law. On the other hand, the volume of gas in the pores is reciprocally proportional to the pressure of the ambient gas according to Boyle's law since the pressure of pores should be almost in balance with the pressure of ambient gas. Therefore, it is considered that the decrease in the average pore diameter and porosity with increase in pressure of hydrogen is caused by more significant influence of pore volumes than that of the solubility of hydrogen.

Under mixed gas of hydrogen and argon of a constant total pressure, the porosity increases and average pore diameter decreases with increasing partial pressure of hydrogen. Hydrogen solubility in the melt of stainless steel increases with increase of the hydrogen pressure according to Sieverts' law. The pores are composed of only hydrogen since argon is not dissolved in the melts. Since the pressure in the pores should be balanced with the total pressure, the pressure in the pores is constant regardless of the variation of the hydrogen partial pressure under a constant total pressure. Therefore, it is considered that the increase in the hydrogen solubility due to the increase of the partial pressure of hydrogen leads to the increase in porosity under a constant total pressure.

Such a pressure dependence of porosity of lotus stainless steel fabricated by the continuous zone melting technique is qualitatively consistent with that of lotus

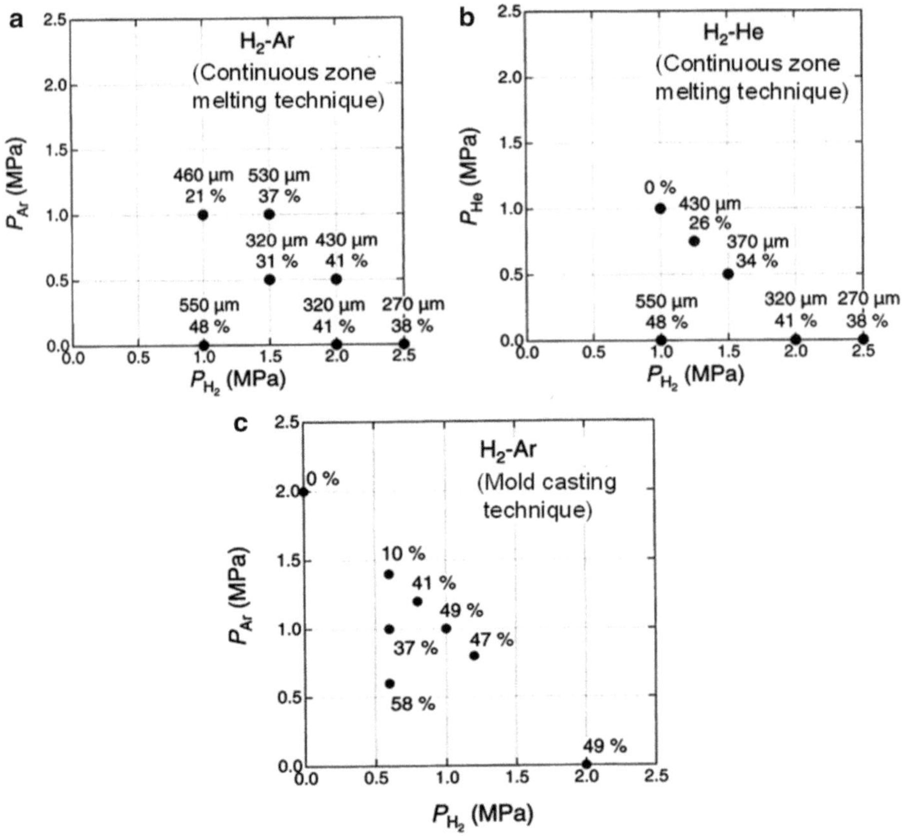


**Fig. 5.5** Average pore diameter (upper number) and porosity (lower number) of the lotus stainless steel fabricated by continuous zone melting technique with gas-blowing cooling under various pressures of mixed gases composed of (**a**) hydrogen and argon or (**b**) hydrogen and helium. These diagrams are for the cases where the transfer velocity is 330 μm $s^{-1}$. (**c**) The porosity of lotus stainless steel fabricated by mold casting technique (Reprinted with permission from [2]. © 2005 The Minerals, Metals & Materials Society and ASM International)

stainless steel fabricated by mold casting technique [3] shown in Fig. 5.5c. However, the porosity by the continuous zone melting technique is a little lower than that by the mold casting technique. It is understood that the hydrogen concentration in the melt zone may not be saturated since the hydrogen in the ambient atmosphere can be dissolved into the melt only while the metal is transferred across the melt zone of at most 10 mm. Therefore, the hydrogen concentration is a little smaller for the continuous zone melting technique. This leads to result in the smaller porosity of the lotus stainless steel fabricated by the continuous zone melting technique than that by the mold casting technique. In Table 5.1, the porosity decreases little by little with increase of the transfer velocity in the case with gas-blow cooling, comparing with the case without the cooling. This is also possibly due to the decrease of hydrogen dissolution into melt with increase of the transfer velocity.

On the other hand, the average pore diameter increases with increasing argon pressure under a constant total pressure. This is possibly related with the heat conductivity of ambient gas. The heat conductivity of hydrogen is $21.18 \times 10^{-2}$ Wm$^{-1}$ K$^{-1}$ (373 K) [4] and that of argon is $2.12 \times 10^{-2}$ Wm$^{-1}$ K$^{-1}$ (373 K) [4], the latter of which is an order of magnitude lower than that of hydrogen. Therefore, the heat conductivity of ambient gas significantly decreases with increasing partial pressure of argon and hence the efficiency of heat transmission from the metal rod to the ambient gas by gas-blow cooling also decreases. Consequently, this leads to the lowering of the cooling rate and hence to the increase in average pore diameter.

Figure 5.5b shows the pressure dependence of average pore diameter and porosity of the lotus stainless steel fabricated under mixed gas of hydrogen and helium. The porosity decreases with increasing helium pressure under a constant total pressure. This trend is almost consistent with the case where mixed gas of hydrogen and argon is used. The average pore diameter also has a trend of increase with increasing helium pressure. The heat conductivity of helium is $17.77 \times 10^{-2}$ Wm$^{-1}$ K$^{-1}$ (373 K) [4], which is a little smaller than that of hydrogen but in the same order of magnitude as hydrogen. Therefore, it is expected that the heat conductivity of ambient gas and hence the cooling rate does not much vary by increasing helium partial pressure under a constant total pressure. Nevertheless, the average pore diameter depends on the helium pressure under a constant total pressure similar to the case of hydrogen-argon gas. The reason for this is not very clear at present. Further investigations are required.

## 5.3 Control of Pore Size by Addition of Oxide Particles

Suematsu et al. proposed a fabrication method of lotus nickel using moisture during the solidification in argon atmosphere [5]. The pore size of the sample fabricated by this method is smaller than that of the sample fabricated in pressurized hydrogen atmosphere (PGM). Such smaller pore formation is attributed to formation of nucleation sites of the oxide. According to the following reaction,

$$n\ \mathrm{H_2O(gas)} + m\ \mathrm{M} \leftrightarrow \mathrm{M}_m\mathrm{O}_n + 2n\ \mathrm{H}, \tag{5.1}$$

the moisture decomposes into hydrogen and metal oxide. The former produces hydrogen pores, while the latter may serve as the nucleation sites for the pores. Although the moisture could be produced by the reverse reaction (1), the moisture ($H_2O$) itself cannot be dissolved into the molten nickel. Therefore, it is considered that the moisture does not contribute to evolve any pores in the solidified nickel. The experimental results by Suematsu et al. suggests that the pore size in lotus nickel can be controlled by the amount and/or particle size of NiO powder [5].

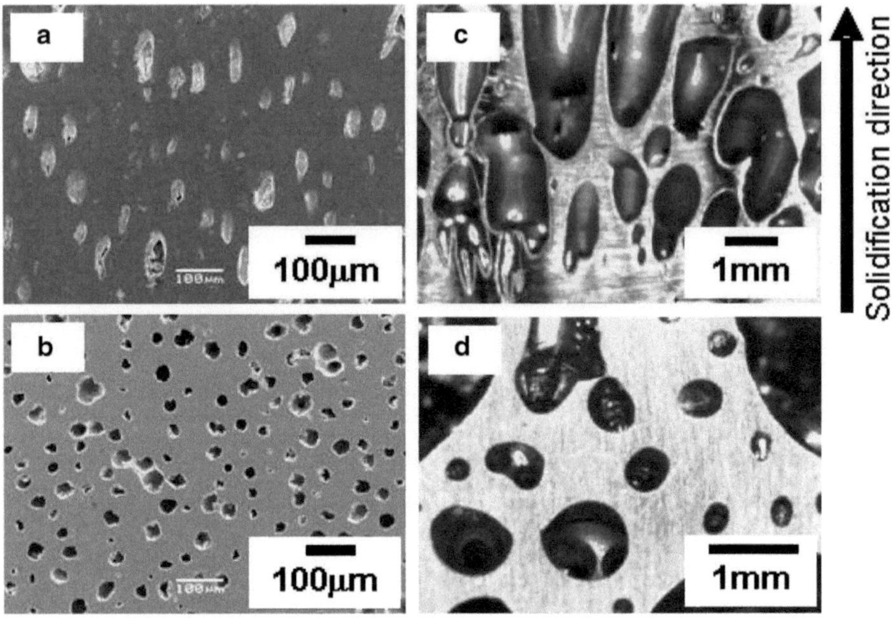


**Fig. 5.6** The longitudinal cross-sectional (**a**) and transversal cross-sectional (**b**) views of the sample fabricated in mixture gas of 0.65 MPa Ar and 0.15 MPa $H_2$ using NiO powder whose weight is 0.5 g and particle size is 7 μm. (**c**) and (**d**) show the longitudinal cross-sectional and transversal cross-sectional views of the sample fabricated in the same atmosphere without NiO powder (Reprinted with permission from [6]. © 2008 Japan Institute of Metals)

Here the effect of NiO powder on the formation of pores was examined by Onishi et al. in order to elucidate the relation among pore size in lotus nickel, the amount, and particle size of NiO powder [6]. The solidification was conducted by mold casting under pressurized gas of argon and hydrogen. 120 g of nickel ingot with 99.9 % purity was heated in alumina crucible by an induction coil and the molten nickel was poured into the mold whose bottom was made of copper and the side was made by molybdenum. The bottom of the mold was cooled by water chiller. An appropriate amount of NiO powder was put on the copper mold before casting.

Figure 5.6a, b shows the longitudinal and transversal cross-sectional views of the sample fabricated in the mixture of gas 0.65 MPa Ar and 0.15 MPa $H_2$ using NiO powder whose weight is 0.5 g and particle size is 7 μm. On the other hand, Fig. 5.6c, d shows longitudinal and transversal cross-sectional views of the sample fabricated under the same atmosphere without NiO powder. The porosity and average pore diameter calculated from Fig. 5.6 are 29 % and 31 μm for (b) and 62 % and 427 μm for (d), respectively. Apparently, the average pore diameter for (b) is smaller than that of (d). On the other hand, the aspect ratio of pores for (a), 2.2, is smaller than that of (c), 4.7. When the melt is solidified, supersaturated hydrogen in the solid diffuses into pores. If the number density of pores increases, the amount

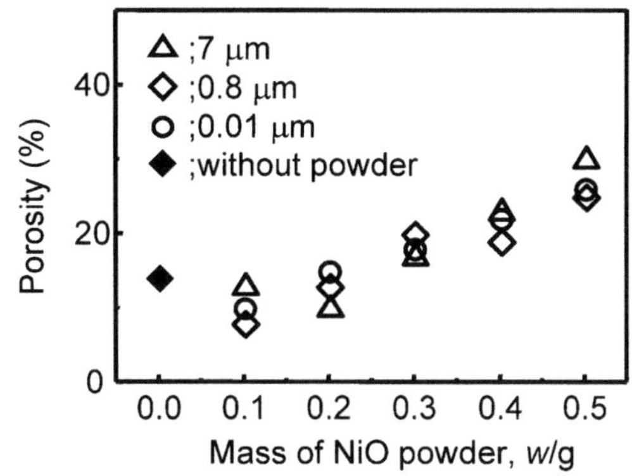


**Fig. 5.7** Dependence of porosity on amount of NiO powder with different sizes of the particles. Lotus nickel was fabricated in mixture gas of 0.85 MPa Ar and 0.15 MPa $H_2$ (Reprinted with permission from [6]. © 2008 Japan Institute of Metals)

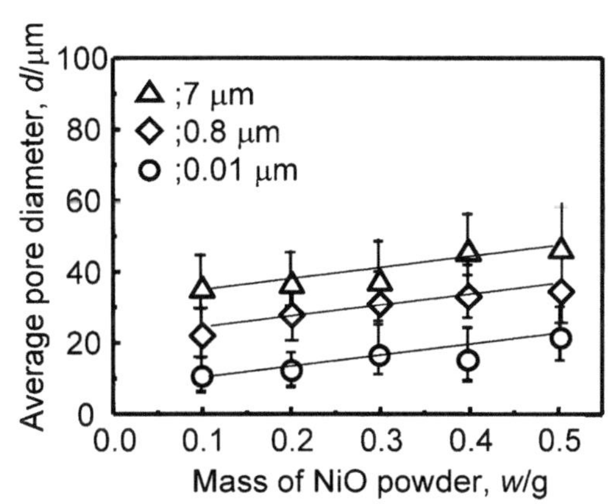


**Fig. 5.8** The average pore diameter of the samples as a function of mass of NiO powder with different sizes of the particles. Lotus nickel was fabricated in mixture gas of 0.85 MPa Ar and 0.15 MPa $H_2$ (Reprinted with permission from [6]. © 2008 Japan Institute of Metals)

of hydrogen which diffuses into each pore from the surrounding region decreases so that the pores cannot grow continuously in the direction of the solidification. Thus, the aspect ratio of the pores in the case of addition of NiO is shorter than that without NiO. Since the pore size decreases and the number of density of pores increases by addition of NiO, it is considered that the oxide powders act as nucleation sites for pores.

Figure 5.7 shows the porosity change with the amount of NiO powder. All samples in Fig. 5.7 were fabricated in the mixture of 0.85 MPa Ar and 0.15 MPa $H_2$. The porosity increases with increasing amount of NiO powder. It was suggested that about half of the hydrogen amount with the solubility gap between solid and liquid is released to the atmosphere not to contribute to the pore formation [7]. However, if the number of the pore nucleation sites increases by addition of excess NiO powder, more insoluble hydrogen may be trapped by the nucleation sites. Thus, higher porosity was observed with increasing mass of NiO powder, as shown in Fig. 5.7. However, no significant effect of the particles size of NiO powder on the porosity was found.

Figure 5.8 shows the average pore diameter of the samples as a function of the mass of NiO powder. The pore diameter decreases with decreasing particle size of

NiO powder. Since the number density of particles significantly increases with decreasing particle size when the same mass of NiO powder is added, the number of pores increases and the resulting pore size decreases as shown in Fig. 5.8. This may be attributed to the increase in the number of hydrogen atoms for formation of the pores by NiO powder addition. On the other hand, the average pore diameter monotonously increases with increasing amount of NiO powder. It is concluded that NiO powder can serve as nucleation sites for the pore formation in the process of the solidification.

## References

1. Hyun SK, Nakajima H (2003) Mater Lett 57:3149–3154
2. Ikeda T, Aoki T, Nakajima H (2005) Metall Mater Trans A 36A:77–86
3. Ikeda T, Tsukamoto M, Nakajima H (2002) Mater Trans 43:2678–2684
4. National Astronomical Observatory (ed) (2002) Chronological scientific tables. Maruzen Co. Ltd, Tokyo, p 404
5. Suematsu Y, Hyun SK, Nakajima H (2004) J Jpn Inst Metals 68:257–261
6. Onishi H, Ueno S, Nakajima H (2008) Mater Trans 49:2670–2672
7. Yamamura S, Shiota H, Murakami K, Nakajima H (2001) Mater Sci Eng A 318:137–143

# Chapter 6
# Details of Fabrication Techniques of Various Lotus Metals and Alloys, Lotus Intermetallic Compounds, Lotus Semiconductors, and Lotus Ceramics

**Abstract** Case studies are shown by different atmospheric gases and by different materials such as intermetallic compounds, semiconductors, ceramics, and magnesium and iron alloys.

**Keywords** Alumina • Aluminum • Ceramics • Czochralski technique • Intermetallic compounds • Iron • Silicon • Solid-state diffusion

## 6.1 Fabrication of Lotus Iron by Nitrogen

So far hydrogen has widely been used as a dissolving gas to fabricate several kinds of porous metals such as lotus metals. However, since the hydrogen gas is inflammable and explosive, its use is not convenient from the industrial point of view. Use of other gases except hydrogen would be desirable. It is well known that nitrogen is an important alloying element widely used to improve corrosion resistance and mechanical properties of steels [1]. Hyun and Nakajima noticed that the temperature dependence of nitrogen solubility in solid and liquid of iron is similar to that of hydrogen [2], which exhibits large nitrogen solubility difference between solid and liquid iron at the melting temperature [3]. Moreover, it is noticed that an invariant reaction of "gas-evolution crystallization reaction" [4] takes place in the Fe–N system, in which the iron melt dissolving nitrogen is solidified to transform into the primary solid solution and nitrogen gas phase. Utilizing the nitrogen solubility difference between liquid and solid, it is possible to fabricate porous iron using safe nitrogen gas instead of hydrogen gas. This new type of porous iron is promising for commercial application.

Figure 6.1 shows the typical optical micrographs of the cross section at the bottom part of the ingot. The overall porosity was measured as functions of the partial pressures of nitrogen and argon. The porosity was found to be affected by the nitrogen and argon pressures, which is similar to the system of hydrogen–argon gas. At a given nitrogen pressure, the porosity decreases with increasing argon

H. Nakajima, *Porous Metals with Directional Pores*,
DOI 10.1007/978-4-431-54017-5_6, © Springer Japan 2013

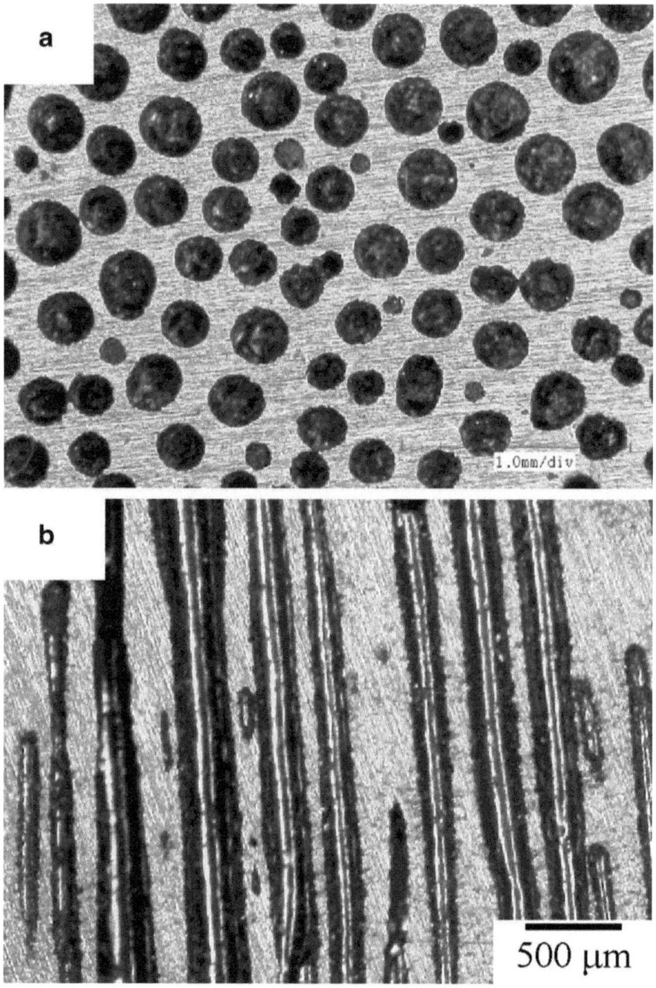


**Fig. 6.1** Optical micrographs of cross sections of lotus iron with 37.7 % porosity. The total pressure 1.5 MPa with the partial pressure of nitrogen of 1.0 MPa and the partial pressure of argon of 0.5 MPa. (**a**) Cross section perpendicular to the pore axis and (**b**) cross section parallel to the pore axis (Reprinted with permission from [2]. © 2002 Japan Institute of Metals)

pressure as shown in Fig. 6.2a. At a given argon pressure, the porosity increases with increasing nitrogen pressure as shown in Fig. 6.2b.

The solubility of nitrogen in iron is proportional to the square root of the nitrogen gas pressure, $\sqrt{P_{N_2}}$, according to Sieverts' law:

$$\left.\begin{aligned} C_N^L &= K^L\sqrt{P_{N_2}} \\ C_N^\delta &= K^\delta\sqrt{P_{N_2}} \end{aligned}\right\}, \tag{6.1}$$

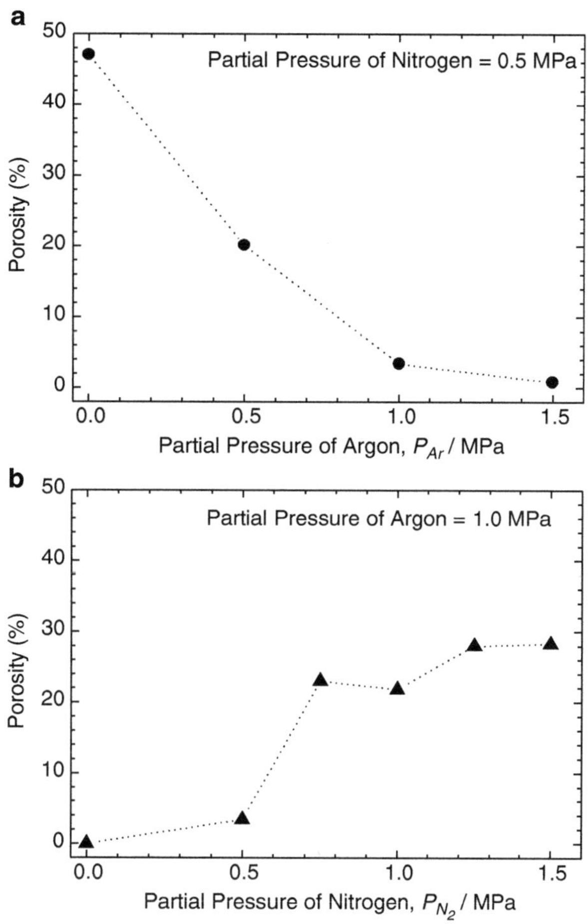


**Fig. 6.2** Dependence of porosity in lotus iron on (**a**) the partial pressure of argon at nitrogen pressure of 0.5 MPa and (**b**) the partial pressure of nitrogen at argon pressure of 1.0 MPa (Reprinted with permission from [2]. © 2002 Japan Institute of Metals)

where $C_{\mathrm{N}}^{\mathrm{L}}$ and $C_{\mathrm{N}}^{\delta}$ are the nitrogen concentration in the liquid and the δ-iron, respectively, and $K^{\mathrm{L}}$ and $K^{\delta}$ are equilibrium constants given by

$$K = \exp\left(-\frac{\Delta G}{RT}\right) = \exp\left(\frac{\Delta S}{R} - \frac{\Delta H}{RT}\right), \tag{6.2}$$

where $\Delta G$ is the Gibbs energy change for the solution of nitrogen, $T$ is an absolute temperature, $R$ is the gas constant, and $\Delta H$ and $\Delta S$ are the standard enthalpy and entropy of solution of nitrogen, respectively. The solubility of nitrogen in the different iron phases is usually evaluated from the slope of the decomposition pressure on logarithmic scale against $1/T$ [5]. The solubility of nitrogen in the molten iron is much higher than that of δ-iron as shown in Fig. 6.3. Therefore, nitrogen in the molten iron is rejected at the solid–liquid interface during solidification and the pores nucleate and grow unidirectionally. Considering the mixture

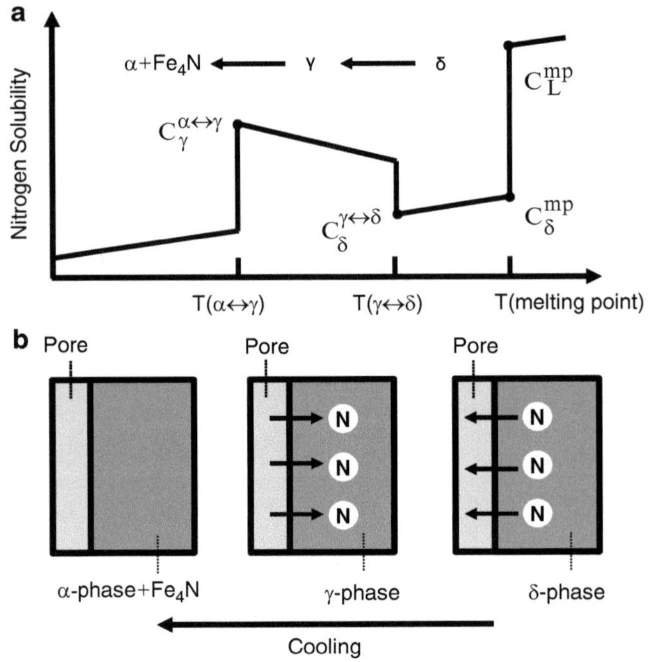


**Fig. 6.3** (**a**) Temperature dependence of nitrogen solubility in iron and (**b**) schematic of nitrogen movement during cooling after solidification in lotus iron fabricated with nitrogen gas (Reprinted with permission from [2]. © 2002 Japan Institute of Metals)

gas of nitrogen and argon, although the solubility of nitrogen in liquid is determined by the partial pressure of nitrogen, the nitrogen solubility near the pores is related with the internal gas pressure $P_i$ for satisfying with the equilibrium between solid phase around the pores and gas phase in the pores. Therefore, Eq. (6.1) can be written as

$$C_N^\delta = K^\delta \sqrt{P_i}. \qquad (6.3)$$

From this equation, it can be seen that the solubility of nitrogen in solid is related with total pressure of nitrogen and argon. With increasing partial pressure of argon at a given total pressure, the nitrogen solubility in liquid decreases but the nitrogen solubility in solid is almost constant.

Figure 6.4 shows the nitrogen concentration in the matrix of lotus iron as a function of the partial pressure of nitrogen when the total gas pressure of nitrogen and argon is kept constant at 2.0 MPa. It was found that the measured nitrogen concentration is in the range from 0.06 to 0.135 mass%. According to the solubility curve of nitrogen in iron [5], the nitrogen solubility in γ-iron decreases with increasing temperature and the maximum solubility is 0.135 mass% at the lowest

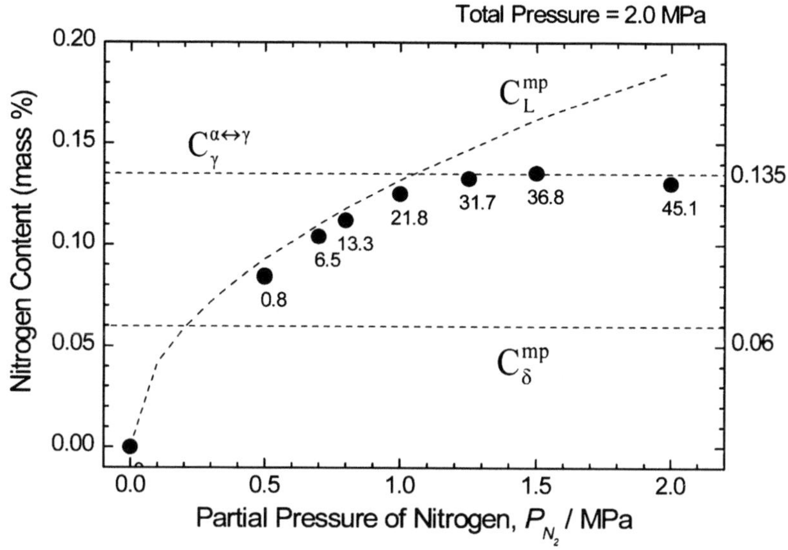


**Fig. 6.4** Nitrogen concentration in the matrix of lotus iron versus the partial pressure of nitrogen gas under constant total pressure of 2.0 MPa (Reprinted with permission from [2]. © 2002 Japan Institute of Metals)

temperature of the γ-phase. On the other hand, in δ-iron, the nitrogen solubility increases with increasing temperature and the maximum solubility is calculated as 0.06 mass% at the temperature just below the melting point of iron. Thus, it is reasonable that the measured nitrogen concentration range as shown in Fig. 6.4 is in between these two limits of the solubility. The reason why the measured nitrogen concentration is much higher than the nitrogen solubility in δ-iron can be explained as follows. As shown in Fig. 6.3 when the molten iron is solidified in the pressurized mixture gas of nitrogen and argon, the pores are formed by the evolution of nitrogen insoluble in the solid iron. In the subsequent cooling process in the solid, the nitrogen solubility in δ-iron gradually decreases with decreasing temperature. Then, the insoluble nitrogen atoms $(C_\delta^{mp} - C_\delta^{\gamma\leftrightarrow\delta})$ diffuse into the pores from δ-iron matrix. However, in further cooling process in γ-iron, the nitrogen solubility abruptly increases at the $\delta \rightarrow \gamma$ phase transformation temperature and continues to increase with decreasing temperature until the $\gamma \rightarrow \alpha$ phase transformation temperature is reached. In this process the nitrogen gas in the pores is dissolved into the γ-iron. Finally, two phases of α(ferrite) phase and $Fe_4N$ compound are formed on the phase transformation from γ-phase to α-phase.

Figure 6.5 shows the microstructure of the specimen containing 0.137 mass% nitrogen in the matrix of quenched porous iron which was fabricated in the mixture gas of $P_{N_2} = 1.25$ MPa and $P_{Ar} = 0.75$ MPa. The proeutectoid ferrite was observed on the grain boundaries and also inside the grains in the as cast porous iron.

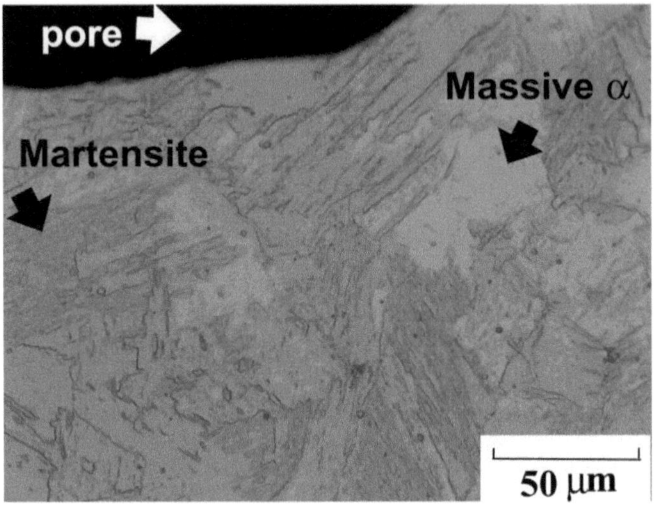


**Fig. 6.5** Optical micrograph of as-quenched lotus iron containing 0.137 mass% nitrogen in the matrix fabricated in the mixture gases of nitrogen 1.25 MPa and argon 0.75 MPa (Reprinted with permission from [2]. © 2002 Japan Institute of Metals)

## 6.2 Fabrication of Lotus Silver by Oxygen

In most combinations of the metals and gas such as hydrogen or nitrogen, the gas solubility in both of liquid and solid phases tends to decrease with decreasing temperature. However, a unique tendency is observed in the silver–oxygen system; the oxygen solubility in liquid silver increases with decreasing temperature [6]. It is interesting to know the effect of such an inverse temperature dependence of solubility on the pore formation and growth during the unidirectional solidification. The lotus-type porous silver was produced by unidirectional solidification through the mold casting technique and the Czochralski technique under pressurized oxygen [7]. Figure 6.6 shows a schematic picture of the apparatus for the Czochralski technique consisting of a melting part and a transfer mechanism of the rod in a high-pressure chamber. First, silver in an alumina crucible was melted by an induction heating in a mixture gas of oxygen and argon under a given pressure. A silver rod set upward was moved down to the surface of the melt, and then the rod was moved upward at the rate of 1.6 mm $min^{-1}$. The dimensions of the resulting silver ingot were about 15 mm in diameter and 100 mm in height.

The shape of pores in porous silver is different from those in other lotus copper and iron. Figure 6.7 shows an example of optical microscopic observation of the structure of lotus-type porous silver fabricated by the mold casting technique, indicating the cross section parallel and perpendicular to the solidification direction. In the cross section parallel to the solidification direction, although characteristic

**Fig. 6.6** Schematic drawing of the apparatus of Czochralski method for fabricating lotus silver (Reprinted with permission from [7]. © 2005 Japan Institute of Metals)

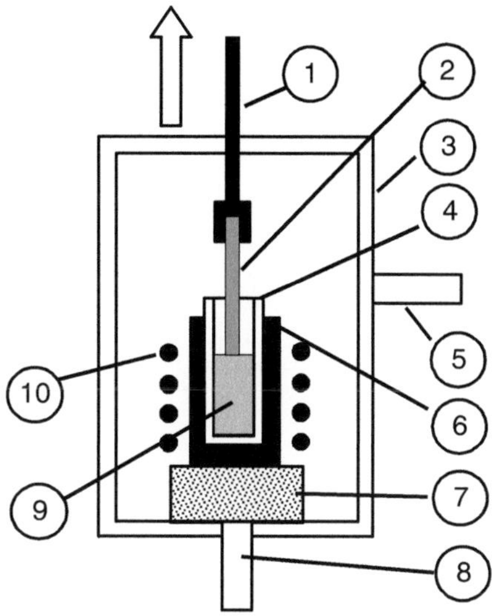


1.Transference mechanism
2.Silver rod
3.High pressure chamber
4.Alumina crucible
5.Inlet and outlet of oxygen and argon
6.Platinum crucible
7.Refractory
8.Tube for evacuation
9.Molten silver
10.Induction heating coil

cylindrical pores were observed to extend to the solidification direction, their shapes are not straight and their diameters are not constant in the direction. Similarly the pore shape is distorted, the distribution density of pores is not homogeneous, and furthermore, the pore size distribution is broad in the cross section of the perpendicular direction compared with other lotus metals. Such particular features to porous silver are discussed on the basis of the binary equilibrium phase diagram. A schematic drawing of the binary equilibrium phase diagram for silver–oxygen system is illustrated in Fig. 6.8, where $T_m$ and $T_n$ mean the holding temperatures of silver melt in oxygen (1,573 K) and the gas-evolution crystallization reaction temperature, respectively. On the assumption that the solidification progresses in equilibrium state, oxygen concentration of the liquid silver increases from A to B with decreasing temperature. Subsequently at $T_n$ the reaction of $L \rightarrow (S) + G$ occurs. In this case the solid phase appears below $T_n$, expecting that the pore shape would be similar to that of other lotus metals. However, the experimental results are not the case. Thus, it is reasonable to consider that the solidification progresses in nonequilibrium state by rapid cooling without

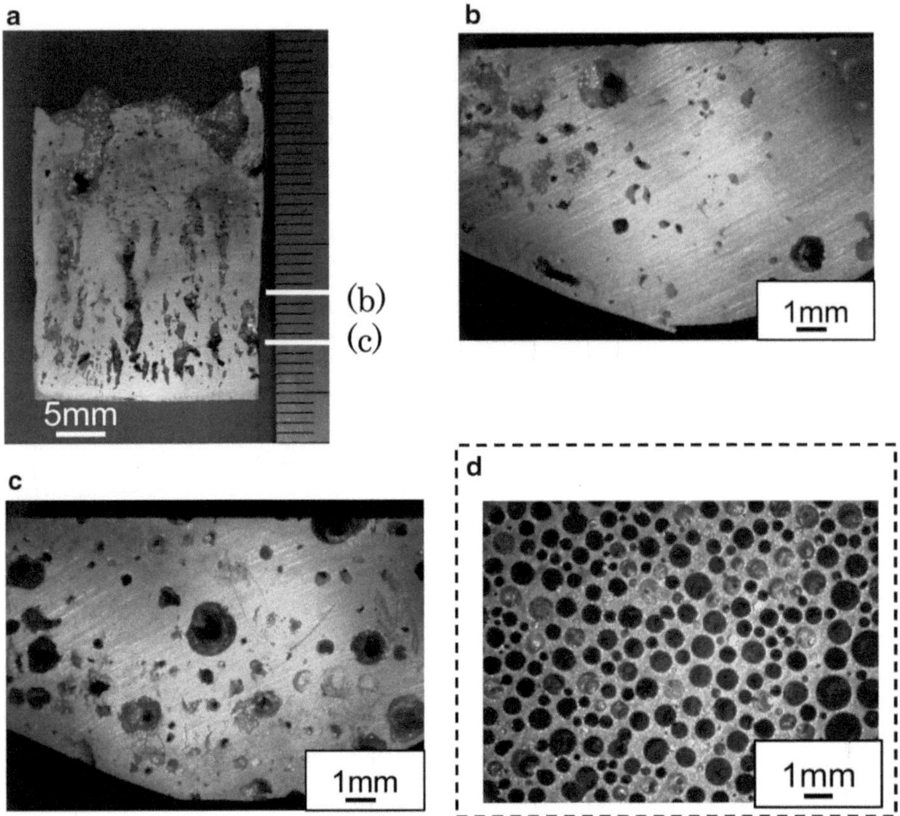


**Fig. 6.7** Cross sections of lotus silver fabricated by the mold casting technique under 0.05 MPa of oxygen and 0.15 MPa of argon. (**a**) Parallel to the solidification direction and (*b*, *c*) perpendicular to the solidification direction at the distance (**b**) 10 mm and (**c**) 5 mm from the copper chiller (the bottom). (**d**) Cross section perpendicular to the solidification direction of lotus copper (Reprinted with permission from [7]. © 2005 Japan Institute of Metals)

oxygen supply from the gas phase to the liquid phase and the oxygen concentration of the liquid phase is constant from A to C. In this case unlike other lotus metals, the solidification progresses through $L + (S)$ phase: the solid phase coexists with the liquid phase. This can prevent the growth of cylindrical pores similar to the case of lotus magnesium alloys [8]. Thus, the solidification in the $L + (S)$ phase could be a part of reason for the distorted pore shape.

Figure 6.9 shows the cross section parallel and perpendicular to the solidification direction of the specimen fabricated by the Czochralski technique. The distribution density of pores fabricated by the Czochralski technique is more homogeneous than that by the mold casting technique. In the Czochralski technique, the liquid phase continuously contacts the gas phase during the solidification. Thus, it is expected that oxygen can easily be supplied from gas phase to the liquid–solid interface and

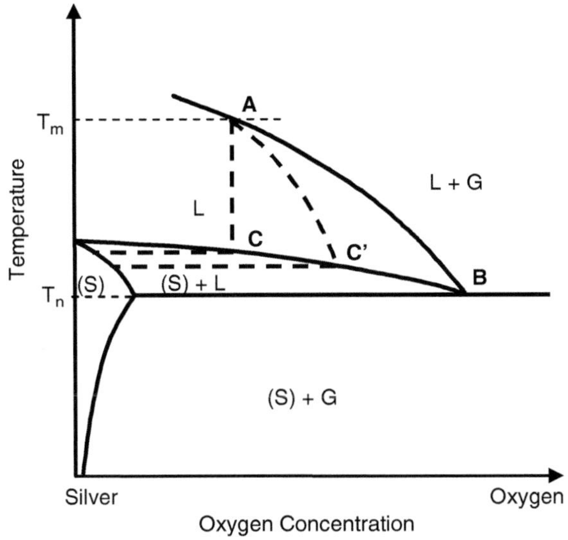


**Fig. 6.8** Silver-rich side of binary equilibrium phase diagram for silver–oxygen system (Reprinted with permission from [7]. © 2005 Japan Institute of Metals)

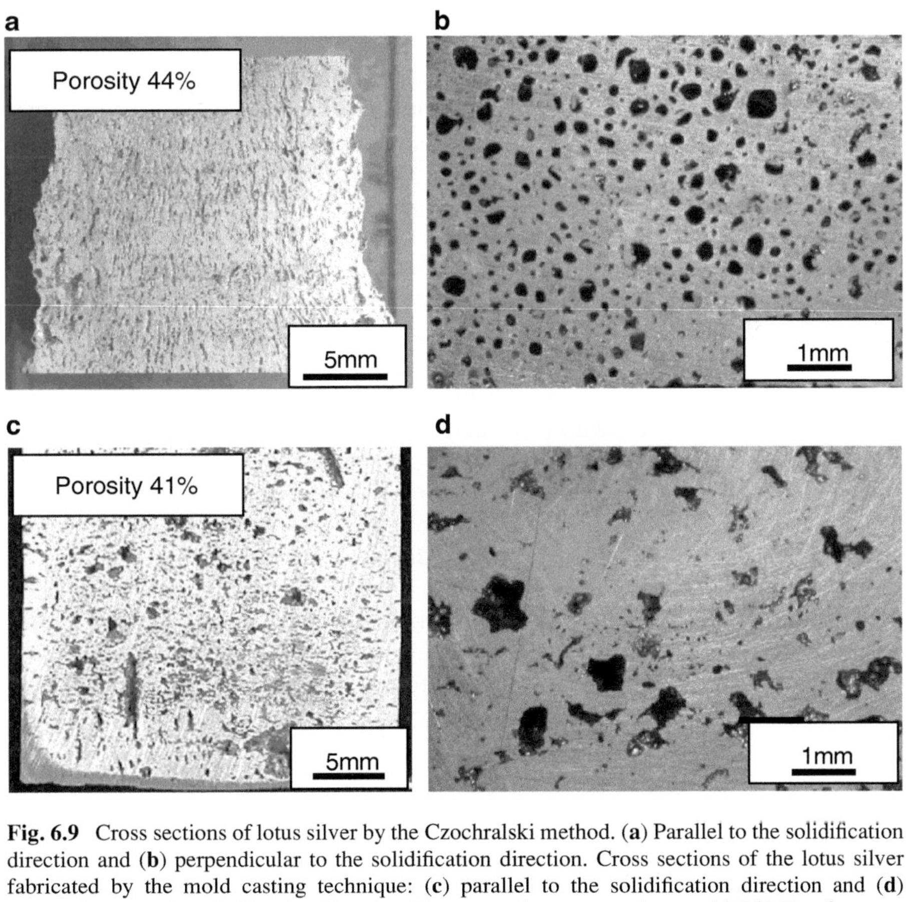


**Fig. 6.9** Cross sections of lotus silver by the Czochralski method. (**a**) Parallel to the solidification direction and (**b**) perpendicular to the solidification direction. Cross sections of the lotus silver fabricated by the mold casting technique: (**c**) parallel to the solidification direction and (**d**) perpendicular to the solidification direction. The atmosphere was a mixture of 0.55 MPa of oxygen and 0.55 MPa of argon (Reprinted with permission from [7]. © 2005 Japan Institute of Metals)

then the porosity also must increase. The distribution of the pore size $\sigma$ and the circularity R at the cross section perpendicular to the solidification direction are calculated as

$$\sigma = \sigma_s / d_a$$
$$R = 4\pi \times A/L^2, \tag{6.4}$$

where $\sigma_s$, $d_a$, $A$, and $L$ are the standard deviation of pore sizes, the average pore diameter, the area of pores, and the circumference length of the pores, respectively. It turned out that the distribution of pore size $\sigma$ decreases and the circularity $R$ increases in the lotus silver made by the Czochralski technique compared with the ingot obtained by the mold casting method. These results support the assumption that the oxygen gas in the pores formed at the liquid–solid interface redissolves to the molten silver with insufficient oxygen concentration in the fabrication by the unidirectional solidification method.

## 6.3 Fabrication of Lotus Intermetallic Compounds

Intermetallic compounds exhibit superior properties of high-temperature strength, oxidation resistance, and anti-corrosion and are expected to be used as high-temperature structural materials and functional materials. If elongated pores aligned in one direction can be evolved in the intermetallic compounds, various applications such as lightweight high-temperature structural materials, high-temperature catalyst, and high-temperature filters can be expected. In particular, aluminide intermetallic compounds have attracted much attention for lightweight high-temperature materials [9–11]. Recently Hyun et al. [12] fabricated lotus-type porous intermetallic compounds of $Ni_3Al$ and TiNi for the first time.

Figure 6.10 shows optical micrographs of the cross-sectional views of lotus-type porous NiAl compounds perpendicular to the solidification direction fabricated by the continuous zone melting technique. The fabrication was carried out at 2.5 MPa hydrogen and with the transfer velocity of 330 μm $s^{-1}$. The pores in lotus Ni-15%Al alloy are round, while those in lotus Ni-28%Al and Ni-31%Al alloys are in distorted morphology. The solidification structures in longitudinal cross section parallel to the solidification direction are shown in Fig. 6.11. The dendritic structure is observed in the Ni-28%Al alloy. When the pores nucleated in apertures of the dendrite grow, the pores may touch to the dendritic arms and cannot grow straight, which result in distorted pore morphology. On the other hand, in Ni-15%Al alloy no dendrite structure is observed so that round-shaped pores observed in the section perpendicular to the solidification direction grow straight in the solidification direction. This is attributed to the narrow two phases' region of mixture of liquid and solid and to the suppressive growth of the dendritic arms.

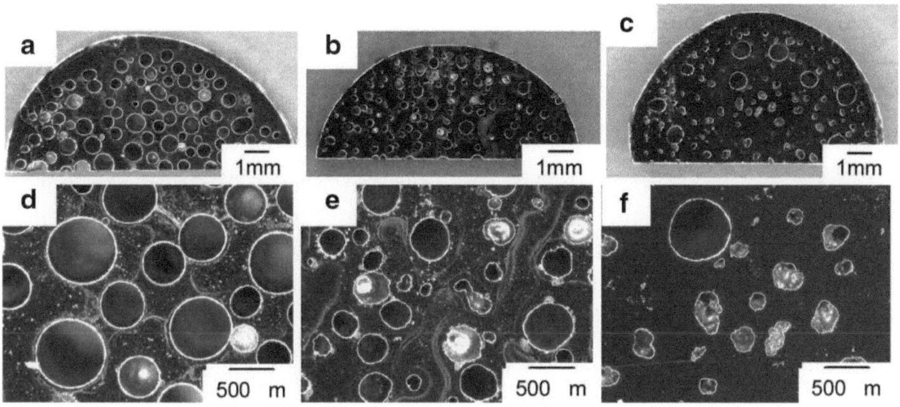


**Fig. 6.10** Optical micrographs of lotus (**a**, **d**) Ni-15%Al, (**b**, **e**) Ni-28%Al, and (**c**, **f**) Ni-31%Al intermetallic compounds in the cross sections perpendicular to the solidification direction fabricated by continuous zone melting technique in 2.5 MPa hydrogen atmosphere (Reprinted with permission from [12]. © 2004 Japan Institute of Metals)

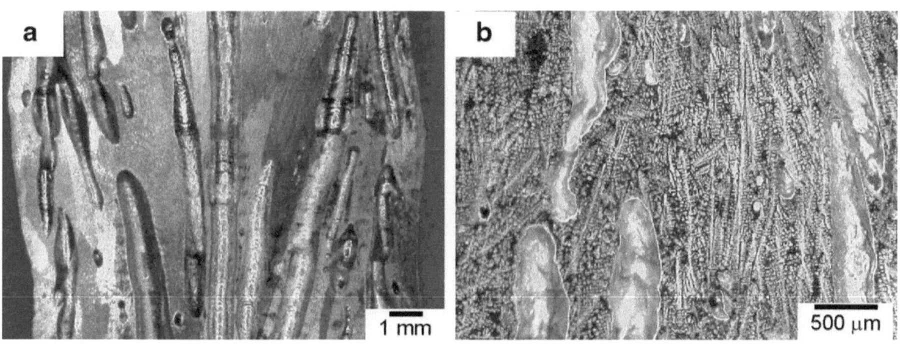


**Fig. 6.11** Optical micrographs of lotus (**a**) Ni-15%Al and (**b**) Ni-28%Al intermetallic compounds in section parallel to the solidification direction (Reprinted with permission from [12]. © 2004 Japan Institute of Metals)

The porosity decreases from 40 to 18 % with increasing Al content from 15 to 31 %. The porosity is closely related with and proportional to the hydrogen concentration rejected at the liquid–solid interface. According to Lange and Schenck [13], the addition of aluminum to nickel suppresses the hydrogen solubility in the liquid. For example, the hydrogen solubility in Ni-25%Al alloy at 1,773 K under 0.1 MPa hydrogen pressure is about 50 % smaller than that in pure nickel. Therefore, the decrease in porosity may be due to the decrease in hydrogen content available to the pore evolution. The average pore diameter decreases with increasing Al content, which is also attributed to the same reasoning.

## 6.4 Fabrication of Lotus Silicon

Porous silicon with nano-order sized pores was fabricated by electrolyte reaction [14–16]. However, porous semiconductors with unidirectional elongated micro-order sized pores have never synthesized so far. Nakahata and Nakajima were succeeded in fabricating a lotus silicon by unidirectional solidification in a pressurized hydrogen atmosphere [17]. Different from the conventional nano-sized porous silicon, the pore size of the lotus silicon ranges from 10 μm to 1 mm in diameter. This type of porous silicon is considered to be novel porous semiconducting material. The lotus silicon with directional cylindrical gas pores was fabricated by unidirectional solidification through mold casting technique of the melt dissolving hydrogen in a pressurized hydrogen atmosphere.

Figure 6.12a shows an overview of lotus silicon fabricated in the hydrogen atmosphere of 0.21 MPa. Figure 6.12b, c shows the optical micrographs of the cross section perpendicular and parallel to the solidification direction, respectively. The pores in the cross section perpendicular to the solidification direction are round-shaped, while those parallel to the solidification direction are long and

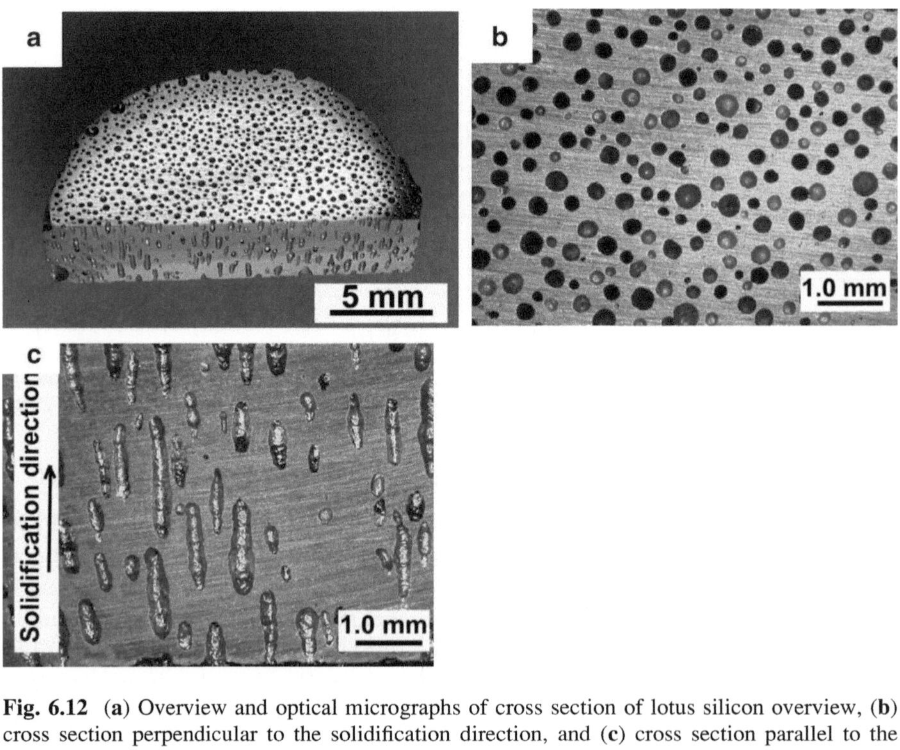


**Fig. 6.12** (**a**) Overview and optical micrographs of cross section of lotus silicon overview, (**b**) cross section perpendicular to the solidification direction, and (**c**) cross section parallel to the solidification direction. The solidification occurred upward as shown in figure (**c**) (Reprinted with permission from [17]. © 2004 Elsevier B.V.)

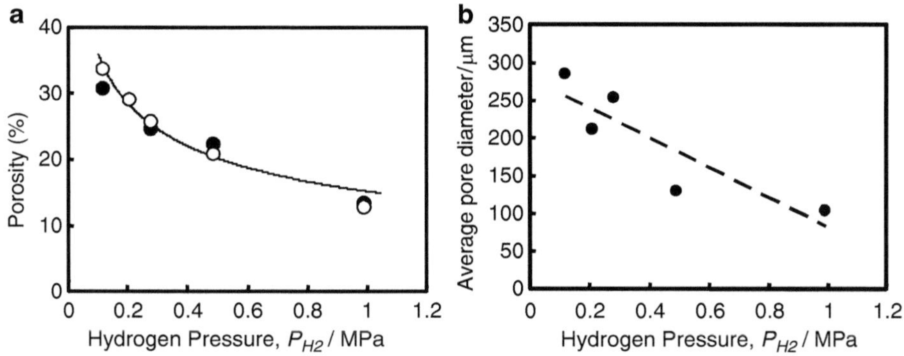


**Fig. 6.13** (**a**) Porosity of lotus silicon as a function of hydrogen pressure: (*filled circle*) evaluated from apparent density of lotus silicon and (*open circle*) determined by the microstructural analysis of the cross sections at the distance 5 mm from the copper chiller. (**b**) Average pore diameter of the lotus silicon as a function of hydrogen pressure. The *closed circles* represent the average pore diameter determined by microstructural analysis of the cross sections at the distance 5 mm from the copper chiller and the *broken line* shows the fitted one by linear regression (Reprinted with permission from [17]. © 2004 Elsevier B.V.)

cylindrical. Figure 6.13a shows hydrogen pressure dependence of the porosity of the lotus silicon. The porosity decreases with increasing hydrogen pressure. Figure 6.13b shows hydrogen pressure dependence of the average pore diameter in the cross section on the sample plane at 5 mm in height from the bottom of the ingot. The pore diameter decreases with increasing hydrogen pressure; the pore size is about 100 μm in diameter at hydrogen pressure of 0.99 MPa. From Sieverts' law and Boyle's law, the following expression is derived:

$$V = \frac{(k_l - k_s) \times P_{H_2}^{1/2} \times R \times T_n}{2(P_{H_2} + 2\sigma/r)}, \tag{6.5}$$

where $k_l$ and $k_s$ are the equilibrium constants of hydrogen dissolving reaction in liquid and solid, respectively. $\sigma$ and $r$ are the surface tension of the melt and pore radius. Since the surface tension of silicon at $T_n$ (1,687 K) is about 0.74 $Nm^{-1}$ [18] and the pore diameter in the range from 103 to 286 μm, the value of $2\sigma/r$ is estimated at most to be 0.03 MPa, which is in fact negligible small in comparison with the corresponding $P_{H_2}$ of 0.99 MPa. Thus, the pore volume is inversely proportional to $P_{H_2}^{1/2}$. The porosity increases by a factor of about 2.7 when the hydrogen pressure decreases from 0.99 to 0.12 MPa, which is in good agreement with the estimation by the above equation; the porosity increases by a factor of $(0.99/0.12)^{1/2} \approx 2.9$. Although the present evaluation is not exact because of neglecting some effect such as temperature gradient, the pore evolution behavior can be explained by supersaturation of hydrogen in the solidified silicon.

## 6.5 Fabrication of Lotus Alumina by Unidirectional Solidification

Porous oxide ceramics have been used as hot gas or molten metal filters and catalyst because of their excellent mechanical properties, high-temperature resistance, and chemical stabilities. Porous ceramics with highly oriented cylindrical pores are convenient to use as a gas filter and as a catalyst carrier. Ishizaki et al[19] suggested that the appropriate pore size for gas filter application is between 10 and several hundred micrometer. Hence, in the development of a gas filter, control of the pore size is an important issue. Several research groups have proposed to the fabrication processes of porous alumina with oriented cylindrical pores. Zhang et al. [20] fabricated porous alumina with unidirectionally aligned continuous pores via the slurry coating of fugitive cotton filter. Porous alumina with approximately 150-μm-diameter pores and 35 % porosity was fabricated using their method. Ding et al. [21] prepared a porous alumina with oriented pores by combining a foaming method with sol–gel technology. A porous alumina with 1-mm-diameter pores and 35 % porosity was fabricated from their method. Isobe et al. [22] prepared porous alumina with oriented cylindrical pores by an extrusion method. In their method, a porous alumina with 14-μm-diameter pores and 35 % porosity was fabricated. Fukasawa et al. [23] have developed a fabrication method for porous alumina with oriented pores using the freeze-drying technique. However, it is difficult to control the porosity and/or pore size of the porous alumina by these methods.

On the other hand, in the field of metals, porous metals with cylindrical pores called as lotus metals were proposed. The porous metals are fabricated by unidirectional solidification under pressurized hydrogen gas, utilizing the hydrogen solubility gap in between liquid and solid. For such metals, the phase diagrams for metal–hydrogen system have been established. However, data of hydrogen gas solubility in oxides are hardly available. Ueno et al. [24] investigated lotus alumina fabricated by unidirectional solidification under pressurized hydrogen gas to elucidate the formation mechanism of pores.

Feed rods were prepared by high-purity alumina powder (99.99 % purity). The powder was mixed with a binder in water and a green rod was prepared by the slip-casting method. After drying in air, calcination was performed at 1,473 K for 7.2 ks in air, and then, a feed rod with 8 mm in diameter and 150 mm in length was obtained. The solidification was carried out using an optical floating zone apparatus under pressurized hydrogen gas where 100%$H_2$ gas, 50%$H_2$-Ar mixture gas, or 10%$H_2$-Ar mixture gas was used as the environmental gas, as illustrated in Fig. 6.14. A xenon lamp was used as an optical source. The light from the xenon lamp was put on the focus by the elliptical mirror to set up the melting zone of the feed rod. The top of the feed rod was hooked on the upper shaft and the bottom was fixed to the lower shaft. The melting system was set up in a quartz tube. During the unidirectional zone melting, the solidification rate and the rod rotation were fixed to 200 mm $h^{-1}$ and 20 rpm, respectively.

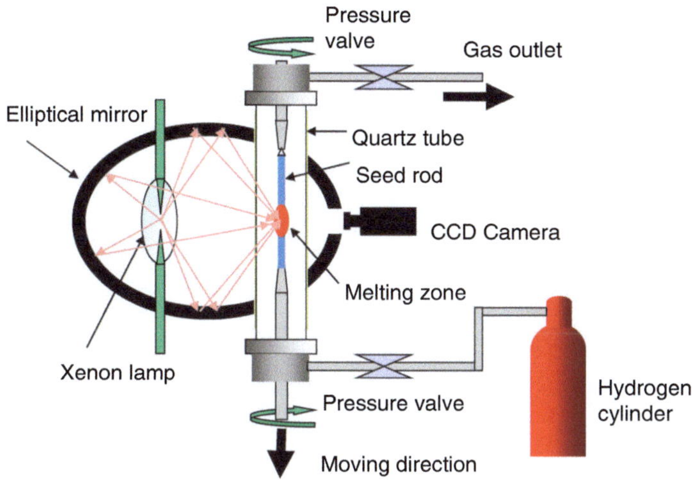


**Fig. 6.14** Schematic drawing of fabrication apparatus for lotus ceramics

Lotus alumina was successfully fabricated by unidirectional solidification under a pressurized hydrogen atmosphere. Figure 6.15a, b shows transversal and longitudinal cross-sectional views of the solidified sample prepared in 100%$H_2$ gas under the pressure of 0.8 MPa. Many cylindrical pores were aligned along the solidification direction. Some large pores were seen in the inner periphery and in the center of the sample. In the longitudinal cross-sectional view, the large pores possessed a spherical shape, which is quite different from the morphology of cylindrical pores. Hence, it is considered that the formation mechanism of these two types of pores is different. For the cylindrical pores, which are the dominant pores found in the figure, the shape of the pores is very similar to that of lotus metals. Thus, this type of pore is formed at the solid–liquid interface during the solidification due to a hydrogen solubility gap between liquid and solid phases. One possible model for the formation of sphere-shaped pores is trapping gas bubbles that are released from the solid–liquid interface to the liquid phase by the solid phase during unidirectional solidification.

Figure 6.16 shows the total pressure dependence of the porosity in the samples prepared in 100%$H_2$, 50%$H_2$-Ar, and 10%$H_2$-Ar gases. For the samples prepared in 100%$H_2$, the porosity once increases with increasing the pressure up to 0.4 MPa and then gradually decreases. A similar tendency was observed for lotus copper. Yamamura et al. suggested that there are two factors controlling the porosity in the formation of lotus metals: Sieverts' law and Boyle's law [25]. It is known that the porosity of lotus metals is proportional to hydrogen partial pressure. However, the pressure inside the pore is driven by environmental pressure according to Boyle's law. Hence, the porosity decreases with increasing total pressure according to Boyles law. Thus, the porosity of lotus materials is determined by two contrary laws related to hydrogen partial pressure and total pressure. Because Ar gas cannot be dissolved into liquid and solid metal phases, Ar gas in the

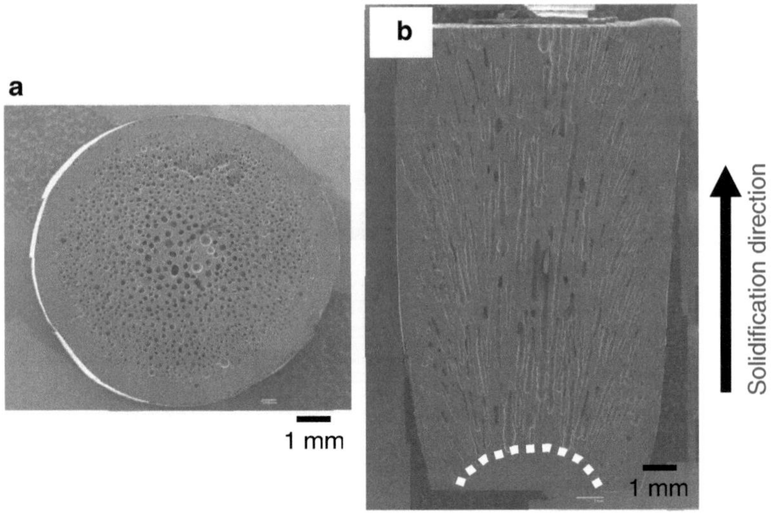


**Fig. 6.15** Transversal cross-sectional and longitudinal cross-sectional views of the solidified sample prepared in mixed gas of 50%$H_2$-50%Ar at 0.8 MPa (Reprinted with permission from [24]. © 2007 The American Ceramic Society)

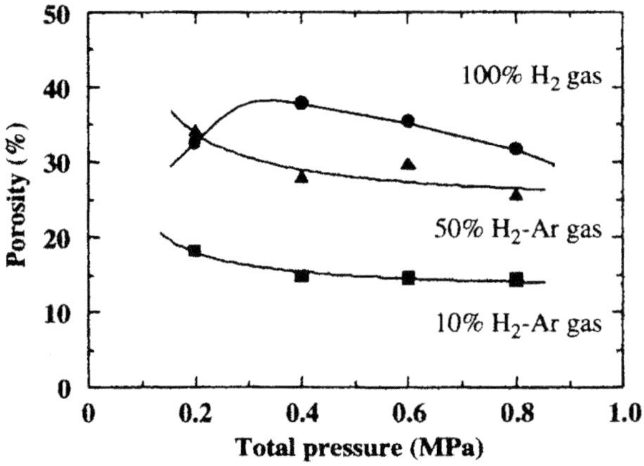


**Fig. 6.16** Porosity changes in the samples prepared in 100%$H_2$, 50%$H_2$-Ar, and 10%$H_2$-Ar gases (Reprinted with permission from [24]. © 2007 The American Ceramic Society)

$H_2$-Ar mixture gas affects only to increase the total pressure so that the effect of Boyle's law became dominant. The same reasoning is also applied to lotus alumina as shown in Fig. 6.16.

Figure 6.17 shows the total pressure dependence of the average pore diameter of the samples. The pore size decreases with increasing total pressure of the gas, which is explained by Boyle's law. On the other hand, since the absolute amount of hydrogen gas increases with increasing hydrogen partial pressure, the pore volume increases with increasing hydrogen partial pressure.

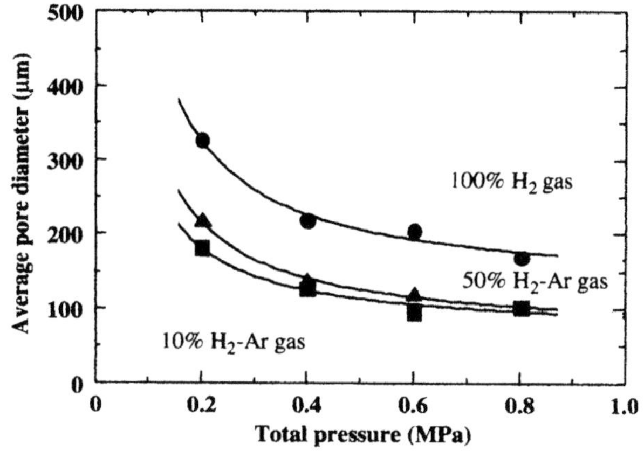


**Fig. 6.17** Longitudinal cross-sectional views of the solidified sample prepared in 50%$H_2$-Ar gas at 0.8 MPa (Reprinted with permission from [24]. © 2007 The American Ceramic Society)

## 6.6 Fabrication of Lotus Brass by Solid-State Diffusion

It is not easy to fabricate lotus brass by unidirectional solidification method in gas atmosphere. This is probably because of its very small hydrogen solubility difference between solid and liquid of brass. Instead of solidification, lotus brass is fabricated by utilizing solid-state diffusion of zinc into porous copper which is made by unidirectional solidification in high pressure of hydrogen atmosphere [26]. The vapor deposition method is adopted; the lotus copper sample was wrapped in a tantalum foil and was annealed together with the zinc piece in a quartz tube with vacuum of $1 \times 10^{-4}$ Pa. During annealing, zinc was evaporated, deposited onto the lotus copper, and diffused into the copper. After annealing, the samples were observed by an optical microscope and their compositions were analyzed by an electron probe microanalysis (EPMA).

Figure 6.18 is the result of the analysis by EPMA on the sample annealed at 953 K for $4.32 \times 10^4$ s. No zinc pieces remained in the quartz tubes after annealing. All of zinc atoms evaporated and diffused into the samples. It was found that the composition was almost constant between two neighboring pores on the cross section parallel to pore direction; the zinc concentration is around 40 at.% and almost uniform composition. Thus, the lotus brass was fabricated by zinc diffusion into lotus copper.

## 6.7 Effect of Microstructure on Pore Morphology in Lotus Magnesium Alloys

Magnesium is considered to be a rather new material as a lightweight structural metal. Usually the magnesium alloys are used for high-mechanical strength and good anti-corrosion resistance. As mentioned before, lotus magnesium can be produced by unidirectional solidification through the mold casting technique [27]. However,

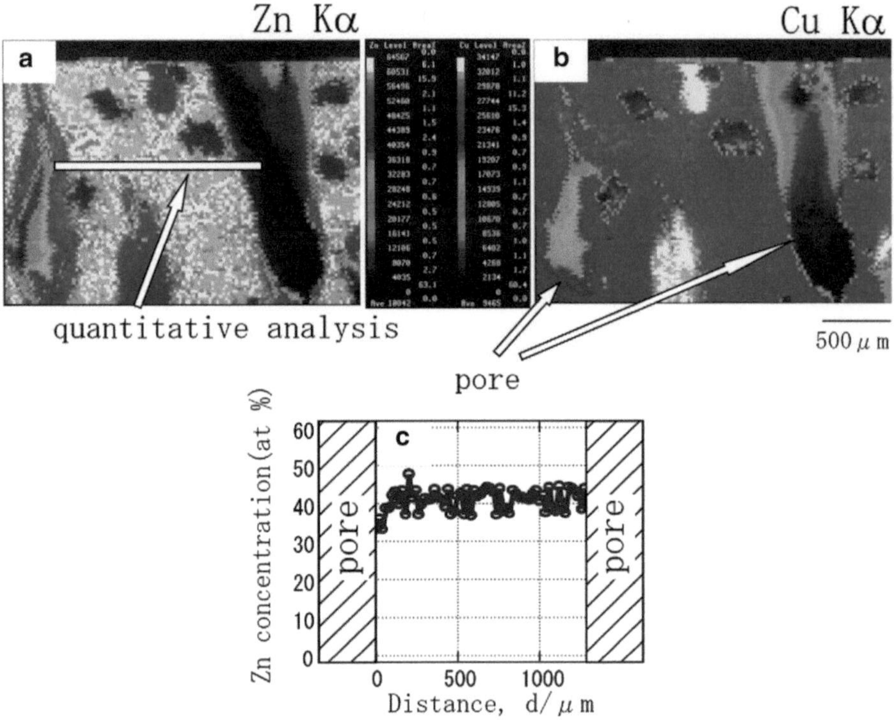


**Fig. 6.18** Results of electron probe microanalysis in the samples annealed at 935 K for $4.32 \times 10^4$ s in a quartz tube with a zinc piece: (**a**) map of Zn Kα intensity, (**b**) map of Cu Kα intensity, and (**c**) composition–distance profiles between two pores (Reprinted with permission from [26]. © 2003 Japan Institute of Metals)

it was found by Hoshiyama et al. [28] that addition of alloying elements significantly changes the pore shape and the porosity. Three kinds of alloys, Mg-9%Al-0.75%Zn (AZ91D), Mg-3Al-1Zn (AZ31B), and Mg-3Al-0.25Zn (AZ91D), the latter of which is diluted with magnesium and is abbreviated as (1/3)AZ91D, were used. Figure 6.19 shows photographs on the longitudinal sections of porous (1/3)AZ91D, AZ31B, and AZ91D alloys. In the lower region near the bottom chiller of (1/3)AZ91D and AZ31B alloys, elongated pore growth is observed, while almost all pores in AZ91D alloy are spherical, not cylindrical. Comparing (1/3)AZ91D with AZ31B the pores in the former alloy are longer than that in the latter, because total concentration of the alloying elements is lower in the former. The number of pores and porosity decrease with increasing concentration of the alloying elements Al, Zn, and Mn. This result is explained in terms of the decrease in hydrogen solubilities in the magnesium alloys due to addition of alloying elements [29]. Figure 6.20 shows Mg-Al equilibrium phase diagram in magnesium-rich side. The temperature interval $\Delta T$ between liquidus and solidus lines at each composition increases by 43.5 K, 52.5 K, and 150 K in order

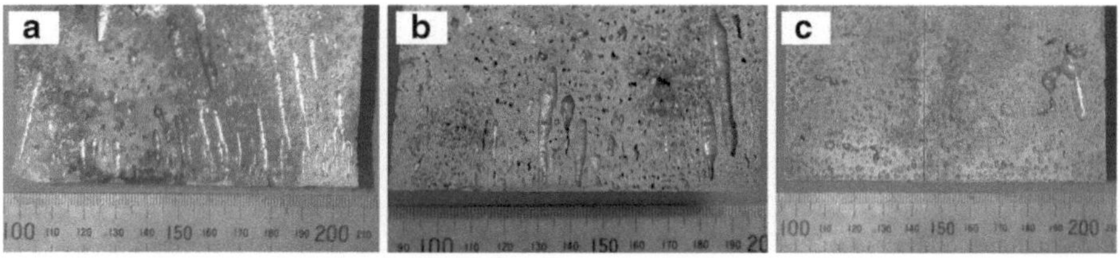


**Fig. 6.19** Photographs on longitudinal sections of lotus magnesium alloys. (**a**) (1/3)AZ91D, (**b**) AZ31B, and (**c**)AZ91D alloys. Copper chiller was used for the bottom plate of the mold (Reprinted with permission from [28]. © 2008 Freund Publishing House Ltd)

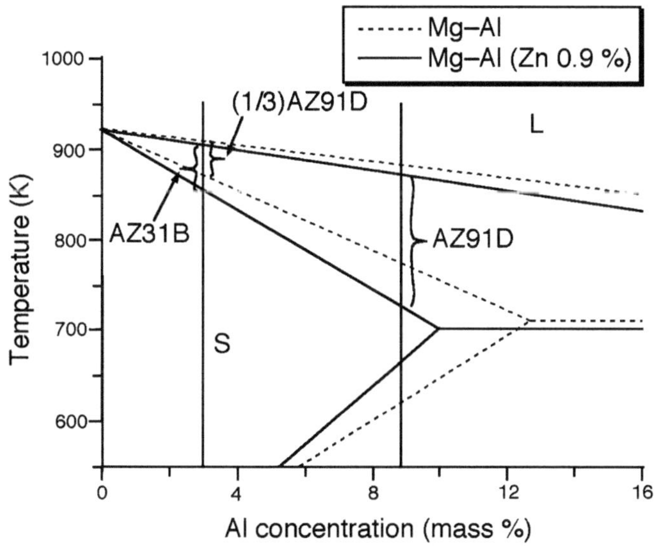


**Fig. 6.20** Magnesium-rich side of Mg-Al equilibrium phase diagram (Reprinted with permission from [28]. © 2008 Freund Publishing House Ltd)

of (1/3)AZ91D, AZ31, and AZ91D, respectively; $\Delta T$ corresponds to increase in coexistence region of solid and liquid phases.

Let us think about pore growth mechanism in such mushy zone. When the alloy is solidified by way of the mushy zone, the ratio of solid phase portion in the mushy zone increases with decreasing temperature and the hydrogen rejected in the solid phase forms the pores. The schematic drawing of wide mushy zone is illustrated in Fig. 6.21a. Even if the pores are nucleated, the solid phase which is precipitated in front of evolved pores prevents the pores from growing toward the solidification direction. On the other hand, in the short mushy zone in Fig. 6.21b, the top of the cylindrical pores can grow without the precipitation of solid phase. Thus, it is concluded that the shorter the mushy zone is, the longer the pores grow.

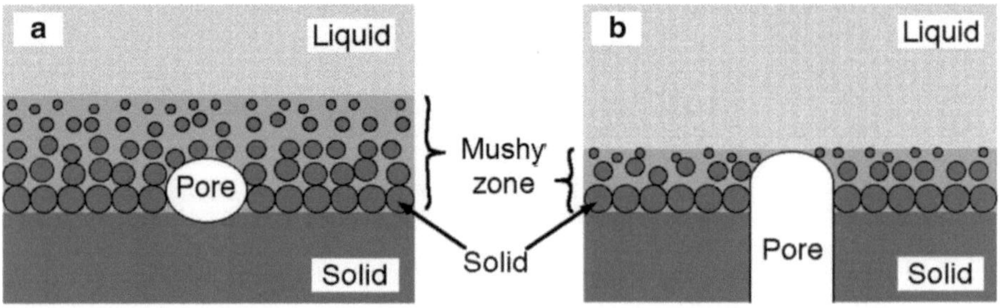


**Fig. 6.21** Effect of the mushy zone on pore growth. Width of the mushy zone is (**a**) wide and (**b**) narrow. The mushy zone affects the pore evolution in unidirectional solidification (Reprinted with permission from [28]. © 2008 Freund Publishing House Ltd)

## 6.8 Fabrication of Lotus Carbon Steel by Continuous Casting Technique

Lotus carbon steel is suitable for lightweight high-strength structural material. For practical purpose, lotus carbon steel plates are expected to be used for movable structural bodies of machine tools because of the effects of reduction of weight and vibration damping, which improve the precision and productivity of machining, the tool life, etc. Although the continuous casting apparatus was used to fabricate lotus copper, magnesium, and aluminum, some modifications of the apparatus were required for fabrication of lotus carbon steel, because there are several problems as follows. Since iron reacts easily with carbon, nitrides, and so on, endurance of the refractory, a ceramic material which is not reactive with iron, was selected as the refractory. In order to improve the endurance of the refractory, a technique for a rigid connection among the crucible, the break ring, and the mold was developed [30]. The second problem was the friction between the mold and the carbon steel, which causes the blockage of melt in the mold during continuous casting. To solve this problem, intermittent motion was adopted, where a cycle of transfer and stop is repeated periodically. This technique can give a vibration to the carbon steel sample in order to deduce the friction. In the following sections, transfer velocity means the average value, which is the transfer distance per unit time over several cycles as shown in Fig. 6.22. By developing this technique, it became possible to fabricate long lotus carbon steel up to 600 mm in length.

Figure 6.23 shows the fabricated lotus carbon steel at 100 mm min$^{-1}$. On the surface, ripple marks are observed periodically. These ripple marks were generated at each cycle of the intermittent motion and correspond to the solid–liquid interface. The ripple marks show that the solid–liquid interface changed from horizontal to concave shape with a depth $h$ during the continuous casting. The change can be explained by the position of interface, which is considered to have dropped during continuous casting, because cooling capacity of the mold was not enough to solidify the total amount of the melt, which entered during one cycle. The length of the

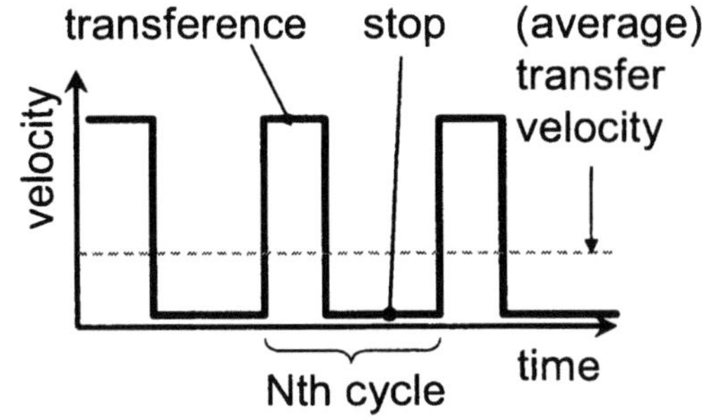


**Fig. 6.22** Intermittent motion of transfer (Reprinted with permission from [30]. © 2007 MIT-Verlag)

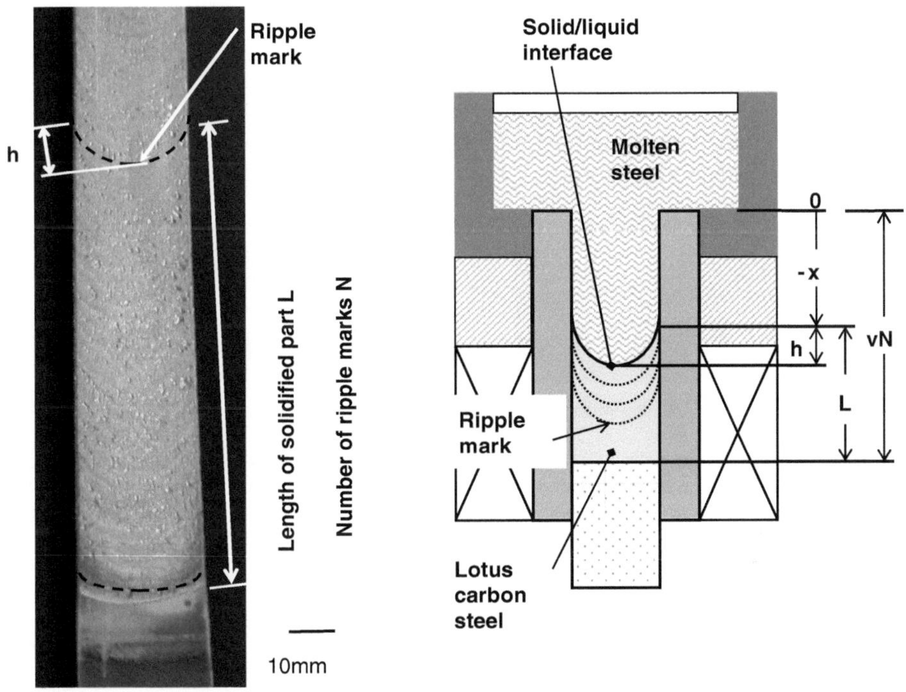


**Fig. 6.23** Fabricated lotus carbon steel. (*left*) *Ripple mark* on the surface of solidified slab of lotus carbon steel. (*right*) Schematic drawing around the solid/liquid interface during continuous casting (Reprinted with permission from [30]. © 2007 MIT-Verlag)

solidified part $L$ until the Nth cycle (see Fig. 6.22) was measured by the ripple mark. Then, the position of the solid/liquid interface can be evaluated by the difference between $L$ and $vN$, which is the moving distance of the dummy bar, where $v$ is a moving distance per a cycle.

Figure 6.24 shows the measured position and depth of the solid–liquid interface at each cycle in the experiment of 100 mm min$^{-1}$. At the beginning of the continuous casting, the position is close to the crucible. The small h-value shows that the interface is nearly horizontal. In this case small pores are distributed.

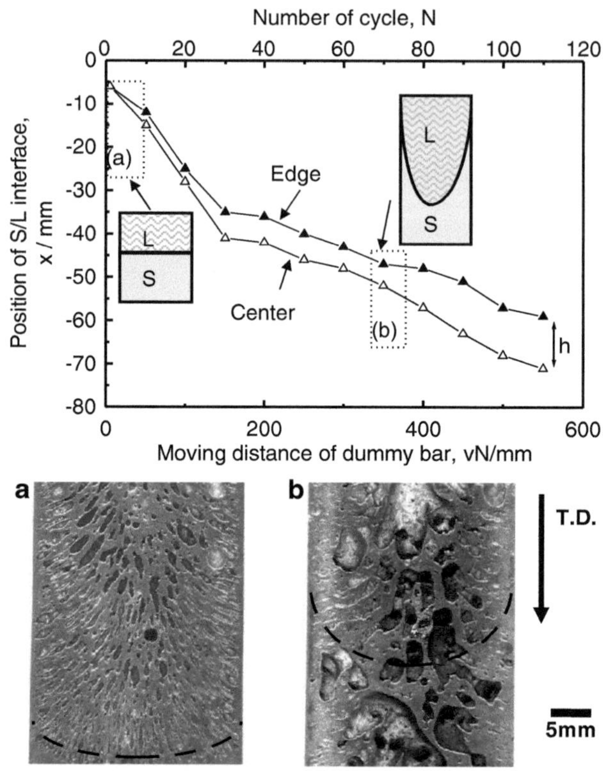


**Fig. 6.24** Position of the solid–liquid interface at each cycle number. The pictures show the cross sections parallel to the transfer direction at the initial transference (**a**) and at about 80 cycles (**b**). The *dotted lines* show the solid–liquid interface (Reprinted with permission from [30]. © 2007 MIT-Verlag)

With increase of the cycle number, the position dropped and the interface became a deep concave shape with a large value of $h$. In this case coarsened irregular pores are observed.

The mechanism of these results is illustrated in Fig. 6.25. In the initial state (a), the melt is cooled mainly by the heat flow in the longitudinal direction, because this is a heating zone. Then the solid–liquid interface is horizontal. But if the solid–liquid interface is located in the cooling zone (b), the heat flow is mainly in the width direction and the interface becomes concave. As the pores grow perpendicular to the solid–liquid interface, the pores merge with each other in the center. Coarsened irregular pores are generated.

From these results the mechanism of pore formation depending on the transfer velocity became clear as shown in Fig. 6.26. Then one can find the condition for small pores distributed homogeneously at around 20 mm $min^{-1}$.

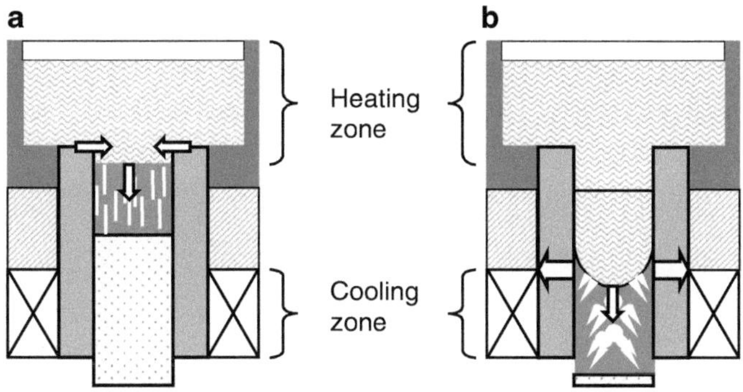


**Fig. 6.25** Mechanism of the descent position of the solid–liquid interface and he change into a concave shape. (**a**) The initial transference and (**b**) after several ten cycles (Reprinted with permission from [30]. © 2007 MIT-Verlag)

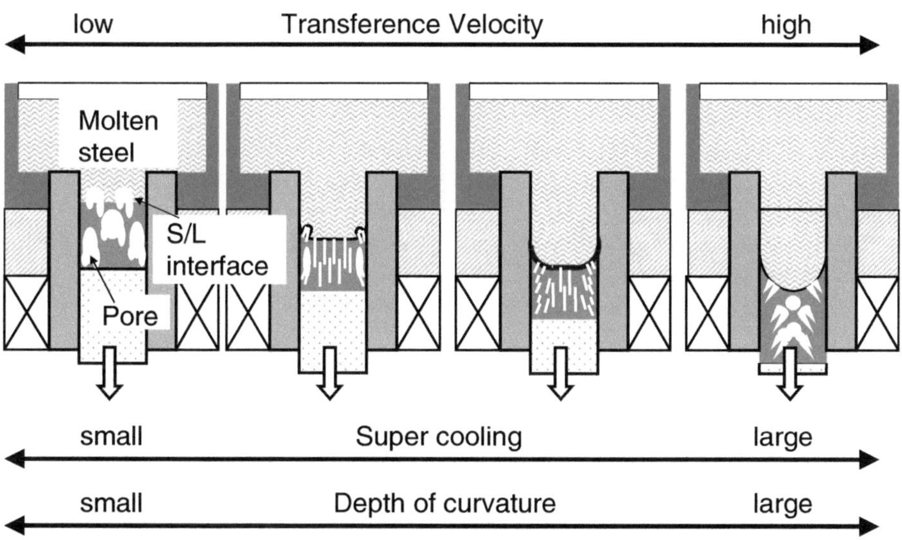


**Fig. 6.26** Schematic illustration of the dependence of pore formation on transfer velocity (Reprinted with permission from [30]. © 2007 MIT-Verlag)

## 6.9 Fabrication of Lotus Aluminum by Continuous Casting Technique

Lotus aluminum possessing slender directional pores aligned in one direction is one of the most promising candidates for lightweight structural materials [31]. Numerous investigations have focused on the fabrication of porous aluminum and its alloys with slender directional pores formed by unidirectional solidification in a hydrogen atmosphere. Particularly, solidification defects have been emphasized for

**Table 6.1** Previously reported porosity and pore size results in lotus aluminum fabricated in hydrogen atmosphere

| Materials (mass%) | Atmosphere ($H_2$/MPa) | Pore diameter (μm) | Porosity (%) | First author (year) [ref] |
|---|---|---|---|---|
| Al (99.8%pure) | 0.1 | | <1.6 | Shinada (1980) [32] |
| Al | 0.05 | 40-60 | 0.2 | Shahani (1985) [33] |
| | 0.1 | 50-60 | 0.9 | |
| | 0.3 | 125-150 | 0.6 | |
| | 0.5 | 150-200 | 1.1 | |
| Al | 0.11 | difficult to produce regular high-porosity Al | | Shapovalov (1993) [34] |
| Al-(2-8)%Fe | 0.11 | $Al_3Fe$ precipitates promote pore formation | | |
| Al | 0.1 | 60 | 5.1 | Zhang (2007) [35] |
| | $0.18H_2$+0.22Ar | 75 | <0.1 | |
| | air | 60 | 0.1 | |
| | air | 60 | 0.3 | |

basic and applied research of a lightweight material. Table 6.1 is a compilation of these works [32–35]. None of the previous works produced porous aluminum with directional pores and a high porosity; the porosity is limited to 5 % at most.

Lotus aluminum can be fabricated utilizing the hydrogen solubility gap between a liquid and a solid at the melting point in the unidirectional solidification process in a hydrogen atmosphere. When the melt dissolving hydrogen is solidified, the insoluble hydrogen is precipitated to evolve directional pores in the solidified region at the solid–liquid interface. Therefore, the following conditions are critical to fabricate highly porous lotus metals.

1. The hydrogen concentration dissolved in the molten metal should exceed several atomic percent.
2. The metal should hold a smaller solid solubility of hydrogen so that the solubility gap between liquid and solid metals becomes larger.

However, the hydrogen solubility in liquid and solid aluminum [36] is only about 2 % of those for magnesium, copper, and transition metals [37, 38]. Consequently, previous researchers considered the fabrication of highly porous lotus aluminum difficult; none of the previous reports successfully fabricated lotus aluminum with a porosity above 5 %.

The influence of solidification conditions (solidification velocity, hydrogen partial pressure, temperature gradient, and melt temperature) on pore formation of lotus aluminum was experimentally and theoretically investigated by Ide et al. [39] to elucidate the pore formation mechanism as well as to control porosity. They fabricated lotus aluminum with slender unidirectional pores using a continuous casting technique under controlled solidification conditions. In particular, it was clarified that the porosity and the pore size are affected significantly by the solidification velocity as the velocity decreases the porosity and pore size by one to two orders of magnitude compared to those for copper, magnesium, and transition metals. Based on this knowledge, lotus aluminum with a porosity as high as 40 % was obtained for the first time.

A continuous casting technique was utilized for unidirectional solidification in a pressurized hydrogen atmosphere under controlled solidification conditions. The temperature of the melt in the crucible was measured using two WRe 5/26-type thermocouples located 5 mm from the bottom of the crucible. Then to solidify the melt in a continuous downward direction, a graphite dummy bar pulled molten aluminum through a cooling mold at a constant solidification velocity. The solidification temperature using two K-type thermocouples inserted in the graphite dummy bar. To ensure the temperature gradient was at a steady state, the temperature was measured 20 mm above the top of the dummy bar. The measured temperature was recorded every second using a temperature recorder (Keyence Co., Ltd.; GR-3500). The temperature gradient was calculated using the obtained cooling curve in the temperature range from 933 to 873 K (temperature–time curve). The temperature gradient was determined using the following equation:

$$G = \frac{\dot{V}}{R}, \tag{6.6}$$

where $G$, $V$, and $R$ are the temperature gradient (K mm$^{-1}$), cooling rate during unidirectional solidification (K s$^{-1}$), and solidification velocity (mm s$^{-1}$), respectively. The temperature gradient was determined from the average of two temperature gradients, whose difference was less than 0.3 K mm$^{-1}$).

The effects of different characteristics on pore morphology were examined. The influence of the solidification rate was probed by melting pure aluminum and unidirectionally solidifying in a mixture of hydrogen 0.25 MPa and argon 0.25 MPa with a constant temperature gradient and melt temperature of 9.7 K mm$^{-1}$ and 1,223 K, respectively, while changing the solidification velocity from 0.5 to 0.9 mm min$^{-1}$. The impact of hydrogen partial pressure was investigated by melting aluminum and unidirectionally solidifying in three different atmospheres: hydrogen 0.5 MPa, argon 0.5 MPa, or mixed gas (0.5 MPa) consisting of hydrogen (0.25 MPa) and argon (0.25 MPa) when the solidification velocity, temperature gradient, and melt temperature were set to 0.9 mm min$^{-1}$, 9.5 K mm$^{-1}$, and 1,223 K, respectively.

Figure 6.27 shows the typical pore morphology perpendicular (upper row) and parallel (lower row) to the solidification direction of lotus aluminum as a function of solidification velocity from 0.5 mm min$^{-1}$ to 0.9 mm min$^{-1}$. Lotus aluminum was fabricated by unidirectional solidification in a mixture of hydrogen (0.25 MPa) and argon (0.25 MPa). Figure 6.27f shows an outer view of lotus aluminum with the highest porosity of nearly 40 %. Unidirectional pores aligned parallel to the solidification direction are observed in the solidified lotus aluminum. The porosity and average pore diameter decrease as the solidification velocity increases.

Figure 6.28 shows the pore morphology in lotus aluminum fabricated by unidirectional solidification at a solidification velocity of 0.9 mm min$^{-1}$ as a function of

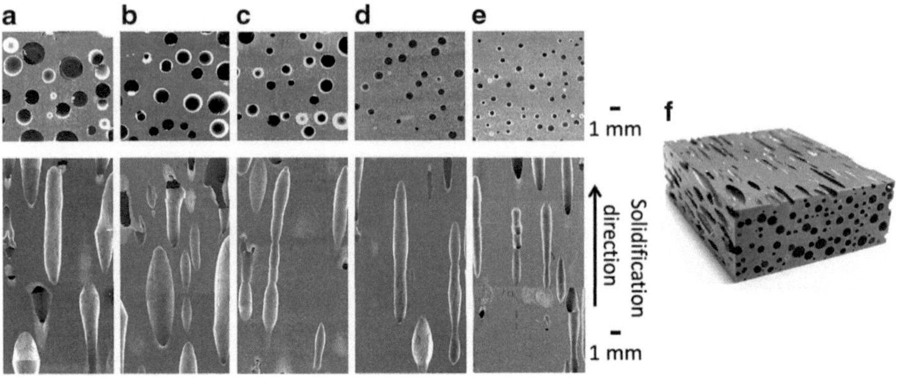


**Fig. 6.27** Cross sections perpendicular (*upper row*) and parallel (*lower row*) to the solidification direction of lotus aluminum. Solidification velocity: (**a**) 0.5 mm $min^{-1}$, (**b**) 0.6 mm $min^{-1}$, (**c**) 0.7 mm $min^{-1}$, (**d**) 0.8 mm $min^{-1}$, and (**e**) 0.9 mm $min^{-1}$. (**f**) Outer view of lotus aluminum with the highest porosity of nearly 40 %

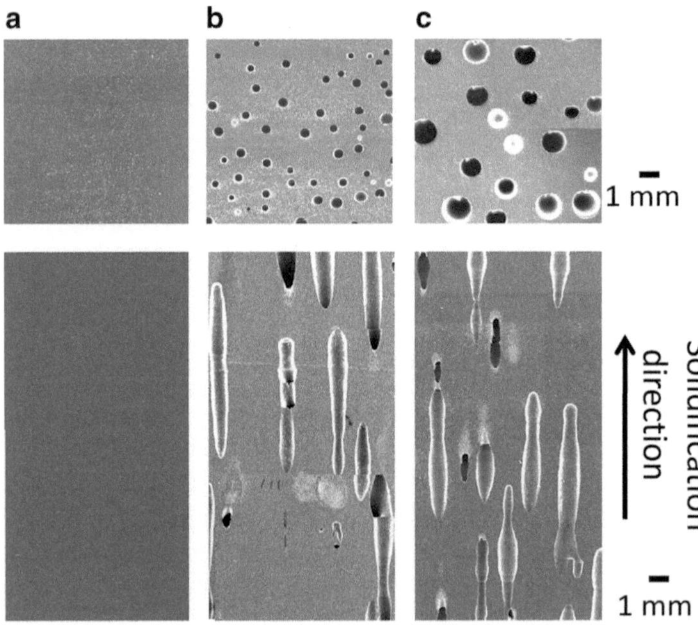


**Fig. 6.28** Cross sections perpendicular (*upper row*) and parallel (*lower row*) to the solidification direction of lotus aluminum fabricated in different atmospheres: (**a**) argon (0.5 MPa), (**b**) mixed gas of hydrogen (0.25 MPa) and argon (0.25 MPa), and (**c**) hydrogen (0.5 MPa). Temperature gradient and melt temperature are 9.5 K $mm^{-1}$ and 1,223 K, respectively

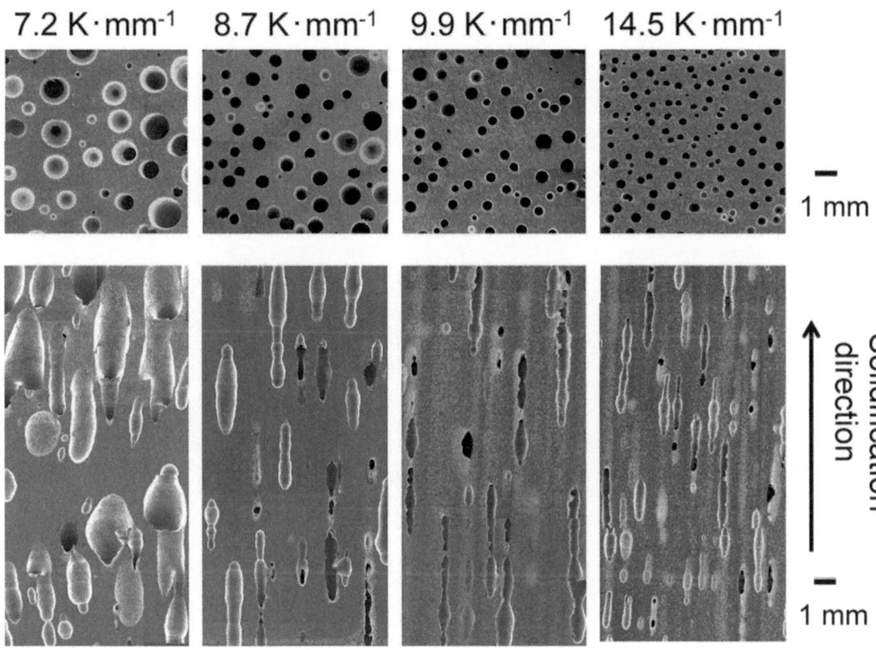


**Fig. 6.29** Cross sections perpendicular (*upper row*) and parallel (*lower row*) to the solidification direction of lotus aluminum fabricated with different temperature gradients at the solid–liquid interface. Lotus aluminum is fabricated in a mixed gas of hydrogen (0.5 MPa) and argon (0.5 MPa) at a melt temperature of 1,373 K and a fixed solidification velocity of 0.5 mm min$^{-1}$

the hydrogen partial pressure; photos in the upper and lower rows are the cross-sectional views perpendicular and parallel to the solidification direction, respectively. The atmospheric total pressure was kept constant at 0.5 MPa, while the partial pressure of hydrogen was either zero, 0.25 MPa, or 0.5 MPa and the partial pressure of argon was 0.5, 0.25 MPa, or zero, respectively. The ingot solidified in an argon atmosphere does not exhibit pores (Fig. 6.28a). In contrast, ingots prepared in a hydrogen atmosphere contain pores (Fig. 6.28b, c). It is apparent that dissolving gas in the melt is responsible for the evolution of pores, because hydrogen but not argon can dissolve in molten aluminum. Porosity and pore diameter increase as the hydrogen partial pressure increases.

Figure 6.29 shows cross-sectional views of the pore morphology perpendicular and parallel to the solidification direction of lotus aluminum fabricated at different temperature gradients. The porosity and average pore diameter decrease as the temperature gradient increases. Figure 6.30 shows the change in pore morphology due to the difference in the melt temperature of lotus aluminum fabricated with a solidification velocity of 0.9 mm min$^{-1}$ in a 0.5 MPa hydrogen atmosphere. The porosity and average pore diameter increase as the melt temperature increases.

As mentioned above, the porosity of lotus aluminum depends on not only hydrogen partial pressure but also the solidification velocity, temperature gradient, and melt

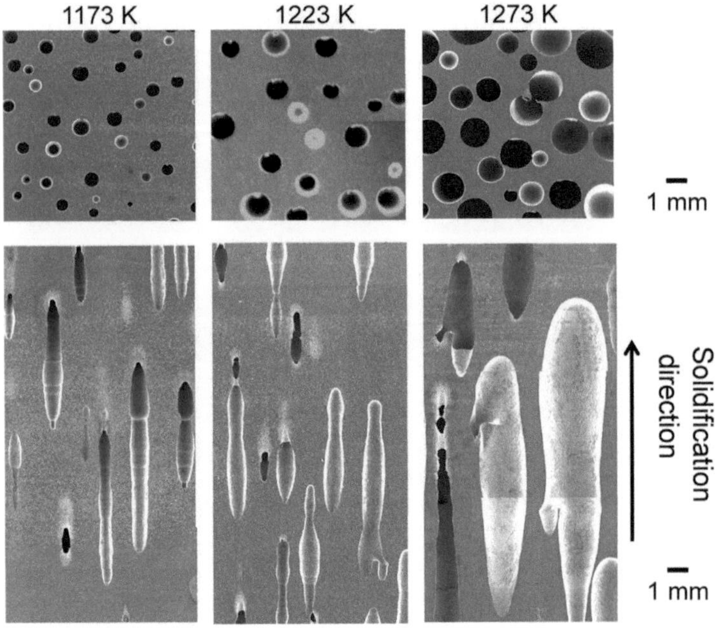


**Fig. 6.30** Dependence of porosity and pore diameter on melt temperature of lotus aluminum fabricated with solidification velocity of 0.9 mm $min^{-1}$ in 0.5 MPa hydrogen. Temperature gradient is $9.5 \pm 0.5$ K $mm^{-1}$

temperature. These dependencies are similar to those of dual-phase formation in eutectic alloys; the flux of solute atoms due to solidification conditions affects the morphology of the phases. This suggests that a change in the hydrogen flux by different solidification conditions significantly affects pore formation in lotus aluminum.

The solubility difference of hydrogen in aluminum is more than one order of magnitude smaller than that in copper, whereas the diffusion coefficients are similar in both aluminum and copper. To promote pore growth in lotus aluminum, hydrogen in aluminum should diffuse over a longer distance than in copper because the solubility of hydrogen is lower in aluminum. Figure 6.31 schematically depicts the effect of a long diffusion distance of hydrogen rejected in a solid in the vicinity of the solid–liquid interface on pore evolution in lotus aluminum. Assuming the diffusion distance of hydrogen increases, the porosity and pore diameter in lotus aluminum increase as the number of hydrogen atoms increases, resulting in the formation and growth of pores. Thus, the mechanism for pore evolution in lotus aluminum may be the opposite of the mechanism for lotus copper and stainless steel where the formation and growth of pores easily occur even for short distant diffusion of hydrogen due to the high content of hydrogen rejected in the solid. Therefore, the condition of a lower solidification velocity is certainly crucial to realize highly porous lotus aluminum.

The porosity and pore diameter decrease as the solidification velocity and temperature gradient increase, but increase as the hydrogen partial pressure and

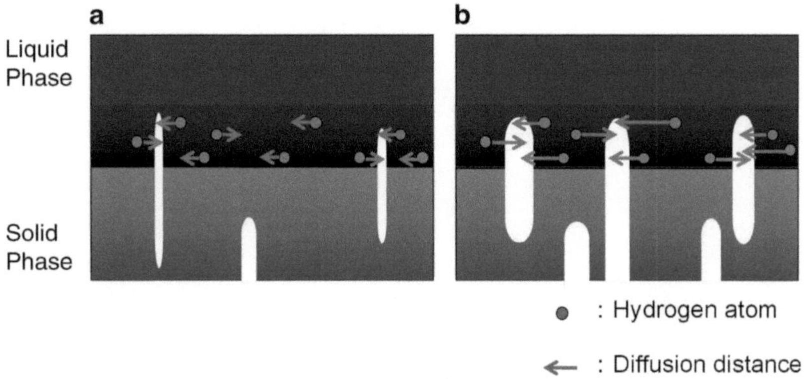


**Fig. 6.31** Two-dimensional model for the growth of unidirectional pores near the liquid–solid interface. Hydrogen rejected in solidified aluminum accumulates in the liquid near the liquid–solid interface. (**a**) When the solidification velocity is fast, hydrogen diffuses only a short distance. Consequently, the impact of hydrogen on growth is small, leading to small pores and a low porosity. (**b**) When the solidification velocity is slow, hydrogen diffuses over a longer distance. Because a large amount of hydrogen contributes to growth, large pores and a high porosity are obtained

melt temperature increase. Previous works have reported the effects of solidification velocity and hydrogen partial pressure on pore formation of lotus metals. However, the influence of solidification velocity on pore formation in lotus aluminum differs from previous reports using other lotus metals such as copper [25] and stainless steel [40]. Similar to the fabrication of lotus copper or stainless steel, the pore diameter decreases as the solidification velocity increases. However, unlike for lotus aluminum, the porosity for lotus copper or stainless steel is independent of solidification velocity. Such a difference is attributed to the hydrogen solubility between the liquid and solid phases. For example, the hydrogen solubility gap between the liquid and solid phases in aluminum ($4.93 \times 10^{-4}$ mol% under $H_2$ 0.1 MPa [36]) is about 40 times smaller than that of copper ($2.09 \times 10^{-2}$ mol% at $H_2$ 0.1 MPa [37]), but there is not a significant difference in the diffusion coefficients [41, 42]. Therefore, the supersaturated hydrogen atoms in aluminum have to migrate a relatively long distance toward the pores to contribute to pore formation and growth. For a fast solidification velocity, the hydrogen atoms in aluminum cannot migrate a sufficient distance to grow large pores. On the other hand, for a slow solidification velocity, the hydrogen atoms in aluminum can migrate a longer distance, allowing more supersaturated hydrogen atoms to contribute to pore formation and growth, producing a larger porosity.

Next, we discuss effects of a hydrogen flux, which depends on the solidification conditions, on pore formation in lotus aluminum. In eutectic alloys, the morphology of a lamellar or rod structure is affected by solidification conditions. Such a change is due to the change in the flux of the solute. For a binary system composed of metal and hydrogen, we consider a dual phase, which consists of a combination of pores and aluminum. Figure 6.32 shows a two-dimensional model for the competitive

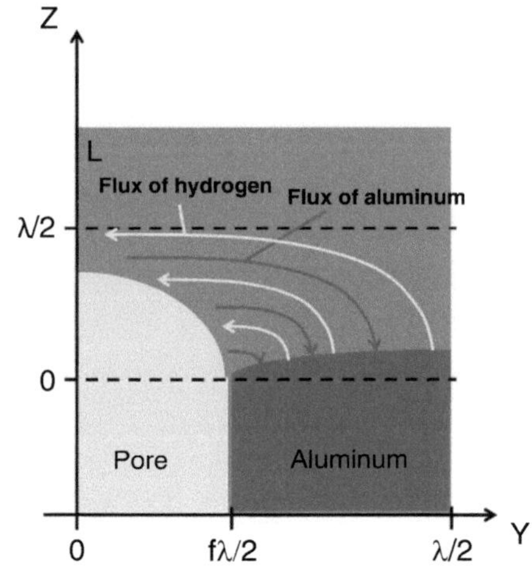


**Fig. 6.32** Schematic drawing of the diffusion flux of hydrogen and aluminum during pore evolution and growth in the solidification front

growth between directional pores and solid aluminum under steady-state conditions. Ide et al. [39] treated the dual phase in two dimensions, and due to symmetry, only half of a lamellae for each phase needs to be considered [43]. As shown in Fig. 6.32, aluminum rejects hydrogen atoms due to the solubility gap in the melt, whereas the pore rejects aluminum atoms during unidirectional solidification. As well as the solute distribution in the eutectic structure, the mass of diffusing hydrogen (aluminum) atoms into pore (aluminum) per unit of time can be obtained by solving the diffusion equation for solute in the diffusion layer at the solidification front. Therefore, when the *z*-axis is parallel to the solidification direction and the *y*-axis is perpendicular to the solidification direction in the $y{-}z$ coordinate system (Fig. 6.32), the distribution of the solute concentration at the solidification front can be expressed by [43]

$$C = C_0 + A\ \exp\left(\frac{-Vz}{D}\right) + B\ \exp\left(\frac{-2\pi z}{\lambda}\right)\cos\left(\frac{-2\pi y}{\lambda}\right), \tag{6.7}$$

where $C_0$ and $D$ are concentration (mol) and diffusion coefficient (m$^2$ s$^{-1}$) of solute in molten metal, respectively, $V$ is the growth velocity (m s$^{-1}$), and $\lambda$ is the distance between periodic phase structure (m). Here $A$ and $B$ are, respectively,

$$A = f\left(C_S^\beta - C_S^\alpha\right) + C_E - C_S^\beta, \tag{6.8}$$

$$B = \frac{f(1-f)V\lambda\left(C_S^\beta - C_S^\alpha\right)}{2D\sin(\pi f)}, \tag{6.9}$$

where $f$ is the volume fraction of the$\alpha$phase, $C_S^\alpha$ and $C_S^\beta$ are the solid solubility limit of $B$ atoms in the $\alpha$ and $\beta$ phases (mol m$^{-3}$), respectively. When $B$ atoms are rejected by the $\alpha$ phase and diffuse into the $\beta$ phase through $y = f\lambda/2$, the flux of $B$ atoms, $J$ (mol s$^{-1}$ m$^{-2}$) is obtained by

$$J = -D\left(\frac{\partial C}{\partial y}\right)_{y=\frac{f\lambda}{2}} = \pi f(1-f)V\left(C_S^\beta - C_S^\alpha\right)\exp\left(-\frac{2\pi z}{\lambda}\right). \tag{6.10}$$

Here, the width of the inspissated solute layer (diffusion layer) is considered to be about $\lambda/2$. When a small distance $dl$ is solidified, the amount of diffusing hydrogen, $C_{\mathrm{dl}}$ (mol m$^{-1}$), into the nearby phase through the small $\Delta l$ is given by

$$J_{\mathrm{dl}} = \int_0^{\lambda/2} Jdz \times \frac{dl}{V} = f(1-f)\left(C_S^\beta - C_S^\alpha\right)\lambda dl, \tag{6.11}$$

where $\alpha$ and $\beta$ are the aluminum and pore phases, respectively, and $A$ and $B$ atoms denote aluminum and hydrogen atoms, respectively. $C_\alpha$ and $C_\beta$ are the hydrogen solubility in solid aluminum and the gas phase (mol m$^{-3}$), respectively. Because solid aluminum is not dissolved in the gas phase, the molar ratio of hydrogen is equal to 1. Therefore, $C_\alpha$ and $C_\beta$ are expressed as

$$C_\alpha = \frac{\eta_{T_n}}{V_{\mathrm{Al}}} \times \sqrt{P_{H_2}}, \tag{6.12}$$

$$C_\beta = \frac{2RT_n}{P} \times \sqrt{P_{H_2}}. \tag{6.13}$$

Here the volume fraction in the equilibrium phase diagram is used as the volume fraction $f$. However, because the porosity of lotus aluminum depends on the solidification conditions, the volume fraction of aluminum phase $f$ is obtained from the experimental results. In the present work, Yamamura's model for mass balance of lotus copper [25] is modified by considering hydrogen diffusion, i.e., the terms indicating the mass of hydrogen are substituted using Eqs. (6.9)–(6.11).

[Initial amount of hydrogen contained in the liquid in the volume element]
+ [Amount of hydrogen contained in the liquid flowing into the volume element to compensate for solidification shrinkage]
– [Amount of hydrogen contained in the liquid flowing out of the volume element due to pore formation]
+ [Amount of hydrogen dissolved in solid copper around the pore with a length of $l + dl$].

In Yamamura's model, the hydrogen mass balance for a unit length as the solidification front advances $dl$ is considered. Because the number of pores per unit length and supplied hydrogen from both sides of the pores have to be considered as the solidification front advances $dl$, the mass of diffusing hydrogen is written as

$$W_H = f(1-f)(C_\beta - C_\alpha)\lambda \times \frac{2}{\lambda} \times \frac{M}{2} dl \tag{6.14}$$

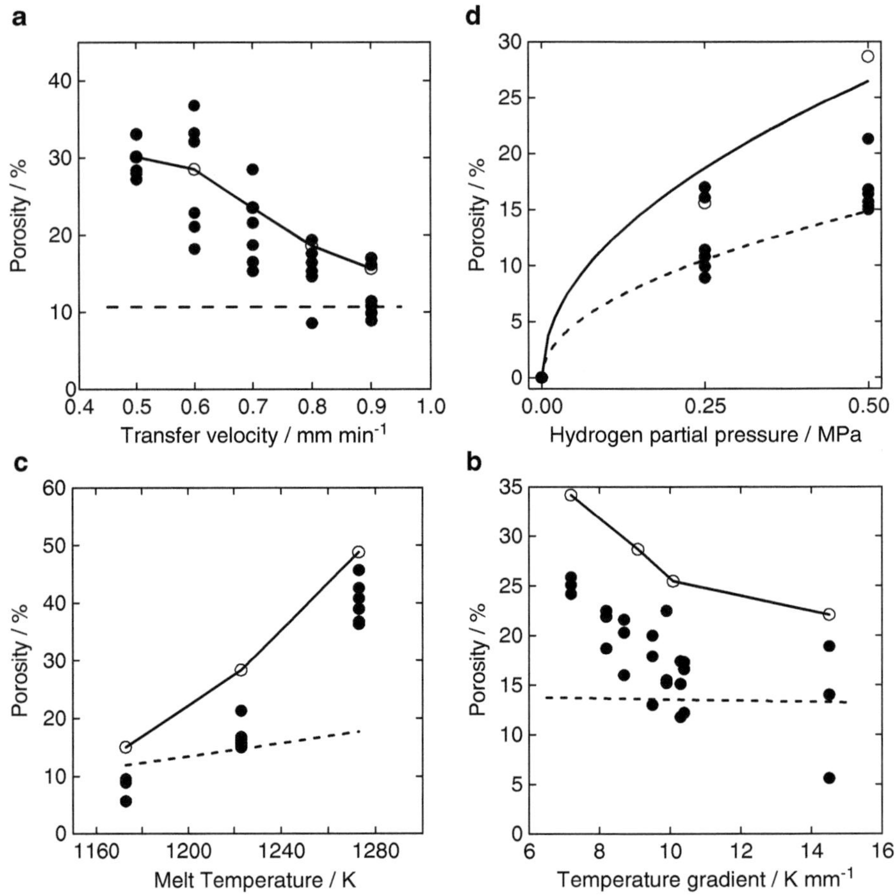


**Fig. 6.33** Dependence of porosity on (**a**) solidification velocity, (**b**) hydrogen partial pressure, (**c**) melt temperature, and (**d**) temperature gradient. *Closed* and *open circles* indicate experimental data and the calculated results using Eq. (6.16), respectively. *Dotted lines* are the predicted results using Yamamura's model

In the present work, $\lambda$ is given by following equation using porosity $\varepsilon$ (%) where pore diameter $d_p$ (m) given by experimental results:

$$\lambda = \frac{d_p}{\varepsilon}. \tag{6.15}$$

Equation (6.14) is substituted in Yamamura's model and the porosity of lotus metal is expressed as

$$\varepsilon = \frac{fM(1-f)(C_\beta - C_\alpha) + (\eta_{T_n} - \eta_{T_n - Gl})\sqrt{P} \times \frac{M}{2V_{Al}}}{\frac{MP}{R(T_n - Gl)} + (\eta_{T_n} - \eta_{T_n - Gl})\sqrt{P} \times \frac{M}{2V_{Al}}} \tag{6.16}$$

Figure 6.33 shows the dependence of porosity on (a) solidification velocity, (b) hydrogen partial pressure, (c) melt temperature, and (d) temperature gradient.

To compare the experimental results and the calculated results predicted using Eq. (6.16) with Yamamura's model [25], closed circles, open circles, and the dotted line denote the experimental results, calculated results using Eq. (6.16), and the results predicted by Yamamura's model, respectively. The calculated porosities using Eq. (6.16) agree well with the experimental results. In Yamamura's model, the porosity was calculated by assuming that all of the rejected hydrogen atoms due to the solubility gap form pores; parameter $a = 1$. Although the dependence of the porosity calculated from Yamamura's model on the hydrogen partial pressure is consistent with the experimental results, the dependence of the porosity for the other factors using Yamamura's model is inconsistent with the experimental results. These observations are reasonable because Yamamura's model does not consider the effect of the solidification condition.

Moreover, even if all of hydrogen atoms rejected in the solidified metal evolve into pores, the maximum porosity calculated by Yamamura's model is less than 15 %, which is far below the maximum experimental porosity of 40 %. This discrepancy suggests that the contribution of hydrogen diffusion in the melt rejected in the solid phase near the solid–liquid interface is critical. Thus, control of solidification conditions such as the solidification velocity, hydrogen partial pressure, temperature gradient, and melt temperature is crucial to increase the porosity of lotus metal with a low hydrogen solubility.

## References

1. Simmons JW (1996) Mater Sci Eng A 207:159–169
2. Hyun SK, Nakajima H (2002) Mater Trans 43:526–531
3. Satir-Kolorz AH, Feichtinger HK (1991) Z Metallkde 82:689–697
4. Levinsky Y (ed) (1997) Pressure dependent phase diagrams of binary alloys. ASM International, Materials Park, p 693
5. Raghaban V (ed) (1987) Phase diagram of ternary iron alloys. Indian Inst Tech, Delhi, p 143
6. Shah ID, Parlee NAD (1967) Trans AIME 239:763–764
7. Nakahata T, Nakajima H (2005) Mater Trans 46:587–592
8. Hoshiyama H, Ikeda T, Murakami K, Nakajima H (2003) J Jpn Inst Metals 67:714–720
9. Hirano T (1990) Acta Metall 38:2667–2671
10. Yamaguchi M, Inui H, Ito K (2000) Acta Mater 48:307–322
11. Deevi SC, Sikka VK (1996) Intermetallics 4:357–375
12. Hyun SK, Ikeda T, Nakajima H (2004) J Jpn Inst Metals 68:39–42
13. Lange KW, Schenck H (1969) Z Metallkd 60:638
14. Turner DR (1958) J Electrochem Soc 105:402–408
15. Unagami T, Seki M (1978) J Electrochem Soc 125:1339–1344
16. Cullis AG, Canham LT (1991) Nature 353:335–338
17. Nakahata T, Nakajima H (2004) Mater Sci Eng A 384:373–376
18. Przyborowski M, Hibiya T, Eguchi M, Egry I (1995) J Cryst Growth 151:60–65
19. Ishizaki K, Komarneni S, Nanko M (1998) Porous materials: process technology and applications, Materials technology series. Kluwer, Dordrecht
20. Zhang GJ, Yang JF, Ohji T (2001) J Am Ceram Soc 84:1395–1397
21. Ding XJ, Zhang JZ, Wang RD, Feng CD (2002) J Eur Ceram Soc 22:411–414
22. Isobe T, Tomita T, Kameshima Y, Nakajima A, Okada K (2006) J Eur Ceram Soc 26:957–960

23. Fukasawa T, Deng ZY, Ando M, Ohji T (2001) J Ceram Soc Jpn 109:1035–1038
24. Ueno S, Lin LM, Nakajima H (2008) J Am Ceram Soc 91:223–226
25. Yamamura S, Shiota H, Murakami K, Nakajima H (2001) Mater Sci Eng A318:137–143
26. Aoki T, Ikeda T, Nakajima H (2003) Mater Trans 44:89–93
27. Ikeda T, Nakajima H (2002) J Jpn Foundry Eng Soc 74:812–816
28. Hoshiyama H, Ikeda T, Nakajima H (2007) High Temp Mater Process 26:303–316
29. Watanabe T (1976) Light Metals 25:167
30. Kashihara M, Yonetani H, Suzuki S, Hyun SK, Kim SY, Kawamura Y, Nakajima H (2007) Porous metals and metallic foams. MIT, Boston, pp 201–204
31. Nakajima H (2010) Proc Jpn Acad Ser B 86:884–899
32. Shinada H, Nishi S (1980) J Jpn Inst Light Metals 30:317–323
33. Shahani H, Fredriksson H (1985) Scand J Metall 14:316–320
34. Shapovalov VI, Timchenko AG (1993) Phys Met Metall 76:335–337
35. Zhang H, Li Y, Liu Y (2007) Acta Metall Sinica 43:11–16
36. Qiu C, Olson GB, Opalka SM, Anton DL (2004) J Phase Equilib Diff 25:520–527
37. Fromm E, Gebhardt E (1976) Gases and carbon in metals. Springer, Berlin
38. San-Martin A, Manchester FD (1987) Bull Alloy Phase Diagrams 8:431–437
39. Ide T, Iio Y, Nakajima H (2012) Metall Mater Trans A 43A:5140–5152
40. Ikeda T, Aoki T, Nakajima H (2005) Metall Mater Trans A36:77–86
41. Papp K, Csetenyi EK (1981) Scr Metall 15:161–164
42. Wright JH, Hocking MG (1972) Metall Trans 3:1749–1753
43. Kurz W, Fisher DJ (1998) Fundamental of solidification. Trans Tech Publications, Switzerland

# Chapter 7
# Mechanical Properties of Lotus Metals and Alloys

**Abstract** Many conventional porous materials possess almost nearly spherical pores and exhibit that those mechanical properties are usually isotropic. Moreover, the shape of pores is not always round and distorted so that the stress concentration takes place easily, which causes to weaken the strength of those materials. In addition, the overall uniformity of the pore size and the porosity is very crucial which may affect the stress concentration; some local strain under the stress is accumulated into the nonuniform region to degrade the materials strength. Thus, the foamed materials, cellular-structured materials, or sintered materials exhibit inferior mechanical properties.

Different from such conventional porous materials, lotus materials have elongated cylindrical pores aligned unidirectionally, and various mechanical properties are remarkably different from those of isotropic porous materials and show significant anisotropic behavior. Although the data on the mechanical properties of such anisotropic porous materials are not so much accumulated compared with the isotropic porous materials, it is at present good enough to review some systematic view of the mechanical properties of the lotus metals.

**Keywords** Compressive strength • Elastic constants • Fatigue strength • Internal friction • Tensile strength

## 7.1 Elastic Properties

Tane et al. measured the anisotropic elastic constants of lotus iron fabricated using the continuous zone melting technique [1, 2], extended the effective-mean-field (EMF) theory so as to take account of the pore-orientation effect on the effective elastic constants, and applied the extended EMF theory to lotus iron to validate the theory as a prediction method of the elastic properties of lotus metals. They prepared two kinds of lotus iron, containing hydrogen pores or nitrogen pores, with various porosities. The specimens were cut out of the ingots and machined into

H. Nakajima, *Porous Metals with Directional Pores*,
DOI 10.1007/978-4-431-54017-5_7, 

the rectangular parallelepipeds with the surfaces normal to the pore (solidification) direction.

When the material possesses orthorhombic symmetry, there are eight independent groups of free vibrations: OD(dilatation), EV(torsion), OX, OY, OZ (shear), and EX, EY, EZ (flexure) [3, 4]. The resonance frequencies of the vibration modes depend on the mass density, dimensions, and elastic constants of the specimen. With the known elastic constants, the resonance frequencies can be calculated using the measured density $\rho$ and dimensions:

$$\rho\omega^2 = \int_V \frac{1}{4} C_{ijkl} \left( \frac{\partial u_i}{\partial x_j} + \frac{\partial u_j}{\partial x_i} \right) \left( \frac{\partial u_k}{\partial x_l} + \frac{\partial u_l}{\partial x_k} \right) dV, \tag{7.1}$$

where $C_{ijkl}$ is an elastic constant tensor in the four-index notation and $\omega = 2\pi f$ is the angular frequency, with $f$ the resonance frequency. Equation (7.1) is computed by the Rayleigh–Ritz method, in which the displacement vector $u$ is expressed by the linear combination of the basis functions $\{\phi\}$:

$$u = (a_p\phi_p, a_q\phi_q, a_r\phi_r). \tag{7.2}$$

Products of the Legendre functions are often chosen for $\{\phi\}$ [3]. It is not straightforward to find the elastic constants from the resonance spectrum. They are determined inversely through iterative calculations. First, the resonance frequencies are measured using two piezoelectric transducers; one sends continuous waves into the specimens and the other detects the signals to pick up resonance peaks. Next, the resonance frequencies are calculated using the assumed elastic constants and are compared with the measured resonance frequencies. If a good agreement between the two spectra is achieved, the assumed values are regarded as the true values. During this procedure, one should know an exact correspondence between the resonance peaks and the vibration modes. For this, one has employed the mode-selective electromagnetic acoustic resonance method (EMAR), which can excite and detect only one vibration group among OX, OY, OZ, and OD vibration groups. Figure 7.1 shows the excitation, as an example, of OX group by EMAR. The resonance frequencies have been measured at room temperature within the frequency range of 50–300 kHz at 0.01-kHz steps.

Figure 7.2 shows the porosity dependence of the Young's moduli $E_\perp$ and $E_\parallel$ and the elastic stiffness $c_{11}$, $c_{33}$, $c_{13}$, $c_{44}$, and $c_{66}$ of lotus iron with hydrogen or nitrogen pores, where $E_\parallel$ and $E_\perp$ indicate Young's moduli in the directions parallel and perpendicular to the pore direction, respectively. As well as the porosity dependence of other lotus metals [5, 6], $E_\parallel$ decreases linearly, while $E_\perp$ drops steeply in the small porosity region. On the other hand, Hyun et al. have found that the mechanical strength of lotus iron prepared in nitrogen atmosphere is much higher than that in hydrogen atmosphere owing to the solid solution hardening by the solute nitrogen [7]. Regarding its elastic properties, however, any significant modifications was not observed [8].

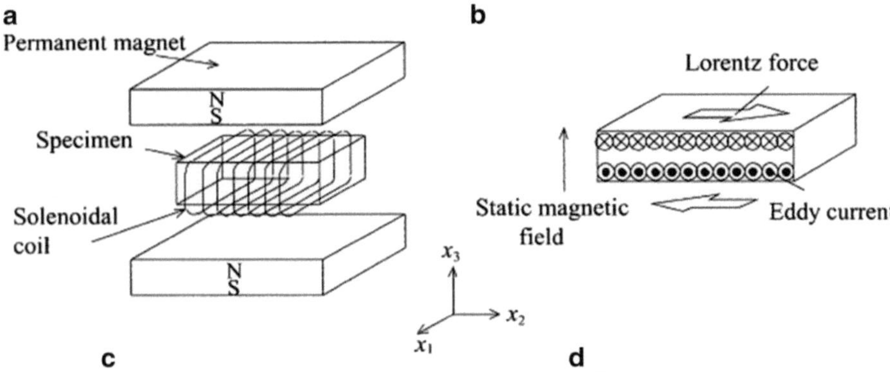


| | $x_1$ | $x_2$ | $x_3$ |
|---|---|---|---|
| $u_1$ | Odd | Odd | Odd |
| $u_2$ | Even | Even | Odd |
| $u_3$ | Even | Odd | Even |

**Fig. 7.1** Excitation of the OX vibration group by the EMAR method. (**a**) A specimen is inserted into the coil and the magnetic field is applied. (**b**) Eddy currents on the specimen surface are induced by RF current in the coil, and the shearing Lorentz force is generated. (**c**) The Lorentz forces deform the specimen as shown in *thick arrows*. The displacement $u_2$ is an even function, for example, against the $x_2$ mirror plane. (**d**) The parity of the displacements for the OX group. The displacement $u_2$ in (**c**) only satisfies the OX parity (Reprinted with permission from [5] © 2002 Elsevier Science Ltd)

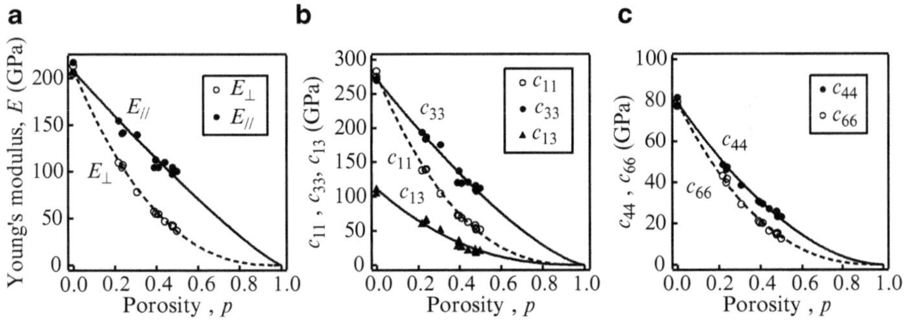


**Fig. 7.2** Porosity dependence of (**a**) two Young's moduli, $E_{\parallel}$ in the direction parallel to the $x_3$-axis and $E_{\perp}$ in the direction perpendicular to $X_3$, and the elastic stiffness coefficients, (**b**) $c_{11}$, $c_{33}$, $c_{13}$, (**c**) $c_{44}$ and $c_{66}$ of lotus iron. The transverse isotropy condition $c_{66} = (c_{11} - c_{12})/2$ holds. Each line is obtained by fitting the equation (7.3) to the measurements (Reprinted with permission from [8] © 2004 Elsevier Ltd)

By extensive researches on the effective physical properties (e.g., the electrical conductivity, yield stress, and elastic modulus) of the porous materials, their porosity dependencies are experimentally found to follow the power-law formula [9–13]:

$$M = M_0(1 - p)^m, \tag{7.3}$$

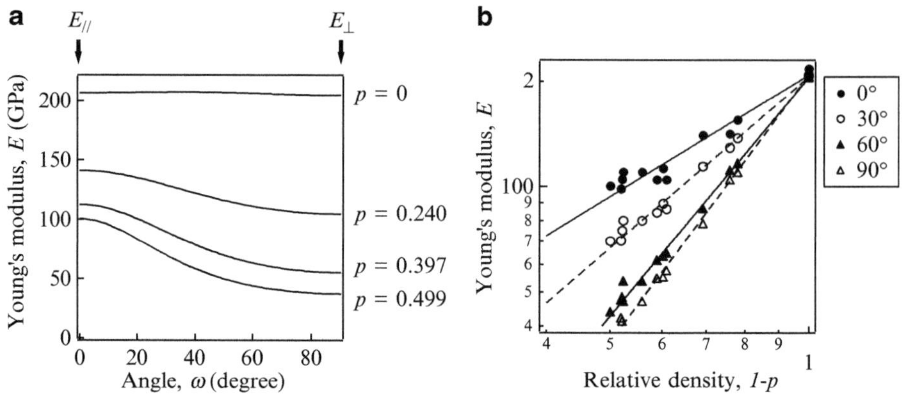


**Fig. 7.3** (**a**) Young's modulus as a function of the angle ω and (**b**) log–log plot of Young's modulus versus porosity in the direction of ω from the $x_3$-axis for lotus iron (Reprinted with permission from [8] © 2004 Elsevier Ltd)

where $M$ and $M_0$ are the physical properties of porous and nonporous material, respectively, and $m$ is the coefficient empirically determined. The solid and broken lines in Fig. 7.2 indicate Eq. (7.3) fitted to the measurement data. Thus, one finds that the measurements lie on them, indicating that Eq. (7.3) holds for the effective elastic constants of the anisotropic porous metals. The values of $m$ for lotus iron are similar to those for lotus magnesium, i.e., they seems to be independent of Poisson's ratio of the isotropic matrix and mainly depend on the shape and alignment of the pores. In contrast, the values of $m$ for lotus copper are slightly different from the others. They appear to depend also on the elastic anisotropy of the matrix.

Since lotus metals are fabricated by using the unidirectional solidification method, preferential orientations in the crystal growth usually appear. This is the case for lotus copper and lotus magnesium [5, 6]. In the case of lotus iron, however, X-ray diffraction proved that the matrix has no texture (not presented here). Hence, despite the fact that the single-crystal iron shows a relatively strong elastic anisotropy $(2c_{44}/(c_{11} - c_{12}) \approx 2.4)$ [12, 13], the nonporous iron has isotropic elastic constants and the two directional Young's moduli of nonporous iron are substantially equal to each other. Although the matrix of lotus magnesium has a texture, it virtually exhibits elastic isotropy because the elastic property of single-crystal magnesium is nearly isotropic [6]. In contrast, the matrix of lotus copper shows elastic anisotropy that originates from a texture consisting of elastically anisotropic crystals [5]. Thus, the texture effects on the matrix metal are observed case by case.

Since all the independent elastic constants were determined, the elastic modulus in an arbitrary direction is available through the coordinate conversion. Figure 7.3a shows the Young's modulus $E$ of lotus iron as a function of the angle $\omega$ from the $x_3$ axis. The Young's modulus monotonically decreases with increase in $\omega$, which simply arises from the increase of the stress concentration around pores. Figure 7.3b shows the log–log plot of Young's modulus of lotus magnesium versus porosity in the direction of $\omega = 0$, 30, 60, and 90°. The solid lines in Fig. 7.3b show fits of Eq. (7.3) to the measurement data. Thus, Young's moduli in arbitrary directions can also be expressed using Eq. (7.3).

## 7.2 Internal Friction

In lotus materials hydrogen exits in a solid solution state in metals and in gaseous state in pores. For the practical use of these materials, it is important to know the behavior of hydrogen, especially transportation paths and desorption temperature. Internal friction (IF) method has been known to be sensitive to impurity atoms and lattice defects. Ota et al. measured IF of lotus copper and found a large IF peak around 700 K [14]. They suggested that hydrogen is responsible for the IF peak, although the detailed mechanism was not known yet. Recently Yoshinari et al. investigated the IF and hydrogen desorption behavior of lotus copper with various pore sizes in order to discuss the mechanism of the influence of hydrogen on internal friction [15].

The specimens for the IF measurement which have a cylindrical shape (2 mm in diameter and 20–30 mm in length) were cut from the lotus ingot by a spark-erosion wire-cutting machine. Two types of specimens were prepared; the cylindrical axis of specimens are selected so as to be parallel or perpendicular to the direction of the lotus pores whose average pore size and porosity are compiled in Table 7.1. The crystal grains in specimens were grown in the transfer direction. The grain size of lotus copper was almost the same as the pore size, while that of nonporous copper was about 0.5 mm. Hereafter lotus specimens with directional pores which are parallel and perpendicular to the specimen axis will be referred to as, e.g., lotus15$\|$ and lotus15$\perp$, respectively. An apparatus of a forced excitation torsion pendulum type was used for IF measurements as illustrated in Fig. 7.4. The measurements were done under the condition s of measuring frequencies of 0.1–10 Hz, temperature range of 300–1,100 K, and a heating rate of 3 K min$^{-1}$. The torsional vibration was excited by supplying a torque to a magnet attached to torsion axis through the external AC magnetic field. The torsion strain was detected by an optical lever system. Both the excitation system and the detection system were connected to a personal computer via DA and AD converters, respectively. The IF, $Q^{-1}$, is detected as $\tan\varphi$, where $\varphi$ is a phase lag between stress and strain (or between

**Table 7.1** Fabrication conditions for prepared specimens

| Specimen | Ar pressure (MPa) | $H_2$ pressure (MPa) | Transfer velocity (mm min$^{-1}$) | Porosity (%) | Average pore size (mm) |
|---|---|---|---|---|---|
| lotus15 | 0.6 | 0.4 | 15 | 34.2 | 0.30 |
| lotus20 | 0.6 | 0.4 | 20 | 35.8 | 0.27 |
| lotus50 | 0.6 | 0.4 | 50 | 35.1 | 0.06 |
| Nonporous | 1.0 | 0.0 | 20 | | |

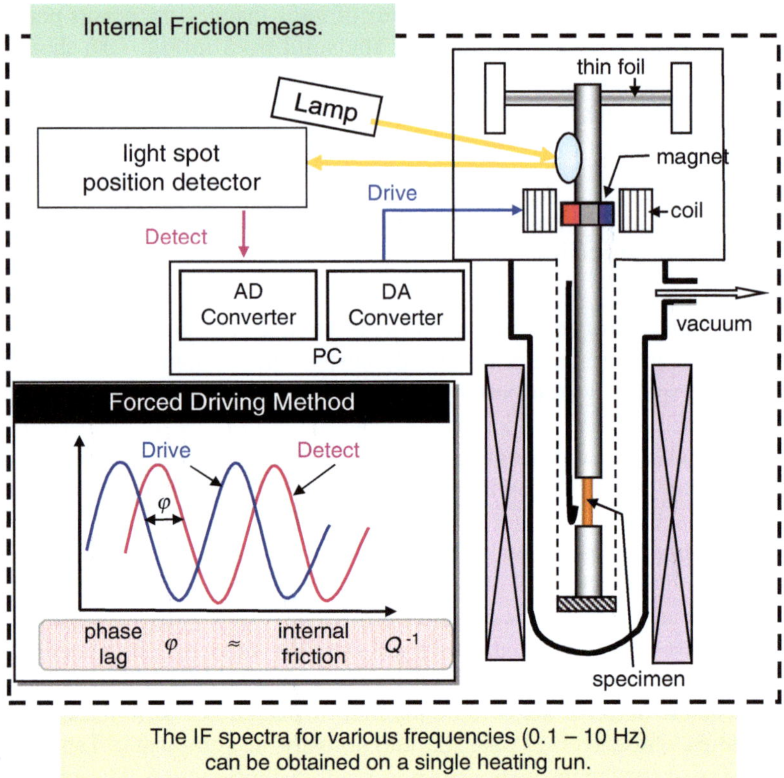


**Fig. 7.4** An apparatus for measurement of internal friction and the internal friction spectra for various frequencies (0.1–10 Hz)

driving force and detected strain). The phase lag was obtained by curve fitting of stress and strain signals. The typical accuracy for $\varphi$ is about $10^{-4}$ (0.005°). The IF spectra for various frequencies were measured on a single heating run. Thermal desorption spectrum (TDS) of hydrogen was measured as follows: hydrogen gas desorbed from specimen was detected by a quadrupole mass spectrometer during constant heating of 12 K $\text{min}^{-1}$ and dynamic evacuating. Spectrum of hydrogen pressure versus temperature was obtained in a range between 300 and 1,000 K.

Figure 7.5 shows the results of IF measurements for lotus15||, lotus20||, and lotus50|| specimens. On the first heating runs for as-prepared specimens, a broad IF peak with a peak temperature of 610–710 K was observed. After heating up to 1,100 K, the specimens were subjected to the second heating measurements (Fig. 7.5b). On the second run, an IF peak appears about 100 K higher than that on the first run. In order to confirm whether this peak shifts from the lower temperatures or it newly appears, measurements were repeated for another lotus50|| specimen with raising the upper limit of the heating as shown in Fig. 7.6. The peak, which appears around 620 K on the first run, decreases and shifts toward lower

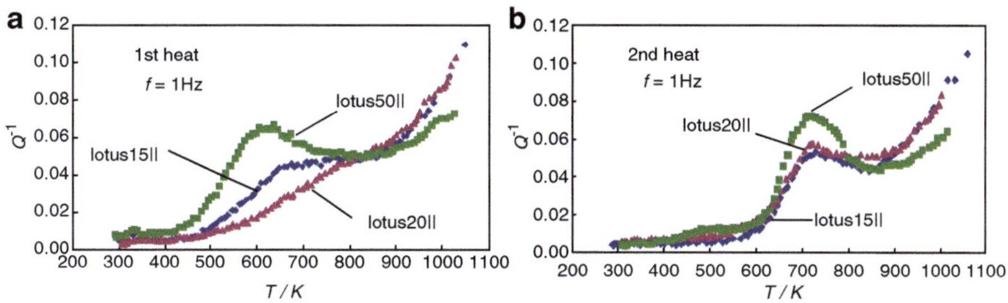


**Fig. 7.5** IF spectra of lotus copper for frequency of 1 Hz. (**a**) 1st and (**b**) 2nd heating runs (Reprinted with permission from [15] © 2012 GSIntervision)

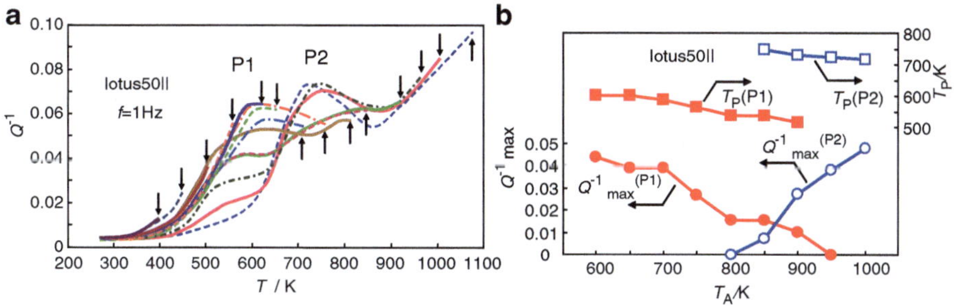


**Fig. 7.6** (**a**) IF spectra of lotus 50|| measured repeatedly by increasing the upper heating limit. (**b**) Variation of peak temperature ($T_p$) for $f = 1$ Hz and peak height ($Q^{-1}{}_{max}$) for P1 and P2 with annealing temperature ($T_A$) (Reprinted with permission from [15] © 2012 GSIntervision)

temperature with the annealing. A new peak appears around 750 K after annealings at temperatures higher than 850 K. It should be noted that the two peaks coexist after the annealings at 850–950 K. Therefore, it is clarified that the two peaks have different origins. In the following, 620 and 750 K peaks are referred as P1 and P2, respectively. In Fig. 7.6b the heights, $Q^{-1}{}_{max}$, and peak temperatures, $T_p$, of P1 and P2 are summarized as a function of the annealing temperature, $T_A$, which is a maximum temperature of the previous heating run. Although $Q^{-1}{}_{max}$ of P2 also decreases with $T_A$, it should be noted that the relation between $Q^{-1}{}_{max}$ and $T_p$ is opposite for P1 and P2. The IF was also measured for specimens in which directional pores lie in the direction perpendicular to that of specimen axis. The results for lotus15⊥ specimen are shown in Fig. 7.7.

The IF of nonporous specimen was also measured for comparison. Figure 7.8 shows the first and second heating measurements for nonporous specimen. It is shown that there is almost no IF peak except for a continuously rising background on the first heating run, while a peak is observed around 670 K on the second run. Since the properties of the peak are similar to P2 observed for lotus specimens, it is considered that it has the same origin as P2.

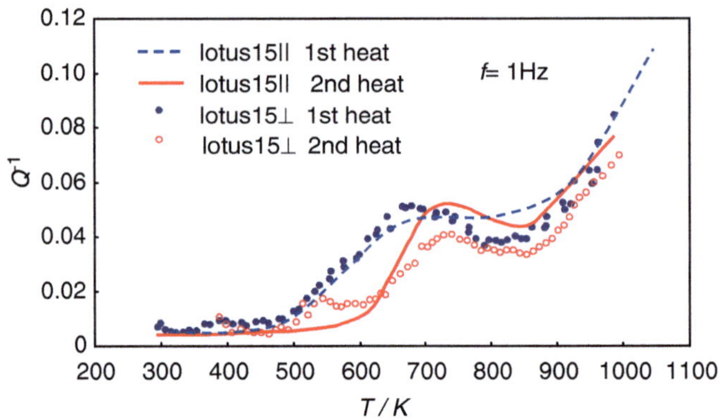


**Fig. 7.7** IF spectra with pore direction perpendicular to specimen axis (lotus15⊥) comparing with parallel pore direction (lotus 15||), $f = 1$ Hz (Reprinted with permission from [15] © 2012 GSIntervision)

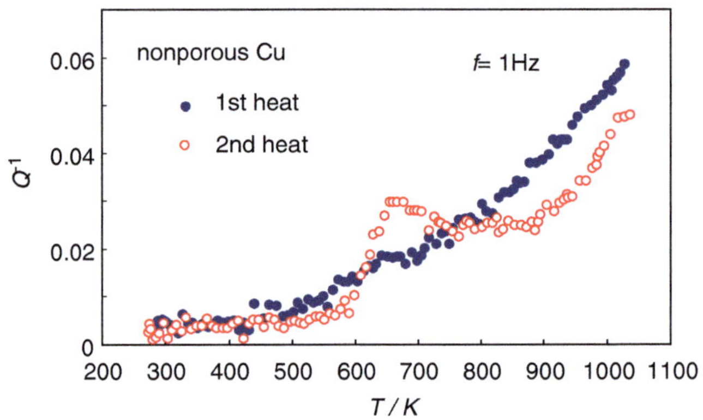


**Fig. 7.8** IF spectra of nonporous copper specimens on first and second heating runs. $f = 1$ Hz (Reprinted with permission from [15] © 2012 GSIntervision)

Figure 7.9 shows hydrogen spectrum for lotus50 specimen. Two peaks are observed; a clear peak is observed around 690 K and a shoulder peak is also observed around 850 K. On the second heating run, the 690 K peak completely disappears, while the 850 K peak still remains although its magnitude somewhat decreases. Therefore, it is considered the 690 K peak corresponds the hydrogen desorption from the specimen and the 850 K peak contains hydrogen release from places other than the specimen. The amount of hydrogen estimated from the 690 K peak was about 40 and 10 at.ppm for lotus50 and lotus20 specimens, respectively. The amount of hydrogen is much smaller than the hydrogen solubility of 470 at. ppm under the specimen fabrication condition (melt temperature 1,523 K and hydrogen pressure 0.4 MPa) [16]. It is considered that a large amount of hydrogen was released during the cooling process of specimen fabrication.

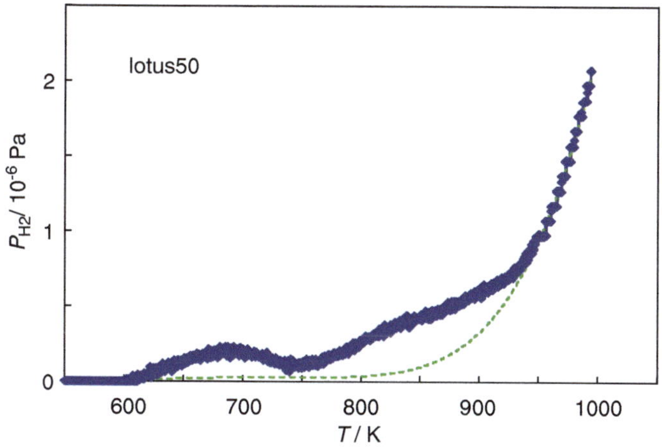


**Fig. 7.9** Hydrogen TDS for lotus 50 specimens. *Dashed line* shows a suitable background (Reprinted with permission from [15] © 2012 GSIntervision)

The origin of P2 may not be related with hydrogen because P2 appears after the hydrogen release from specimens and P2 is also observed for nonporous specimen which is fabricated under atmosphere without hydrogen. It is well known for aluminum and copper that a large IF peak is observed for a temperature range of 0.4–0.6 $T_m$, where $T_m$ is the melting temperature. The mechanism of the peak was believed to be the grain boundary (GB) sliding, because no IF peak is observed for single-crystalline specimens and the peak has been referred to as the GB peak or Kê peak [17]. Although it is not clear whether P2 is caused by GB sliding or dislocation movements, anelastic strain which is responsible for P1 may be the same as that of P2 since the peak height of P1 is compatible to P2 and the magnitudes of them are complementary as shown in Fig. 7.6b. In the case of CW peak observed in bcc metals, the existence of hydrogen retards the dislocation motion and the CW peak appears at higher temperatures than the original peak. It is considered for the case of P1 that the existence of hydrogen helps the motion of the defects which is responsible for the P2 relaxation. As a result, P1 appears at lower temperatures than P2. It is a new idea that the hydrogen affects the so-called GB relaxation and it may be helpful for the elucidation of its mechanism. Furthermore, it is interesting to use lotus specimens for such a purpose since anisotropic GBs and hydrogen atoms simultaneously exist in them.

## 7.3 Tensile Strength

Little data have been available on the mechanical properties of this type of lotus metals. Wolla and Trovenzano [18] and Simone and Gibson [19] measured the tensile strength of porous copper with pore orientation parallel to the tensile direction. However, their data exhibited large scatter, which may be attributed to

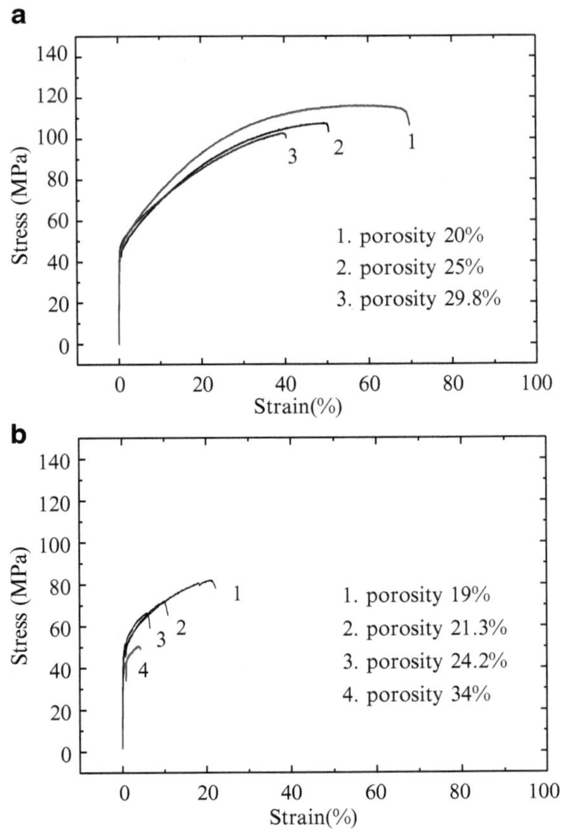


**Fig. 7.10** Stress–strain curves for the specimens with the pore orientation (**a**) parallel and (**b**) perpendicular to the tensile direction of lotus copper (Reprinted with permission from [20] © 2001 Elsevier Science Ltd)

the microstructural variations both within each specimen and among specimens. Furthermore, no investigations were carried out on the anisotropy in the mechanical properties of the porous metals with elongated cylindrical pores. Hyun et al. [20] undertook to measure the ultimate tensile strength and the yield strength of the lotus copper with elongated cylindrical pores and to elucidate the uniaxial tensile behavior of lotus copper with the pore orientation parallel to and perpendicular to the tensile direction. Moreover, the fracture surfaces of the selected tensile test specimens were investigated to obtain information concerning the fracture mode of these materials. Typical stress–strain curves for the lotus copper specimens are shown in Fig. 7.10. The curves show linear elastic behavior at small strains, followed by yield and strain hardening up to the peak stress. The ductility of the specimens decreases with increasing porosity. The ultimate tensile strength of specimens with cylindrical pore orientation parallel to the tensile direction is plotted against the porosity in Fig. 7.11. The data points for the ultimate tensile strengths lie on a straight line which passes through the point of 0 MPa at the porosity of 100 %; the specific ultimate tensile strength does not change by the pore existence. This fact indicates that the pores whose axes are aligned parallel

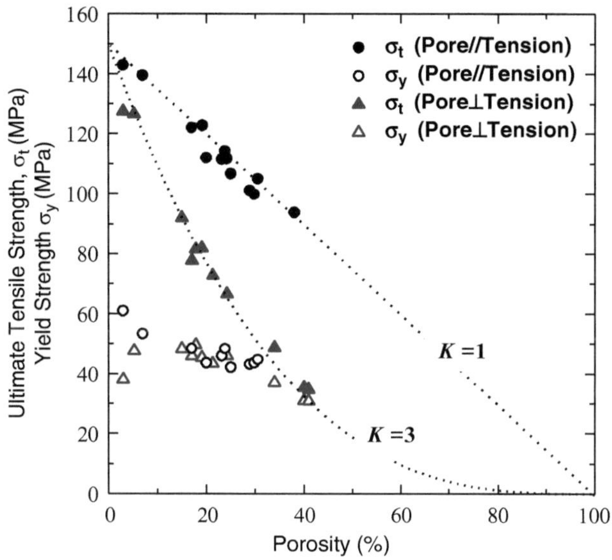


**Fig. 7.11** Ultimate tensile strength and yield strength of lotus copper in the direction parallel and perpendicular to pore axis versus porosity (Reprinted with permission from [20] © 2001 Elsevier Science Ltd)

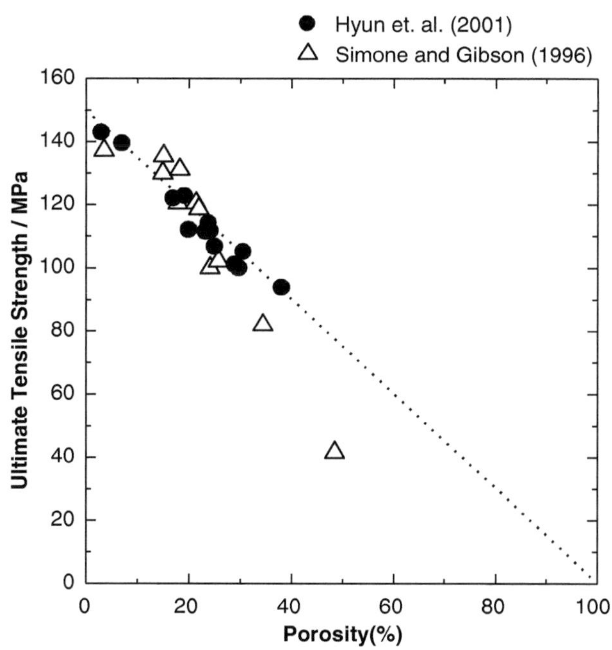


**Fig. 7.12** Comparison of data obtained from two different groups for ultimate tensile strength of lotus copper in the direction parallel to the pore axis (Reprinted with permission from [20] © 2001 Elsevier Science Ltd)

to the tensile direction cause little stress concentration in the tensile specimens. Simple rule of mixture of the empty pores and the solid body can be applied to these specimens.

As shown in Fig. 7.12, the ultimate tensile strength obtained by Hyun et al. [20] agrees well with the data by Simone and Gibson [19] at a low porosity levels, but the strength is considerably higher in the higher porosity range. This difference may be due to the production method of tensile specimens.

On the other hand, the ultimate strengths of specimens with cylindrical pores perpendicular to the tensile direction are also plotted against the porosity in Fig. 7.11

The ultimate tensile strength of the specimen with pores perpendicular to the tensile direction is much lower than that of specimen with the pores parallel to the tensile direction at a given porosity.

Correlation between the ultimate tensile strength and the porosity of sintered porous materials was studied extensively [21–23]. The measurements were limited in the low porosity levels from 0 to 40 %, because the strength for porosity higher than about 60 % becomes almost 0 MPa. It is considered that both load-bearing areas model [21] and stress concentration model [23] are reasonable to determine the strength of porous materials, since load-bearing area is reduced with increasing porosity and causes stress concentration around the pores.

### 7.3.1 *Ultimate Tensile Strength*

Balshin [22] suggested a relation of the form

$$\sigma = \sigma_0(1-p)^K, \tag{7.4}$$

to describe the variation of strength with the porosity, where $K$ is the empirical constant which depends on the materials and the fabrication method. Recently Baccaccini et al. [23] suggested that the empirical constant $K$ is related with the stress concentration around pores in the porous materials. The stress concentration factor of the pores depends on the pore geometry and orientation with respect to the direction of applied stress and can be expressed by

$$K = \frac{\sigma_{\max}}{\sigma}, \tag{7.5}$$

where $\sigma_{\max}$ is the maximum value of the stress, which is given by the following equation using the radius parameters, $a$ and $b$, of an ellipse-shaped pore in the tensile specimen as shown in Fig. 7.13 [24].

$$\sigma_{\max} = \sigma\left(1 + 2\frac{a}{b}\right). \tag{7.6}$$

For the specimen with the cylindrical pores aligned parallel to the tensile direction, since $b$ becomes infinite, the value of $K$ approaches unity, where no stress concentration takes place. Thus, Eq. (7.6) can be simply rewritten as

$$\sigma = \sigma_0(1-p). \tag{7.7}$$

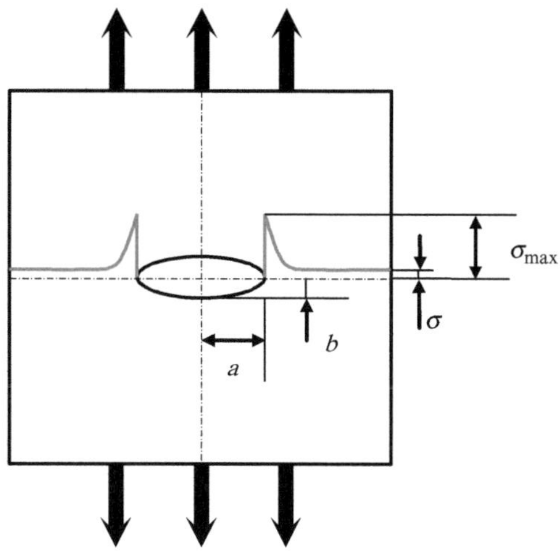


**Fig. 7.13** Stress concentration in the vicinity of elliptical pore in the infinitely large panel. The pore is located in the center. The stress is concentrated in both edges of the pore (Reprinted with permission from [20] © 2001 Elsevier Science Ltd)

For the specimen with cylindrical pores oriented perpendicular to the tensile direction, the equality $a = b$ can be held, so that the value of $K$ approaches 3. Then, the stress can be expressed as

$$\sigma = \sigma_0(1 - p)^3. \tag{7.8}$$

The ultimate tensile strength evaluated from Eqs. (7.7) and (7.8) is shown in Fig. 7.11, where it is found that the experimental results are in good agreement with the dotted lines estimated from Eqs. (7.7) and (7.8).

Such agreement suggests that (1) the ultimate tensile strength for porous copper with cylindrical pores parallel to the tensile direction exhibits no stress concentration and (2) the ultimate tensile strength with the pore orientation perpendicular to the tensile direction can be quantitatively explained in terms of stress concentration. Such anisotropic behavior for the tensile strength is applicable not only to lotus copper but also to other lotus metals and alloys. Thus, the anisotropy is a general trend.

Although the mechanical properties of porous materials have been an object of study for a long time, there is little attention to the correlation between the yield strength and the porosity. Some authors [25, 26] reported that two types of yielding occurred. At low porosity level the yield strength linearly decreased with increasing porosity, and the yield strength over the critical porosity level drastically decreased with increasing porosity. Experimental results shown in Fig. 7.11 are in agreement with their results. Although it is difficult to explain perfectly the reason for this, they suggested the following mechanism. As seen in Fig. 7.14, if a pore is in the tensile plate, the stress concentrates at the edge of the pore. Therefore, the stress level of this region is higher than that of matrix. Before the stress level reaches the yield strength, at first plastic deformation locally occurs in the region of stress concentration. Plastic deformation is accompanied by plastic hardening; therefore, this

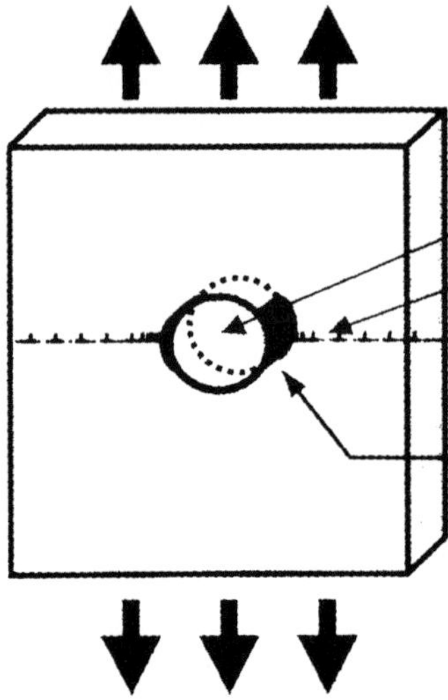


**Fig. 7.14** Formation of barrier to dislocation movement in the region of stress concentration (Reprinted with permission from [20] © 2001 Elsevier Science Ltd)

region becomes barrier such as Cottrell–Lomer locks [27, 28], against moving dislocations in the matrix until the cross slip makes progress. Consequently, the yield strength of this sample will not rapidly decrease with increasing porosity at low porosity levels.

Here, we have to remark on abnormal tensile properties of lotus iron fabricated by nitrogen. Hyun et al. [7] fabricated the lotus iron, not only by hydrogen gas but also by nitrogen gas. They found extraordinary merit of the superior mechanical strength of lotus iron. Typical stress–strain curves of the lotus iron specimens fabricated by hydrogen or nitrogen gas are shown in Fig. 7.15. The tensile direction is parallel to the direction of the pore axes. The curves show a linear elastic behavior at small strains, followed by yield and strain hardening up to the peak stress. Drastic difference in the strength of the specimens fabricated with hydrogen and nitrogen was observed even at similar porosity; the ultimate tensile strength of the specimen fabricated by nitrogen is about twice higher than that by hydrogen. According to the chemical analysis, hydrogen of 27.7 mass ppm is contained in the specimens fabricated under hydrogen atmosphere, which is not considered to affect the strength of iron, since the tensile data extrapolated to the zero porosity as shown in Fig. 7.11 are good fit to the result of nonporous pure iron which hardly contains hydrogen. On the other hand, nitrogen as much as 0.0873 mass% is contained in the specimens fabricated with nitrogen atmosphere. Thus, such remarkable enhancement of the tensile strength is attributed to solid solution strengthening due to solute nitrogen atoms.

The ultimate tensile strength of specimens with cylindrical pore orientation parallel to the tensile direction is much higher than that perpendicular as shown in Fig. 7.16. Such anisotropic strength was interpreted by the previous work on lotus copper [2] mentioned above. The pores whose axes are aligned parallel to the

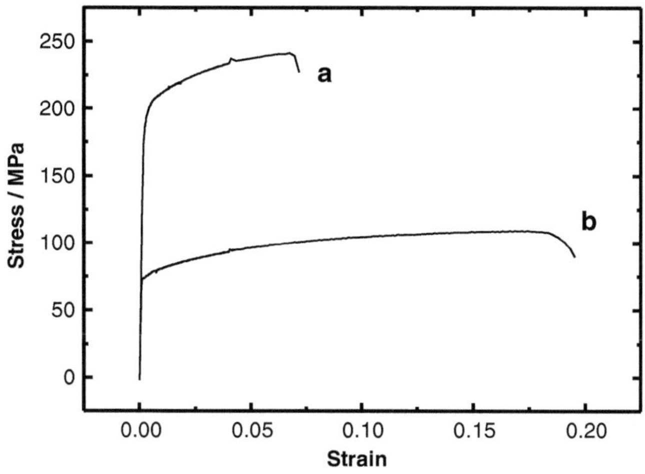


**Fig. 7.15** Stress–strain curves for lotus iron with 50.1 ± 2.6 % porosity fabricated by (*a*) nitrogen or (*b*) hydrogen gas. The tensile direction is parallel to the direction of the pore axis (Reprinted with permission from [7] © 2003 Elsevier Ltd)

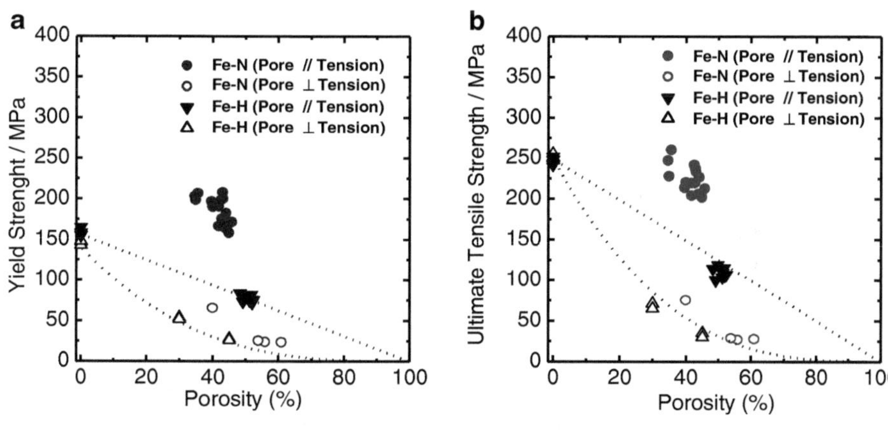


**Fig. 7.16** Porosity dependence of (**a**) yield strength and (**b**) ultimate tensile strength of lotus iron fabricated with hydrogen and nitrogen in the direction parallel and perpendicular to pore growth orientation (Reprinted with permission from [7] © 2003 Elsevier Ltd)

tensile direction cause little stress concentration; the specific tensile strength is constant in spite of the pore existence in the lotus iron. On the other hand, the ultimate tensile strength with the pore orientation perpendicular to the tensile direction can be quantitatively explained in terms of the stress concentration; the stress is accumulated in the vicinity of the pores.

### *7.3.2 Investigation of Tensile Deformation by Acoustic Emission Method*

Anisotropic tensile deformation of lotus copper was investigated with a particular focus on crack formation during deformation. The anisotropy of microscopic deformation was analyzed by an acoustic emission (AE) method that can detect crack formation. The deformed microstructures were observed using a scanning electron microscope (SEM). By correlating the observations pertaining to macroscopic deformation with those pertaining to microscopic deformation, the anisotropic tensile deformation mechanism was examined [29].

A broadband AE sensor was placed on the specimen during the tensile tests, and the AE signals generated by the deformation were used to analyze crack formation. The detected signals were recorded on a personal computer through a discriminator and oscilloscope. Since understanding the crack formation mechanism was central to this study, only the burst AE signals relating to crack formation were analyzed, although some continuous AE signals were also detected. Figure 7.17 shows the amplitudes of burst AE signals detected during the deformation of lotus copper and the corresponding stress–strain curves for parallel loading (Fig. 7.17a) and perpendicular loading (Fig. 7.17b). It is apparent that the elongation and peak stress (ultimate tensile strength) for perpendicular loading are much smaller than those for parallel loading; this result is consistent with that of the previous report mentioned in 7.3.1. Many burst AE signals are detected after yielding to fracture regardless of the loading direction.

In contrast to lotus copper, burst AE signals are not detected when tensile loadings are applied in directions parallel and perpendicular to the solidification direction in nonporous copper as shown in Fig. 7.17c. This implies that the burst AE signals detected in lotus copper are correlated with the porosity. Furthermore, there is no other source of burst AE signals except for crack formation when the normal plastic deformation of ductile copper takes place. Thus, the burst AE signals associated with lotus copper can be attributed to pore-related crack formation. The cumulative AE counts for the parallel and perpendicular tensile loadings illustrate the anisotropy in crack formation (Fig. 7.18). The slopes of cumulative AE counts for perpendicular loading are larger than those for parallel loading, which means that cracks are formed more easily under perpendicular loading.

The change in the cross section, $\Delta S$, during the parallel and perpendicular tensile loadings for lotus copper is shown in Fig. 7.19. $\Delta S$ for nonporous copper is also shown for comparison and is calculated assuming a plastic Poisson's ration of 0.5. The absolute value of $\Delta S$ for perpendicular loading is smaller than that for parallel loading, while the absolute values of $\Delta S$ in lotus copper are much smaller than those in nonporous copper. It is noted that the cross section for perpendicular loading hardly changes despite the occurrence of a large tensile deformation exceeding 20 % strain. These results indicate that the plastic Poisson's ratios of lotus copper are much smaller than the value of 0.5 corresponding to nonporous copper; the porosity strongly decreases the plastic Poisson's ratio.

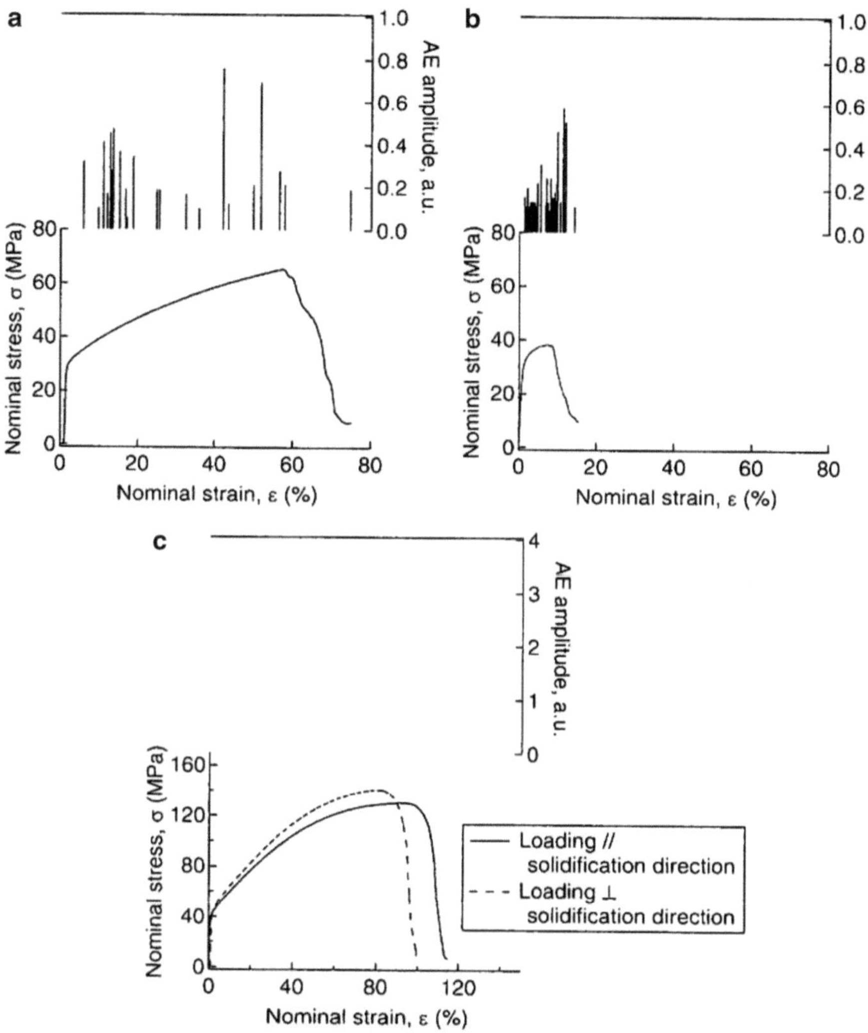


**Fig. 7.17** Amplitude of burst AE signals detected during deformation and the corresponding stress–strain curves for porous copper under (**a**) parallel and (**b**) perpendicular loadings. (**c**) Amplitude of burst AE signals detected during deformation and the corresponding stress–strain curves when tensile loadings parallel and perpendicular to the solidification direction are applied to nonporous copper (Reprinted with permission from [29] © 2010 Materials Research Society)

In addition to other tensile properties, the apparent work-hardening coefficients of lotus copper exhibit anisotropy. The apparent work-hardening coefficients, $n$, were evaluated by fitting the equation $\sigma = K\varepsilon^n$, where $K$ is the strength coefficient, to the true-stress–true-strain curves calculated on the basis of changes in cross section. The apparent work-hardening coefficient for perpendicular loading is

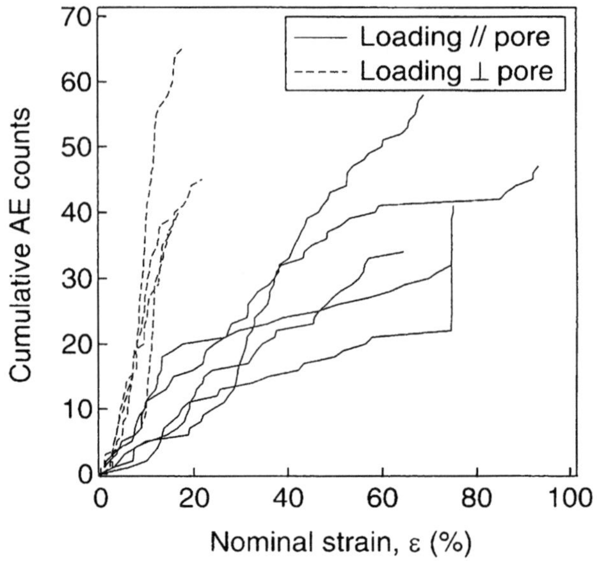


**Fig. 7.18** Cumulative AE counts as a function of nominal strain in the case of parallel and perpendicular loadings (Reprinted with permission from [29] © 2010 Materials Research Society)

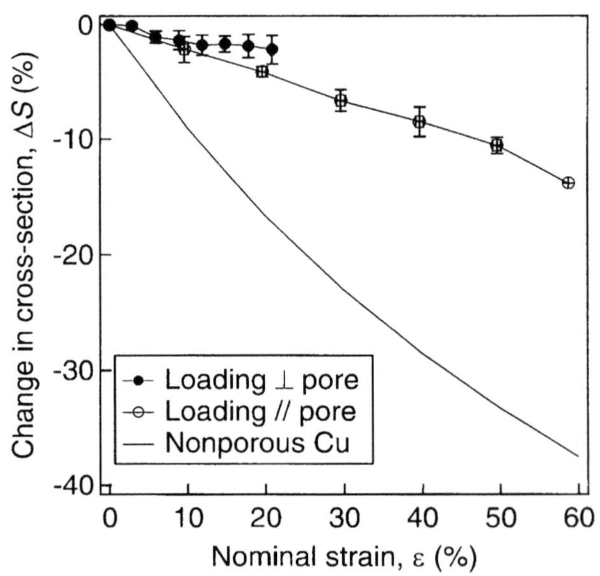


**Fig. 7.19** Change in cross section as a function of nominal strain in the case of parallel and perpendicular loadings. The change in cross section of nonporous copper is shown for comparison. The *plotted values* correspond to the average of five measurements, and the error bars are standard deviation values (Reprinted with permission from [29] © 2010 Materials Research Society)

$0.13 \pm 0.01$, which is smaller than the value of $0.34 \pm 0.04$ for parallel loading. Because the apparent work-hardening coefficients of lotus metals increase with the degree of plastic deformation with regard to applied strain, it is deduced that perpendicular loading produces comparatively less plastic deformation. The reduced plastic deformation is probably the result of the localization of deformation around the pores.

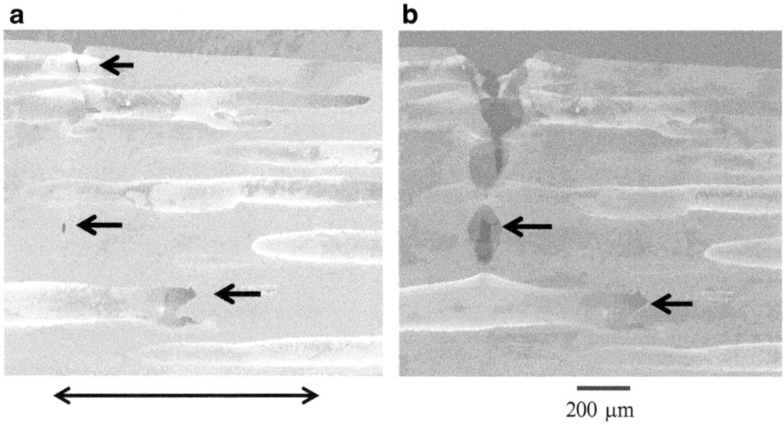


**Fig. 7.20** SEM images of deformed porous structure in the case of parallel loading. The applied nominal strains are (**a**) 31.5 % and (**b**) 41.0 %, which correspond to conditions just before the peak stress and just after the peak stress in the stress–strain curve, respectively (Reprinted with permission from [29] © 2010 Materials Research Society)

Figure 7.20a shows the SEM image of a deformed lotus structure following parallel loading, where the deformation is applied up to just before the point of peak stress in the stress–strain curve. The cracks whose propagation direction is perpendicular to the pore direction are formed as indicated by the arrows. The cracks become remarkably large just after the peak stress, as shown in Fig. 7.20b. Not one crack but many cracks are formed in the vicinity of the peak stress, and they are distributed in the specimen as indicated by the arrows in Fig. 7.21a, b. Figure 7.22a shows the SEM images of a deformed lotus structure just before the peak stress for perpendicular loading. A small crack is formed around the pores as indicated by the arrow. Note that it is more difficult to find cracks on the specimen surface under perpendicular loading than under parallel loading. Just after the peak stress, cracks grow and connect with others as shown in Fig. 7.22b. Figure 7.23a denotes the deformed microstructures on the fracture surfaces under parallel loading. Serpentine glides are formed on the entire observed surface as indicated by the arrows, where the serpentine glide is a sharp and well-defined interwoven pattern [30]. This indicates that slip deformation occurs homogeneously and that the final fracture is caused by glide plane decohesion. In contrast to parallel loading, the deformed microstructures depend on the local positioning in the case of perpendicular loading as shown in Fig. 7.23b. The semibrittle surface is formed near the final fracture region, while slip lines are formed in regions away from the final fracture region. In general, a metal tends to deform in a brittle fashion under high-stress triaxiality [31]. This finding indicates that deformation occurs locally under high-stress triaxiality in the case of perpendicular loading.

The tensile elongation and absolute value of change in the cross section of the specimen are much smaller for perpendicular loading than the corresponding values in the case of tensile deformation of nonporous copper. Under perpendicular

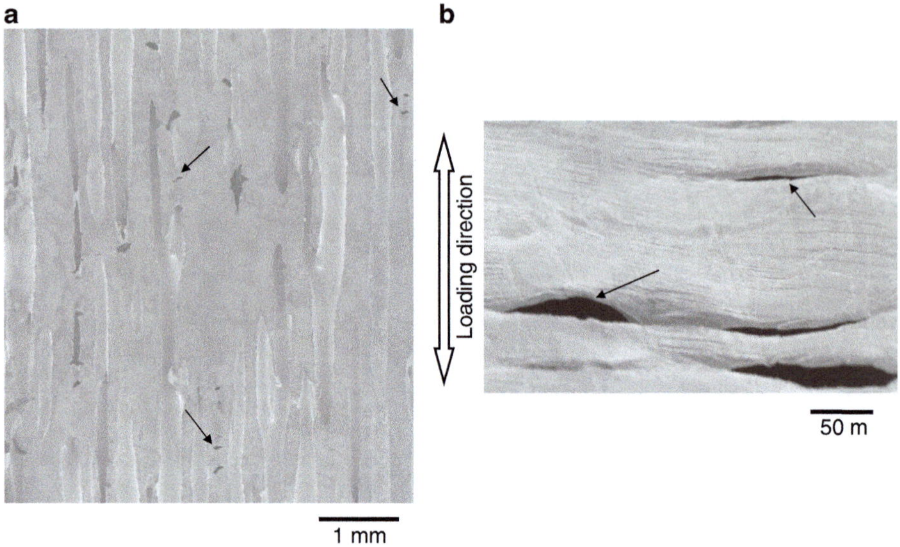


Fig. 7.21 (a) SEM images of deformed porous structure in the case of parallel loading and (b) magnified version of (a). The applied strain is 51.0 %, which corresponds to a strain in the vicinity of peak stress (Reprinted with permission from [29] © 2010 Materials Research Society)

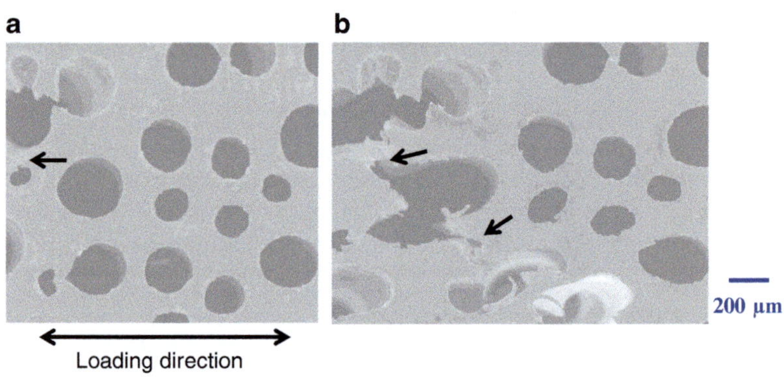


Fig. 7.22 SEM images of deformed porous structure in the case of perpendicular loading. The applied nominal strains are (a) 7.0 % and (b) 9.0 %, which correspond to conditions just before the peak stress and just after the peak stress, respectively (Reprinted with permission from [29] © 2010 Materials Research Society)

loading, lotus copper macroscopically behaves a semibrittle material, although nonporous copper is ductile as shown in Fig. 7.24. Under parallel loading, lotus copper behaves in an intermediate manner between deformation under perpendicular loading and that of nonporous copper as shown in Fig. 7.24. These deformation behaviors of lotus copper are qualitatively consistent with that of a ductile metal, which tends to deform as a brittle material under high-stress triaxiality [31].

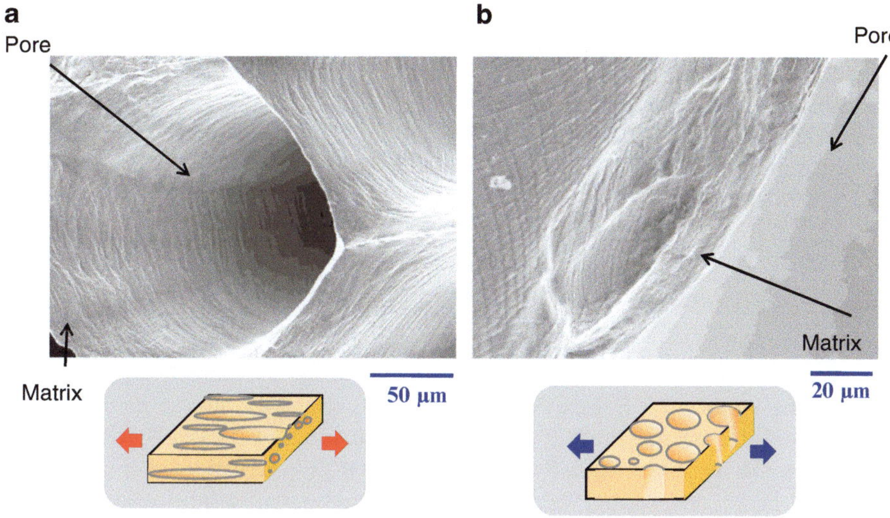


**Fig. 7.23** Deformed microstructure on the fracture surfaces for (**a**) parallel and (**b**) perpendicular loadings (Reprinted with permission from [29] © 2010 Materials Research Society)

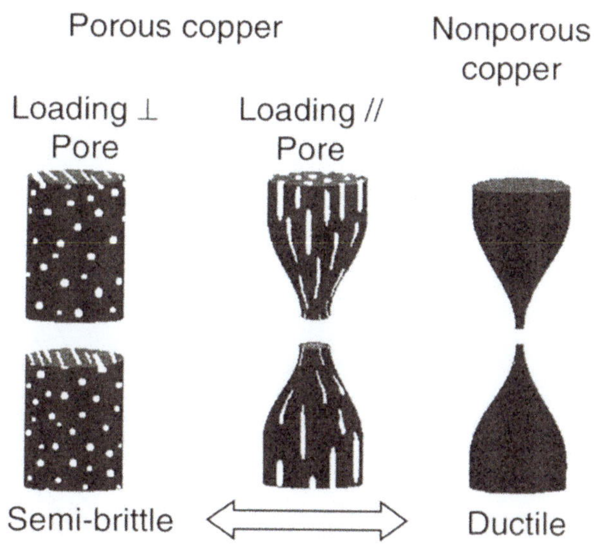


**Fig. 7.24** Macroscopic deformations of porous copper and nonporous copper (Reprinted with permission from [29] © 2010 Materials Research Society)

Crack formation and growth affect tensile deformation in lotus copper, closed-cell metal foams [32, 33], and open-cell metal foams [34] in the same way, being independent of porous structure. Furthermore, in all three materials, the cracks are preferentially formed in weak regions of the matrix. When lotus copper is subjected to parallel loading, reduced stress concentration and stress triaxiality occur around pores as compared with the cases of other porous metals. Thus, crack formation is more difficult, which results in the superior tensile strength and elongation. Therefore, it is important to make the porous structure so homogeneous that the crack formation is suppressed.

## 7.4 Compressive Strength

### *7.4.1 Compressive Yield Strength*

Compressive properties of foamed metals have been extensively investigated. In the strain–stress curves, there is a plateau region, where the stress remains nearly constant with increasing strain [35, 36]. Energy absorption is closely related with this region. On the other hand, Simone and Gibson [37] investigated compressive properties of porous copper fabricated by unidirectional solidification, but no investigation on anisotropy for uniaxial compressive properties was carried out. From the point of view, Hyun and Nakajima [38] investigated the uniaxial compressive properties of lotus copper with elongated cylindrical pores parallel and perpendicular to the compressive direction.

Compressive stress–strain curves for the lotus copper specimens with different porosities and compressive direction are shown in Fig. 7.25. The slope of the stress–strain curves decreases with increasing porosity and depends on compressive direction with respect to the pore growth direction at a given porosity. At low strain level, the stress of the specimen with cylindrical pores parallel to the compressive direction is higher than that of specimen with pores perpendicular. However, this

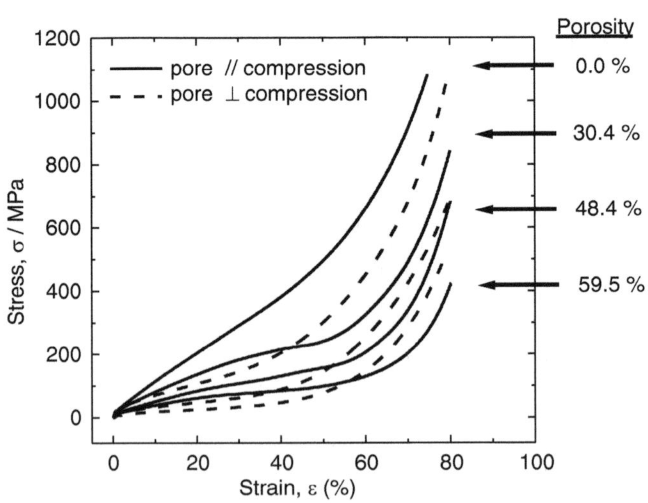


**Fig. 7.25** Compressive stress–strain curves of lotus copper and nonporous copper with different porosities. The compression directions are parallel and perpendicular to the pore growth direction (Reprinted with permission from [38] © 2002 Elsevier Science Ltd)

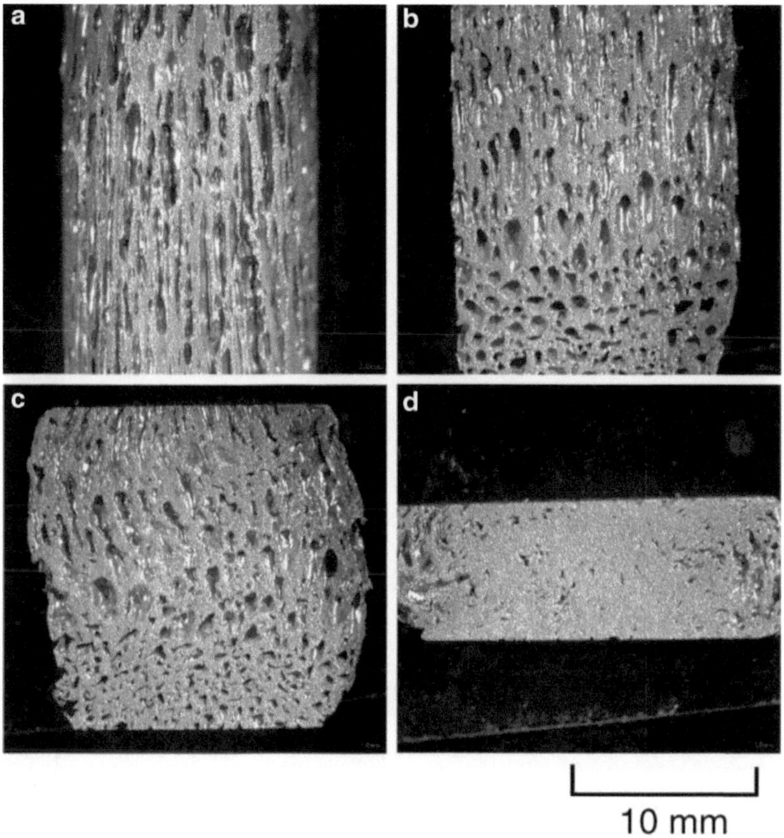


**Fig. 7.26** Optical micrographs of cross section of lotus copper with 59.5 % porosity in the direction parallel to pore axis after compression test: strain (**a**) 0 %, (**b**) 30 %, (**c**) 50 % and (**d**) 80 % (Reprinted with permission from [38] © 2002 Elsevier Science Ltd)

tendency is reversed with increasing strain. For understanding this behavior, two reasonable possibilities were taken into consideration [38].

First, while the stress concentration occurs around the pores in the specimen with pores perpendicular to the compressive direction, there is little stress concentration around the pores in the specimen with the cylindrical pore parallel to the compressive direction. The stress concentration is important factor for determination on the strength of porous metals [20, 23]. The pores perpendicular to the compressive direction are easily deformed at lower stress.

Next, different types of deformation such as buckling occur in the specimens depending on compressive direction with respect to the pore axis. The buckling in a specimen may result from loading misalignment and the specimen can be deformed at lower stresses. The macrographs of the lotus copper with 59.5 % porosity compressed by 0, 30, 50, and 80 % in the direction parallel and perpendicular to the pore axis are shown in Figs. 7.26 and 7.27, respectively. In

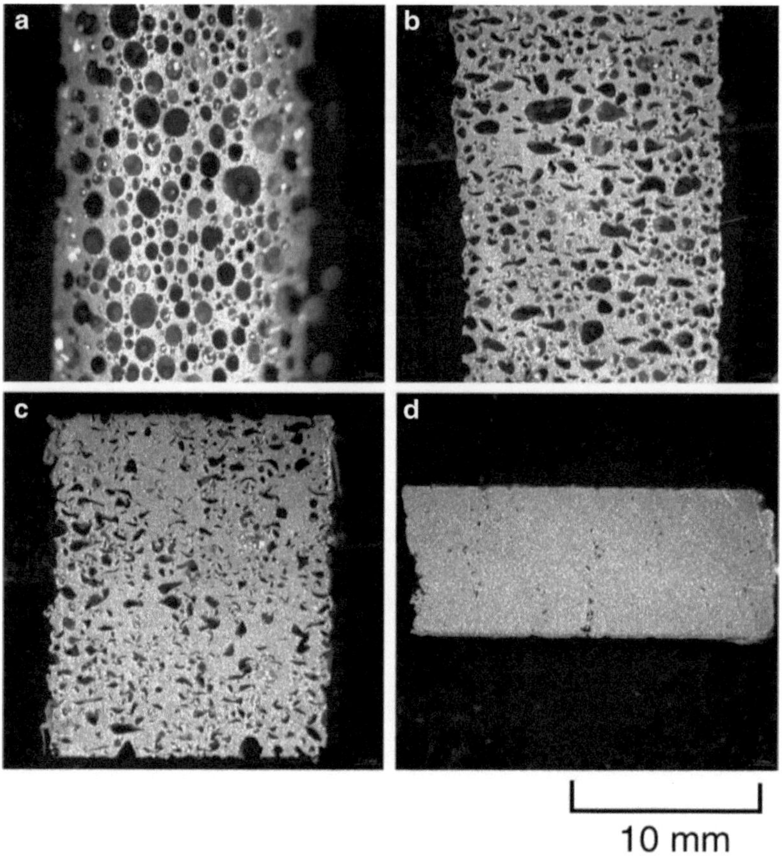


**Fig. 7.27** Optical micrographs of cross sections of lotus copper with 59.5 % porosity in the direction perpendicular to pore axis after compression test: strain (**a**) 0 %, (**b**) 30 %, (**c**) 50 %, and (**d**) 80 % (Reprinted with permission from [38] © 2002 Elsevier Science Ltd)

the parallel case, two types of buckling are observed in the specimen during compression. One is the macro-buckling observed outside of the specimens as shown in Fig. 7.26c. Another is the micro-buckling observed around the pores. Figure 7.28 shows the microstructure around the pores before and after 50 % height reduction, in which the copper ligaments between pores after compression is buckled like zigzag.

However, in the perpendicular case, macro-buckling is not observed outside of specimens during compression, while the micro-buckling is found inside of the specimens. Furthermore, the volume fraction of pores after 50 % compression in the perpendicular case is lower than that in the parallel case. Moreover, the slope of stress–strain curve in the perpendicular case is rapidly increased at 50 % compressive strain as shown in Fig. 7.25. It can be explained that the pores in the perpendicular case are easily crushed during compression due to the stress concentration around the pores. Consequently, such anisotropy of the slope of the

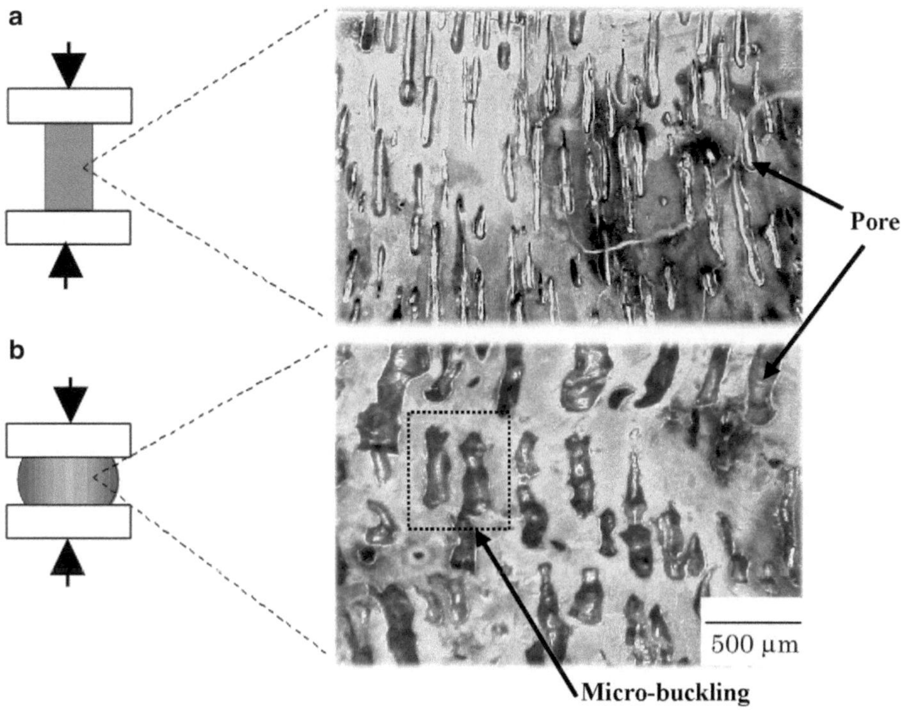


**Fig. 7.28** Optical micrographs of cross section of lotus copper with 59.5 % porosity in the direction parallel to pore axis before and after compression test: (**a**) $\varepsilon = 0$ and (**b**) $\varepsilon = 50$ %, respectively (Reprinted with permission from [38] © 2002 Elsevier Science Ltd)

stress–strain curves may be attributed to the stress concentration, the buckling, and the volume fraction of pores.

### 7.4.2 Energy Absorption

Schematic stress–strain curve of the lotus-type porous copper is shown in Fig. 7.29. The curve exhibits three regions: (I) elastic region characterized by elastic modulus; (II) stress-plateau region characterized by a shallow slope corresponding to the plastic yielding, buckling, and bending of cell ligaments; and (III) densification region characterized by a relative steep slope. From the compression test data, the energy absorption during the compression can be calculated by integrating the area under the stress–strain curve [38]:

$$W = \int_0^{\varepsilon} \sigma(\varepsilon) d\varepsilon, \tag{7.9}$$

where $W$ is the absorbed energy per unit volume and $\varepsilon$ is the strain.

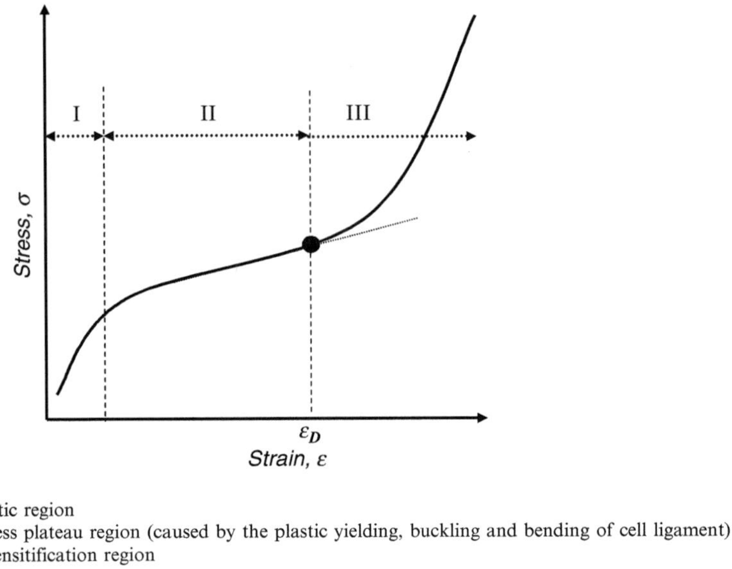


**Fig. 7.29** Schematic compressive stress–strain curve of lotus metal. (*I*) elastic region, (*II*) stress-plateau region (caused by the plastic yielding, buckling and bending of cell ligament), and (*III*) densification region (Reprinted with permission from [38] © 2002 Elsevier Science Ltd)

**Table 7.2** Densification strain and absorption energy of lotus copper

| Porosity (%) | Apparent density (Mg m$^{-3}$) | Compressive direction with pore axis | Densification strain (%) | Stress (MPa) | Absorbed energy per volume (MJ m$^{-3}$) |
|---|---|---|---|---|---|
| 30.4 | 6.25 | Parallel | 48.4 | 234 | 70.8 |
| | | Perpendicular | 29.8 | 144 | 25.7 |
| 48.4 | 4.64 | Parallel | 51.2 | 159 | 47.1 |
| | | Perpendicular | 33.8 | 70.5 | 14.4 |
| 59.5 | 3.57 | Parallel | 46.3 | 91.5 | 27.8 |
| | | Perpendicular | 33.0 | 36.1 | 7.2 |

Table 7.2 compiles the absorbed energy up to densification strain, which is defined as the strain corresponding to the end of the stress plateau of lotus copper. The stress and the absorbed energy at densification strain decrease with increasing porosity. At a given porosity the absorbed energy of the specimen in the parallel case is higher than that of the specimen in the perpendicular case, because the lotus copper with pores parallel to the compressive direction has higher strength and ductility than that with pores perpendicular to the compressive direction. It can be seen that the energy absorption capacity of lotus copper with the pores parallel to the compressive direction is much higher than that perpendicular to the compressive direction at a given porosity.

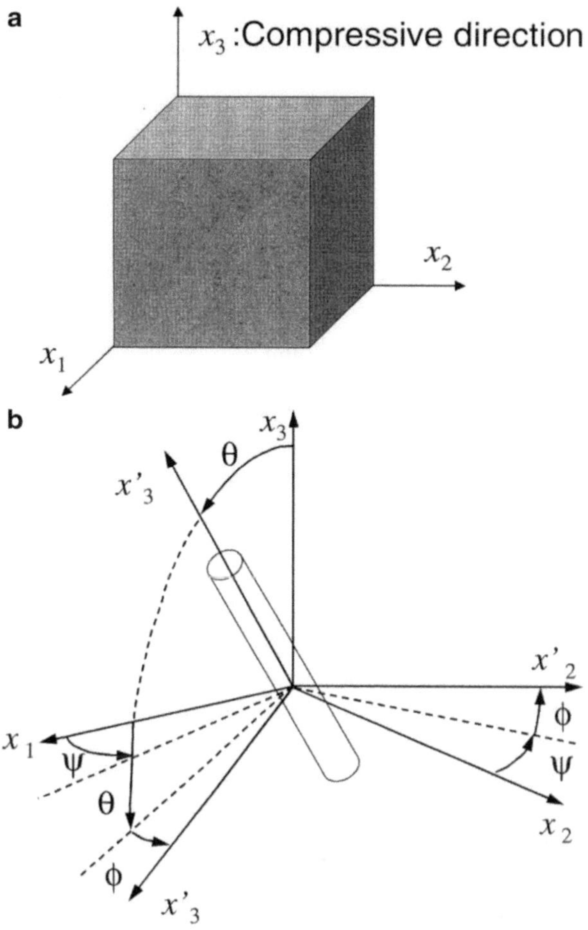


**Fig. 7.30** Schematic illustration showing (**a**) specimen coordinate system and (**b**) three Euler angles, $\psi$, $\theta$, and $\varphi$ defined between specimen and pore coordinate system. The longitudinal axis of the pore is parallel to the $x'_3$ axis (Reprinted with permission from [39] © 2006 Materials Research Society)

### *7.4.3 Effect of Pore Orientation on Compressive Properties*

Ide et al. investigated the compressive properties of lotus stainless steel not only in the direction parallel and perpendicular to the pore direction but also in other directions. The results of compressive tests on lotus stainless steel were described and the yield behavior by using micromechanical mean-field theory was discussed [39].

Nonporous and porous stainless steels (SUS304L) were fabricated by a continuous zone melting technique. Cubic specimens ($5 \times 5 \times 5$ mm$^3$ or $8 \times 8 \times 8$ mm$^3$) with various values of $\theta$ were cut out from the nonporous and porous ingots, where $\theta$ is the angle between the pore direction ($x'_3$ axis) and compressive direction ($x_3$ axis) as shown in Fig. 7.30. The pores inside the specimens are invisible so that they assume that the angle of inside pores is the same as that of pores on the surface and determine $\theta$ by analyzing the pores only on the surface. The angle of the $i$th pore inside a porous specimen, $\theta^{(i)}$, is given by

$$\theta^{(i)} = \tan^{-1}\sqrt{\tan^2 {\theta_1}^{(i)} + \tan^2{\theta_2}^{(i)}}, \tag{7.10}$$

where ${\theta_1}^{(i)}$ is the projection angle between the pore direction and the $x_3$ axis on the $x_2$–$x_3$ plane and ${\theta_2}^{(i)}$ is the projection angle between the pore direction and the $x_3$ axis on the $x_1$–$x_3$ plane. However, since the angles cannot be measured simultaneously for a pore on the same surface, they determined the two projection angles by averaging ${\theta_1}^{(i)}$ and ${\theta_2}^{(j)}$ on the respective surfaces: $\vec{\theta}_1 = \left(\frac{1}{n_1}\right)\left|{\theta_1}^{(i)}\right|$ and $\vec{\theta}_2 = \left(\frac{1}{n_2}\right)\left|{\theta_2}^{(j)}\right|$, where $n_1$ and $n_2$ denote the number of pores on the $x_2$–$x_3$ and $x_1$–$x_3$ planes, respectively. Note that the absolute values are used to take account of the misorientation of pores. The angle $\theta$ is thus given by

$$\theta = \tan^{-1}\sqrt{\tan\bar{\theta}_1 + \tan\bar{\theta}_2}. \tag{7.11}$$

Figure 7.31a shows the compressive stress–strain curves of nonporous stainless steel fabricated through unidirectional solidification, where $\sigma_{||}$ and $\sigma_{\perp}$, respectively, denote the compression stress in the directions parallel and perpendicular to the solidification direction. The compression behavior does not exhibit anisotropy despite the presence of the texture. Figure 7.31b shows the stress–strain curves of lotus stainless steel for compression in the direction of $\theta$. The compressive stress in the direction of $3° < \theta < 15°$ and $58° < \theta < 87°$ is insensitive to $\theta$. In all $\theta$, the compressive stress steeply increases in the high-strain range, and its slope tends to increase with increased $\theta$. Figure 7.31c shows the stress–strain curves of stainless steel subjected to compression in the direction parallel and perpendicular to the pore direction. The compressive stress in both directions decreases with increased porosity. In the low-strain range, $\sigma_{||}$ is higher than $\sigma_{\perp}$. However, this magnitude relation is reversed in the high-strain range. The crossover points of $\sigma_{||}$ and $\sigma_{\perp}$ shift to higher strain with increase in porosity.

Figure 7.32a shows the yield stress (0.2 % proof stress) of lotus stainless steel in the direction of $\theta$, $\sigma_{y\theta}$. The yield stress $\sigma_{y\theta}$ monotonically decreases with increasing of $\theta$. Figure 7.32b shows the porosity dependence of the yield stress of lotus stainless steel. The yield stress $\sigma_{y||}$ decreases almost linearly with increased porosity, while $\sigma_{y\perp}$ rapidly decreases. Thus, $\sigma_{y\perp}$ is lower than $\sigma_{y||}$ over the whole range of porosity. The yield stress of lotus stainless steel is expressed by the power-law equation [20] $\sigma_y = \sigma_{y0}(1-p)^m$. The values of $m$ were found to be 1.0 and 2.4 for the direction parallel and perpendicular to the pore direction, respectively.

The compressive behavior of nonporous stainless steel does not show anistropy; the macroscopic deformation modes is thought to originate from the anisotropic pore structure. Although average compressive strain inside pores, $\varepsilon_\mathrm{p}$, and that inside the matrix, $\varepsilon_\mathrm{M}$, cannot be easily measured, the magnitude relationship between

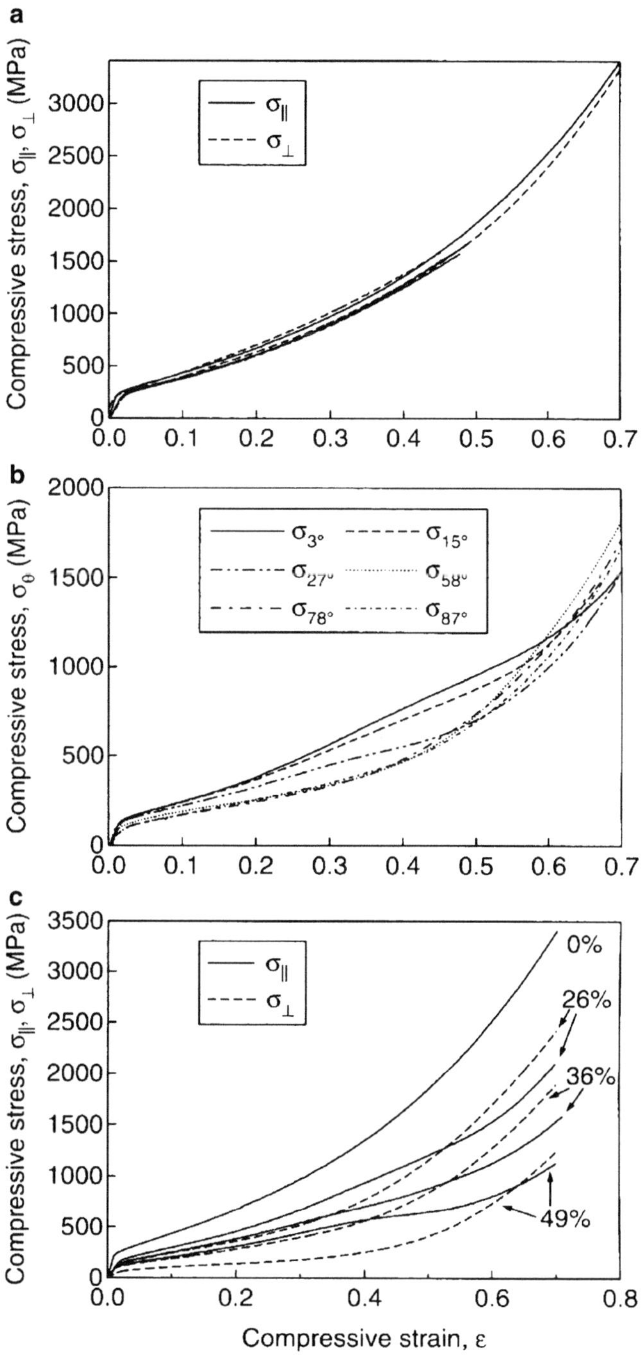


**Fig. 7.31** (**a**) Stress–strain curves of nonporous stainless steel fabricated through unidirectional solidification for compression in the direction parallel and perpendicular to the solidification direction. (**b**) Stress–strain curves of lotus stainless steel ($p = 39$ %) for compression in the direction of $\theta$. (**c**) Stress–strain curves of lotus stainless steel with different porosities for compression in the direction parallel and perpendicular to the pore direction (Reprinted with permission from [39] © 2006 Materials Research Society)

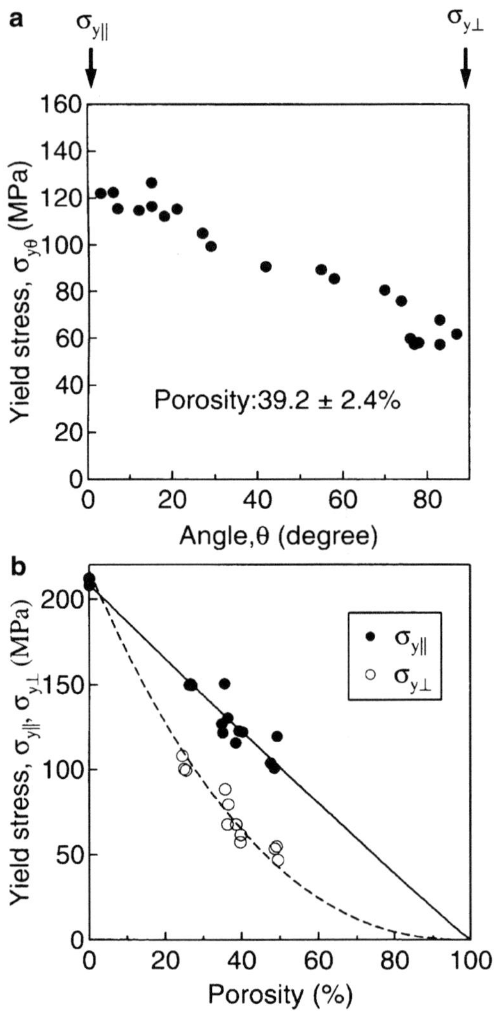


**Fig. 7.32** Yield stress of lotus stainless steel measured in the compression tests: (**a**) Yield stress in the direction of $\theta$. (**b**) Porosity dependence of yield stress in the directions parallel and perpendicular to the pore direction. The plots represent the measurement result, and the lines represent fitting of the power-law equation mentioned above to the measurements (Reprinted with permission from [39] © 2006 Materials Research Society)

them can be calculated according to Qiu and Weng's theory on the elastoplastic behavior of composite materials [40], in which the ductility of the matrix is taken into account by its secant moduli. For compression in the direction perpendicular to the pore direction, it is concluded from the theory that $\varepsilon_p$ is larger than $\varepsilon_M$; the deformation rate of pores with regard to the macroscopic compressive strain is larger than that of the matrix. Thus, in the perpendicular case, porous specimens are densified with increased macroscopic comressive strain, and specimens macroscopocally deforn only in the compressive direction. On the other hand, for compression in the pore direction, the Q–W theory concludes $\varepsilon_p$ almost equals $\varepsilon_M$; the deformation rate of the pore with regard to the macroscopic compressive strain is almost the same as that of the matrix. Therefore, it is apparent that the pores may

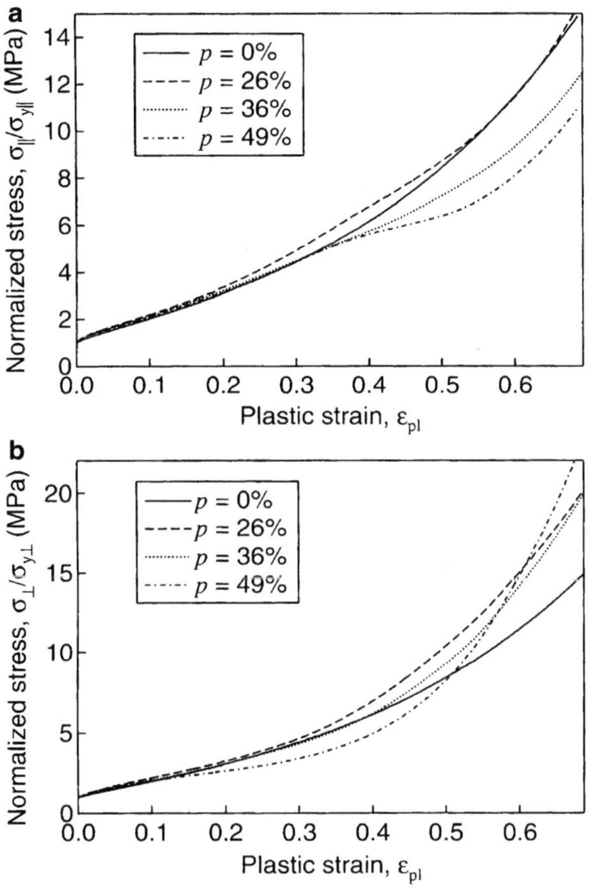


**Fig. 7.33** Normalized stress–plastic strain curves of lotus stainless steel, where the compressive stress of lotus stainless steel is normalized with its yield stress: (**a**) compression in the direction parallel to the pore direction and (**b**) compression in the direction perpendicular to the pore direction (Reprinted with permission from [39] © 2006 Materials Research Society)

remain, even in the high-strain range, and barreling may occur as well as in the case of nonporous materials.

A plateau-stress region exists in conventional porous metals; in this region, first, local weak regions deform and harden, and thus deformation in these regions does not proceed further. Subsequently, other weak regions deform and harden. Densification proceeds by iteration of this phenomenon; pores collapse not simultaneously but sequentially. Figure 7.33 shows the normalized stress–plastic strain $\varepsilon_{\rm p}$ and that inside the matrix, $\varepsilon_{\rm M}$, curves. $\sigma_{||}/\sigma_{y||}$ and $\sigma_{\perp}/\sigma_{y\perp}$ denote the normalized stress in the directions parellel and perpendicular to the pore direction. $\varepsilon_{\rm pl}$ denotes the plastic strain.

For $p = 49$ %, $\sigma_{\perp}/\sigma_{y\perp}$ is lower than that for $p = 0$ % when $\varepsilon_{\rm pl} = 0.15$–$0.5$, while for $p = 26$ % and 36 %, $\sigma_{\perp}/\sigma_{y\perp}$ agrees with that for $p = 0$ %. It is surmised that for $p = 49$ %, the specimen is densified by sequential pore collapse in the same way as for conventional porous materials, which results in the stress-increase rate of $p = 49$ % being smaller than that of the nonporous materials. On the other hand, for

$p = 26$ and 36 %, the rates of stress increase in porous materials, similar to those of nonporous materials. This is because the collapse of the pores appears to occur simultaneously, this densification mode differs from that of conventional porous materials.

For $p = 26$ %, $\sigma_{||}/\sigma_{y||}$ agrees quite closely with that for $p = 0$ % over the whole strain range. This suggests that the specimen is not densified even in the high-strain range, since stress concentration does not occur around the pores; $\varepsilon_p$ almost quals $\varepsilon_M$. $\sigma_{||}/\sigma_{y||}$ is roughly equal to that for $p = 36$ and 49 %, $\sigma_{||}/\sigma_{y||}$ almost agreew with that for $p = 0$ % when $\varepsilon_{pl} < 0.3$, while $\sigma_{||}/\sigma_{y||}$ is smaller than that for $p = 0$ % when $\varepsilon_{pl} \geq 0.3$. This is because the pores bend with barreling when the porosity is high, which results in stress concentration around the pores. The pores then begin to be densified, after the plateau-stress region appears. The deformation modes therefore depend not only on the compressive direction but also on porosity. The rapid stress increase in the latter range of compression also depends on the compressive deformation mode and yield stress.

The yield behavior of lotus stainless steel was dicussed by using micromechanical mean-field theory based on Qiu and Weng's concept [40]. On the assumption that a macroscopic yield is experimentally observed when local region **R** yields, the yield function for x ∈ **R** is expressed as

$$f_P^R = \left(\sum_{\text{eq}}^{(R)}\right)^2 - Y^2, \tag{7.12}$$

$$\left(\sum_{\text{eq}}^{(R)}\right)^2 = (\bar{\sigma}_{\text{eq}})^2 + R\left(\sigma'_{\text{eq}}\right)^2 = (\bar{\sigma}_{\text{eq}})^2 + R\left[\left(\sum\nolimits_{\text{eq}}\right)^2 - (\bar{\sigma}_{\text{eq}})^2\right], \tag{7.13}$$

where $Y$ is the yield stress of the matrix, $\sum_{\text{eq}}$ is the macroscopic effective stress in the matrix, and $\sum_{\text{eq}}^{(\text{R})}$ is the effective stress for x ∈ **R**. The term $R$ is a parameter introduced by Tane et al. and can take into account the microyield around pores; $R$ depends on the local-yield region for macroscopic yield to appear, and $R > 1$ means that local yield is dominant for macroscopic yield [8]. The average and the deviation of local effective stress, $(\bar{\sigma}_{\text{eq}})^2$ and $(\sigma'_{\text{eq}})^2$, are defined as

$$\begin{aligned}\left(\sum\nolimits_{\text{eq}}\right)^2 &= \frac{1}{2}\left\{(\bar{\sigma}_2 - \bar{\sigma}_3)^2 + (\bar{\sigma}_3 - \bar{\sigma}_1)^2 + \cdots + 6(\bar{\sigma}_6)^2\right\} \\ &\quad + \frac{1}{2V_{\text{M}}}\int_{V_{\text{M}}}\left\{\left[\sigma'_2(x) - \sigma'_3(x)\right]^2 + \left[\sigma'_3(x) - \sigma'_1(x)\right]^2 + \cdots + 6\left[\sigma'_6(x)\right]^2\right\}dV \\ &= (\bar{\sigma}_{\text{eq}})^2 + (\sigma'_{\text{eq}})^2,\end{aligned} \tag{7.14}$$

where $(\bar{\sigma}_{\text{eq}})^2$ and $(\sigma'_{\text{eq}})^2$ depend on the average stress and the stress variance of the matrix, respectively; $V_{\text{M}}$ is the volume of the matrix. The numerical-

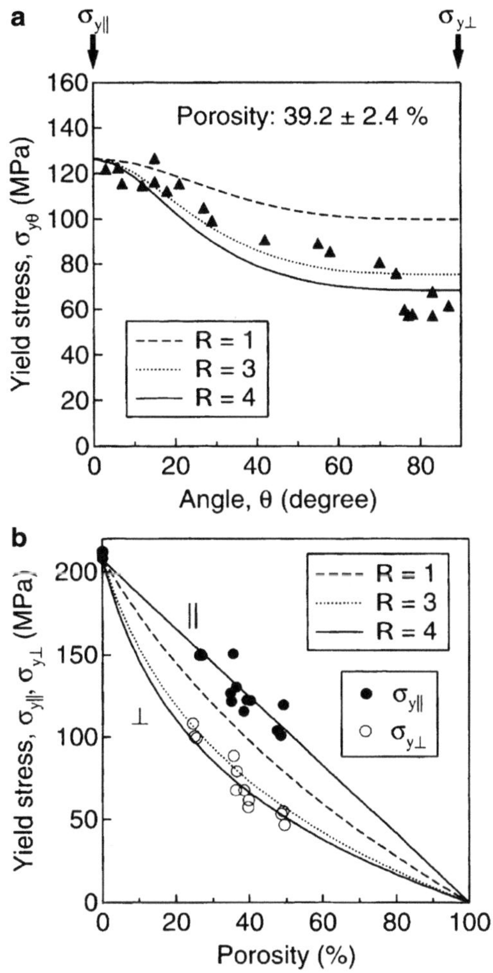


**Fig. 7.34** Yield stress of lotus stainless steel calculated by micromechanical mean-field-theory: (**a**) yield stress in the direction of $\theta$ from the longitudinal axis of the pore, (**b**) porosity dependence of yield stress in the direction parallel and perpendicular to the pore direction. The plots and lines represent the measurement and calculation results, respectively (Reprinted with permission from [39] © 2006 Materials Research Society)

differentiation method based on Eshelby's inclusion theory and Mori–Tanaka's mean-field theory was used to calculate the yield stress by using Eqs. (7.12) and (7.13). Figure 7.34a shows the yield stress of lotus stainless steel in the direction of $\theta$, where the plots denote the measurements and lines denote the calculations for $R = 1,3$, and 4, and $a_3/a_1 = \infty$. For $R = 1$, the calculations are higher than the measurements, since the local yield around the pores is not adequately taken into account. On the other hand, the calculations agree with the measurements for $R = 3$ and 4, because the local yield is adequately taken into account. Figure 7.34b shows the porosity dependence of $\sigma_{y\parallel}$ and $\sigma_{y\perp}$ for $R = 1$, 3, and 4. The yield stress $\sigma_{y\parallel}$ is independent of R, because stress deviation does not appear for compression in the pore direction. The yield stress $\sigma_{y\perp}$ depends on $R$, and $\sigma_{y\perp}$ decreases with increased $R$. This is because high-stress deviation appears on the loading in the direction perpendicular to the pore direction. The calculations for $R = 4$ provide a better fit to the measurements on the whole.

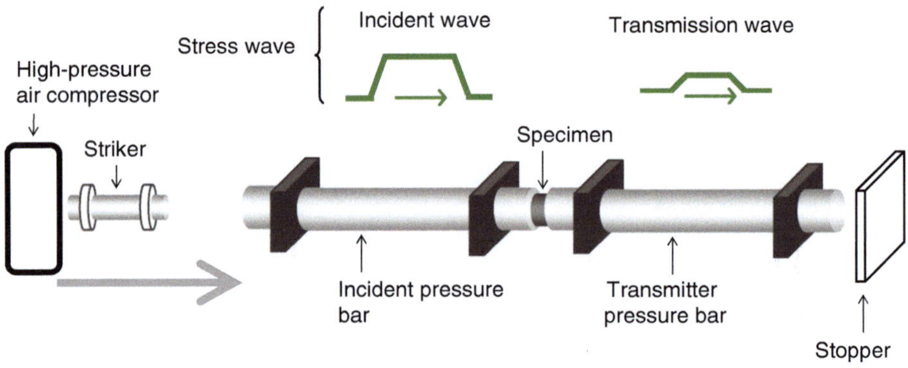


**Fig. 7.35** Apparatus for high strain rate compression test by split Hopkinson pressure bar method

### *7.4.4 Strain Rate Dependence of Compressive Behavior*

Lotus metals are promising not only as functional materials such as sound absorbers but also as lightweight structural materials. There are also high expectations of lotus metals as impact energy absorbers. Because lotus metals have considerably better mechanical strength than metal foams, it is expected that those also exhibit high-energy absorption. However, the high strain rate compression behavior of lotus metals has yet to be investigated. Here high strain rate compression behavior and energy absorption characteristics of lotus iron are described. High strain rate (~$10^3$ $s^{-1}$) compression tests in the directions parallel and perpendicular to the pore direction were done with the split Hopkinson pressure bar (SHPB) method [41, 42]. Low and middle strain rate compression tests were also carried out with a universal Instron testing machine. The apparatus of SHPB mainly consists of striker, incident pressure bar, transmitter bar, and high-pressure air guns shown in Fig. 7.35. The striker bar was made of a SUS304 stainless steel for 20 mm in diameter and 700 mm in length. The incident and transmitter pressure bars are made of 30 mm diameter S45C steel bars with the lengths of 1,600 and 1,100 mm, respectively. The absorbed energy was calculated by Eq. (7.15), while the efficiency of the absorbed energy, $\eta$, is calculated by

$$\eta = \frac{W}{\sigma_d \cdot \varepsilon_d}, \tag{7.15}$$

where $\sigma_d$ is the stress at the densification strain $\varepsilon_d$. Figure 7.36a shows the low, middle, and high strain rate nominal stress–nominal plastic strain curves of porous iron when the compressive direction is parallel to the pore direction [43]. In the low and middle strain rate compression, the shapes of stress–strain curves are almost the same between the low and middle strain rates: (1) The flow stress monotonically increases with increasing strain up to $\varepsilon_p \approx 20$ %, (2) the deformation proceeds with a low work-hardening rate in the range of $\varepsilon_p \approx 20$–$40$ %,

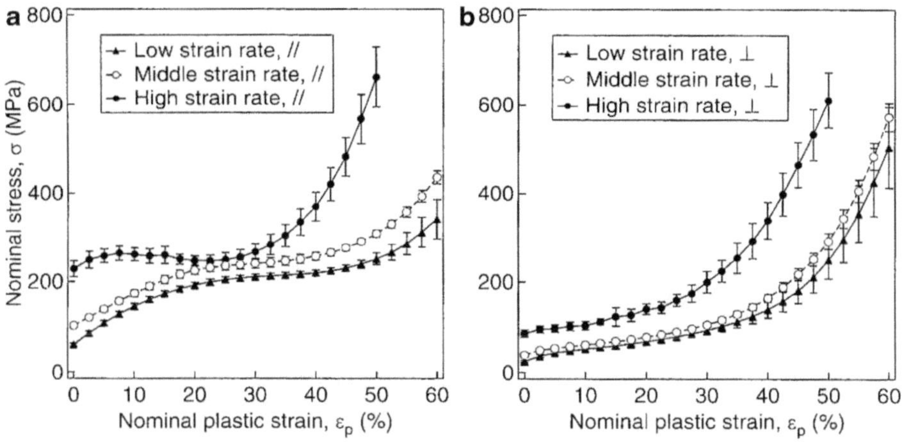


**Fig. 7.36** Low, middle, and high strain rate nominal stress–nominal plastic strain curves of porous iron in directions (**a**) parallel and (**b**) perpendicular to the pore direction. All specimens were prepared in the same hydrogen atmosphere; averages of porosity are in range of 47.9–49.7 %. Specimen heights H are 6, 6, and 3 mm for low, middle, and high strain rates, respectively (Reprinted with permission from [43] © 2010 Materials Research Society)

where the work-hardening rate of porous metals depends on the intrinsic work hardening of matrix iron and the change in the porous structure as the deformation progresses, and (3) the densification region, where stress steeply increases with increasing strain, appears above $\varepsilon_p \approx 47.5$ %. On the other hand, the shape of stress–strain curve in the high strain rate compression is different from those in the other two strain rate compressions. The flow stress slightly increases with increasing strain up to $\varepsilon_p \approx 7.5$ %. Then, the flow stress slightly decreases with increasing strain in the range of $\varepsilon_p \approx 7.5$–20 %. As a result, the flow stress shows hardly any increases up to $\varepsilon_p \approx 30$ %; a clear plateau-stress region appears. After the plateau-stress region, the densification region appears. In three strain rates, the flow stress increases with increasing strain rate up to $\varepsilon_p \approx 20$ % as in the nonporous iron. However, the difference among three strain rates becomes small for $\varepsilon_p \approx 20$–30 % as a result of the appearance of the clear plateau-stress region in the high rate compression.

Figure 7.36b shows low, middle, and high strain rate stress–strain curves of lotus iron in the compressive direction perpendicular to the pore direction. Comparison with Fig. 7.36a indicates that the flow stress in the perpendicular direction is lower than that in the parallel direction in the initial stage of the stress–strain curves. However, the magnitude of the relationship is reversed in the latter stage of the stress–strain curves in low and middle strain rates. In contrast to the parallel direction, the shapes of stress–strain curves in the perpendicular direction are similar for all three strain rates: (1) the flow stress monotonically increases with increasing strain and (2) the densification region appears in the latter stage of stress–strain curves. The flow stress increases with increasing strain rate in the entire strain range as in the case of nonporous iron.

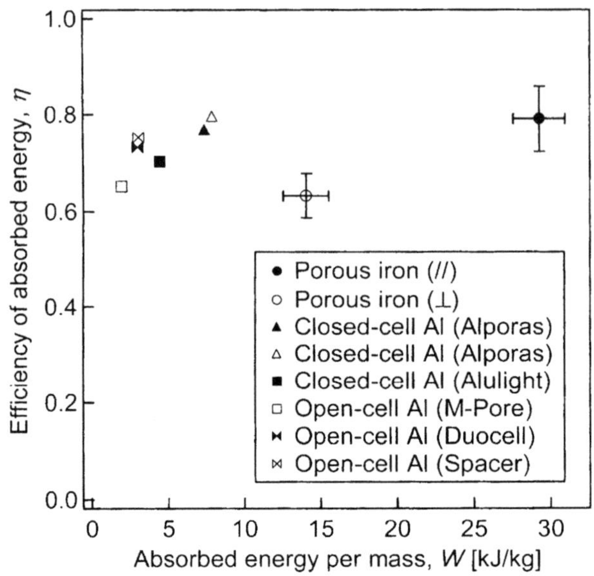


**Fig. 7.37** Efficiency of absorbed energy and absorbed energy per mass in the high strain rate compression in various porous metals: porous iron in directions parallel and perpendicular to pore direction, closed-cell aluminum alloy foams (Alporas), closed-cell aluminum alloy foams (Alulight), open-cell aluminum alloy foam (M-Pore), open-cell aluminum alloy foam (Duocell), and open-cell aluminum alloy foam (Spacer) prepared by the spacer method [43] (Reprinted with permission from [43] © 2010 Materials Research Society)

Figure 7.37 compares the efficiency of absorbed energy $\eta$ and absorbed energy per mass $W$ in high strain rate compression among various porous metals. The absorbed energy $W$ and its efficiency $\eta$ of lotus iron exhibit significant anisotropy; $W$ and $\eta$ in the parallel direction are higher than those in the perpendicular direction. $W$ of lotus iron in both directions are higher than those of metal foams. It is of particular interest that the absorbed energy in the parallel direction is much higher than it is in metal foams. On the other hand, $\eta$ of lotus iron in both directions is comparable with that of metal foams; in fact, the efficiency in the parallel direction is comparable with that of closed-cell aluminum alloy foams (Alporas), which has the highest efficiency among all of the metal foams.

Figure 7.38a shows SEM images of microstructure on the inside wall of a pore before deformation and those after (b) low, (c) middle, and (d) high strain rate compression of approximately 5 % plastic strain is applied along the pore direction. In the deformed microstructure after the high strain rate compression, the deformation bands are observed as indicated by the arrows. However, the deformed microstructure after low and middle strain rate compression is almost the same as the microstructure before compression, because the applied plastic strain is so small that visible slip lines are not formed. Figure 7.39 shows SEM images of the deformation on the inside walls of pores after (a) low, (b) middle, and (c) high

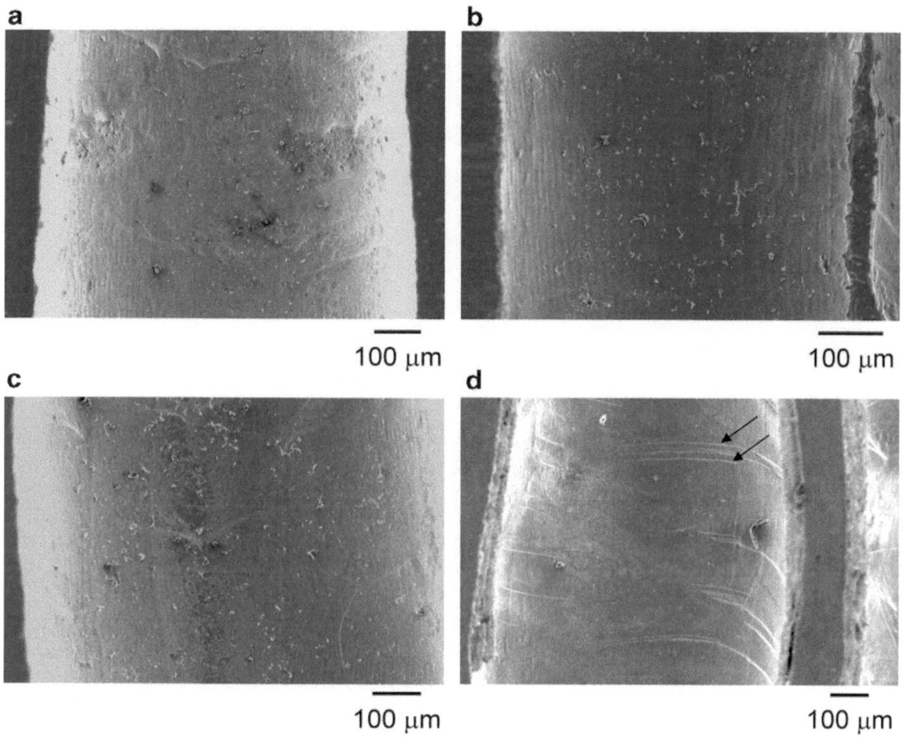


**Fig. 7.38** (**a**) Microstructure on inside wall of pore before deformation. Deformation on inside walls of pores after (**b**) low, (**c**) middle, and (**d**) high strain is applied along the pore direction (Reprinted with permission from [43] © 2010 Materials Research Society)

strain rate compression. The microscopic deformation behaviors are clearly different among three strain rates. In the low strain rate compression, fine slip lines are formed as indicated by the arrows. In the middle rate compression, the slip lines are thicker than those in the low strain compression as indicated by the arrows. In the high strain rate compression, large deformation bands are formed, which are further apart than the slip lines in other two levels of strain rate compression. Comparison between Figs. 7.38d and 7.39c indicates that deformation bands grow with an increase in applied strain from 5 to 35 %.

The strain rate-dependent microscopic deformation behavior can be explained by the strain rate-dependent assistance of thermal activation for dislocations to move. In the low strain rate compression of $\varepsilon_p \approx 35$ %, conventional fine slip lines are formed. In the middle strain rate, dislocation becomes increasingly difficult to move because the assistance from the thermal activation decreases. As a result, slip deformation becomes concentrated on particular slip planes, forming thick slip lines as shown in Fig. 7.39b. In the case of high strain rate compression, dislocations become increasingly difficult to move. Thus, deformation occurs only along limited planes, resulting in the formation of large deformation bands

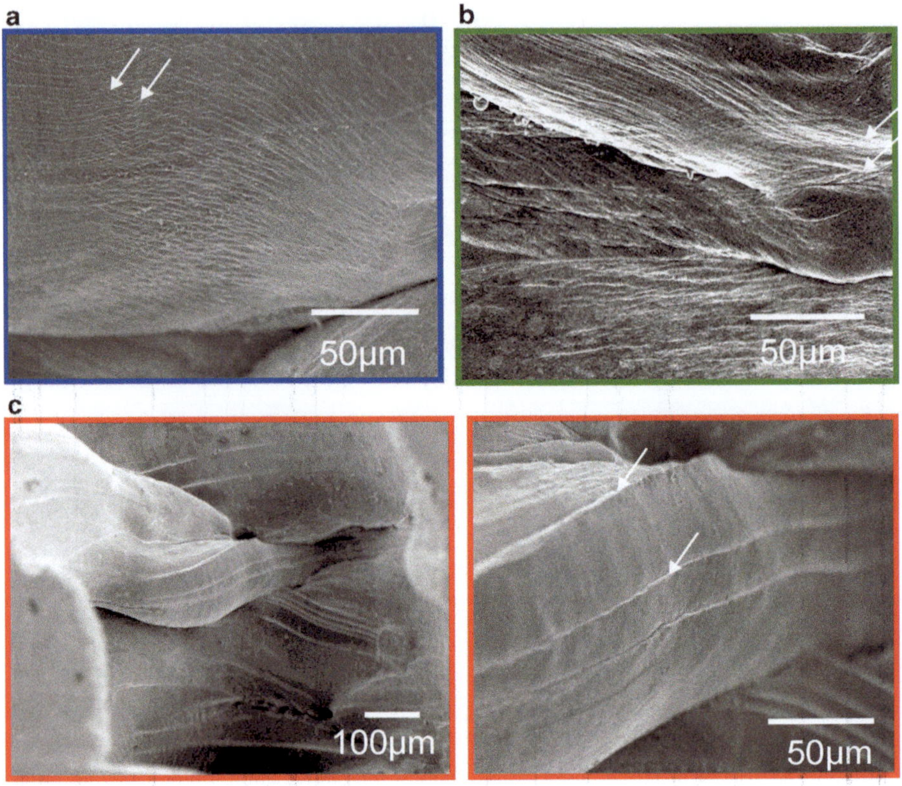


**Fig. 7.39** Deformation on inside walls of pores after (**a**) low, (**b**) middle, and (**c**) high strain rate compression. Compression of approximately 35 % plastic strain is applied along the pore direction (Reprinted with permission from [43] © 2010 Materials Research Society)

as shown in Fig. 7.39c. Similar deformation behavior dependent on a strain rate probably occurs not only on the observed inside wall of pores but also all around the matrix iron. In the early stage of compression, the microscopic deformation behavior is probably similar to those in the middle stage of compression of $\varepsilon_p \approx 35\ \%$, although slip lines cannot be observed in the low and middle strain rate compression because of the small applied strain. Thus, the effects of strain rate on the both microscopic deformation behavior and initial yield stress can be explained by the thermal activation of dislocations.

### *7.4.5 Compression Deformation Behavior of Lotus (γ-TiAl)*

Gamma TiAl(γ-TiAl) has been known to be one of the most promising candidates for high-temperature materials owing to its high strength at elevated temperature and lower density compared with those of NiAl and $Ni_3Al$. Intermetallic compounds

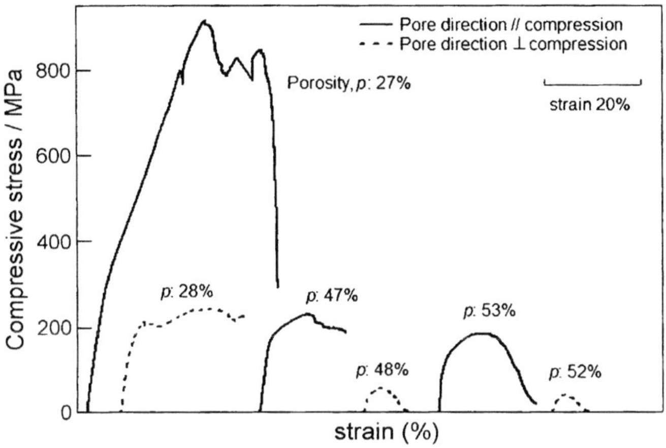


**Fig. 7.40** Compressive stress–strain curves of lotus TiAl with various porosities (Reprinted with permission from [44] © 2009 Elsevier B.V.)

including TiAl are expected to lightweight structural materials at elevated temperature. However, no investigation of TiAl was carried out. Ide et al. investigated fabrication and compressive behavior of γ-TiAl [44].

Raw ingot whose chemical composition was $Ti_{48}Al_{52}$ was prepared by skull melting. Rods of 10 mm in diameter and 100 mm in length were cut from the raw ingot with a spark-erosion wire- cutting machine. The TiAl rod was unidirectionally solidified utilizing a continuous zone melting technique in a chamber pressurized by mixtures of hydrogen and helium gas of 2.5 MPa.

Figure 7.40 shows the stress–strain curves of lotus TiAl with various porosities. In both (parallel and perpendicular) directions, the compressive stress and the strain at the peak strength decrease with increasing porosity. For lotus metals with ductile matrix, the plateau-stress region appears owing to the gradual densification of the specimen [38]. However, for lotus TiAl, the plateau-stress region does not appear in both directions. For lotus TiAl with the similar porosity, the compressive stress in the loading direction parallel to the elongated pore direction (parallel direction) is higher than that in the loading direction perpendicular to the elongated pore direction (perpendicular direction), and the strain at the peak strength in parallel direction is larger than that in perpendicular direction. It is reported that the polycrystalline $Ti_{48}Al_{52}$ shows 4 % plastic strain at room temperature [45]. Therefore, plastic strain at room temperature in lotus TiAl significantly increases compared with plastic strain in nonporous TiAl.

Figure 7.41 shows the porosity dependence of the 0.2 % proof stress of lotus TiAl, where 0.2 % proof stress of nonporous TiAl was referred to as that of polycrystalline $Ti_{48}Al_{52}$ with the same grain size (about 300 μm):332 MPa [46]. In both directions, 0.2 % proof stress decreases with increasing porosity. While 0.2 % proof stress rapidly decreases with increasing porosity in the perpendicular direction, it

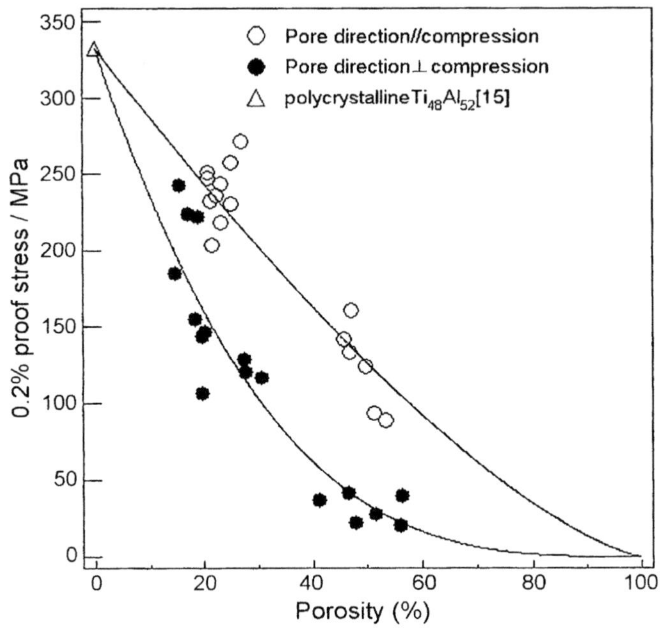


**Fig. 7.41** Porosity dependence of the 0.2 % proof stress of lotus TiAl. Symbols and lines represent the measurements and fittings of the power-law equation to the measurements, respectively (Reprinted with permission from [44] © 2009 Elsevier B.V.)

gradually decreases with increasing porosity in the parallel direction. The porosity dependence of the yield stress (0.2 % proof stress) of lotus metals is expressed using the following equation:

$$\sigma_\gamma^p = \sigma_\gamma^0\left(1 - \frac{p}{100}\right)^K, \tag{7.16}$$

where $\sigma_\gamma^p$ and $\sigma_\gamma^0$ are yield stress of lotus and nonporous material, respectively, and $K$ is a constant determined experimentally. With the experimentally obtained $K = 1.4$ in the parallel direction and $K = 3.3$ in the perpendicular direction, the porosity dependence of the 0.2 % proof stress of lotus TiAl is expressed by the curve using Eq. (7.16).

The anisotropic compressive properties of lotus TiAl are due to anisotropic porous structure. However, the porosity dependence of the 0.2 % proof stress of lotus TiAl differs from that of lotus metals with ductile matrix. Particularly, in parallel direction, the $K$ value obtained from Eq. (7.16) is higher than that of lotus metals with ductile matrix (approximately $K = 1$), i.e., the 0.2 % proof stress of lotus TiAl decreases more with increasing porosity compared with that of lotus metals with ductile matrix. Figure 7.42 shows periodic unloading–reloading

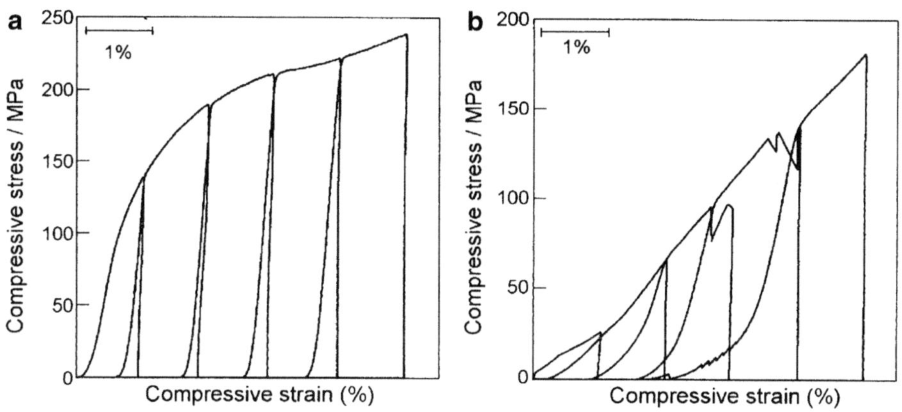


**Fig. 7.42** Cyclic stress–strain curves of lotus TiAl. Porosities of specimens are, respectively, (**a**) 9 16 and (**b**) 17 %. Loading directions are (**a**) parallel and (**b**) perpendicular to the elongated pore direction (Reprinted with permission from [44] © 2009 Elsevier B.V.)

compressive stress–strain curves of lotus TiAl. The loading directions are (a) parallel and (b) perpendicular to the elongated pore direction. When the stress is unloaded, the strain remains in the specimen even in the very low stress level, which is corresponding to the origins that compressive stress approximately linearly increases with increasing compressive strain in parallel and perpendicular directions. Figure 7.43 shows the optical micrographs of cross section of lotus TiAl after periodic unloading–reloading compression tests. The loading direction is perpendicular to the elongated pore direction.

Deformation and cracks locally exist around the pores and between the pores as shown in Fig. 7.43b. Since the deformation and cracking are caused by structural defect in porous metal, it is thought that those can also occur in lotus metal with ductile matrix. However, the specific strength in parallel direction in lotus metal with ductile matrix holds constant despite the presence of pores, and pore collapse does not affect the stress–strain response at early stage of compressive deformation. Therefore, 0.2 % proof stress of lotus metals with ductile matrix is not sensitive to local large stress concentration. On the other hand, for lotus TiAl, it is considered that the lack of ductility of the matrix is the cause of a stress–strain response which is affected by local large stress concentration. The strain locally increases with increasing stress and the macroscopic strain is rapidly increased by local strain and crack caused by local large stress concentration, depending on ductility of matrix. The region causing local deformation and fracture increases with increasing porosity. Therefore, the 0.2 % proof strength of lotus TiAl decreases more significantly with increasing porosity, compared with lotus metal with ductile matrix.

The effect of the brittleness of matrix on compressive properties significantly appears in high-strain region. Figure 7.44 shows optical micrographs of cross

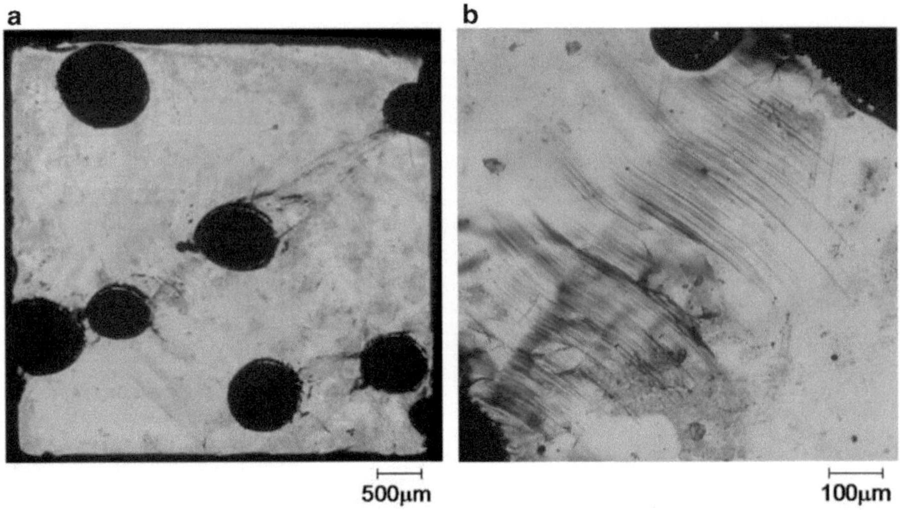


**Fig. 7.43** Optical micrographs of cross section of lotus TiAl, which is fabricated in mixture gas of 1.0 MPa hydrogen and 1.5 MPa helium, after cyclic compression test at (**a**) low and (**b**) high magnifications. Porosity and average pore diameter of specimen are 17 % and 635 μm, respectively. Loading direction is perpendicular to the elongated pore direction (Reprinted with permission from [44] © 2009 Elsevier B.V.)

section of lotus TiAl during compression; the loading direction is perpendicular to elongated pore direction. At the strain 1.5 % as shown in Fig. 7.44a, initial deformation occurs around the pores. Then, deformation and initial cracks are observed around the pores (Fig. 7.44b), which progress between locally neighbor pores (Fig. 7.44c, d). When the loading direction is parallel to elongated pore direction, the local deformation occurs and progresses around the pores (Fig. 7.45a, b) as in perpendicular compression. The compressive stress of the specimen which apparently yielded increases with increasing strain without fracture of specimen. Therefore, the increase of plastic strain in lotus TiAl compared with plastic strain in nonporous TiAl is caused not by the improvement in ductility of matrix but by successive deformation and cracking. For lotus TiAl in both compressive directions, the deformation occurs around the pores and progresses only in local region between pores. The fracture of specimen is caused by buckling of cell wall without the deformation of overall specimen owing to barreling or uniaxial deformation, which is observed in compressive deformation in lotus metals with ductile matrix. However, the densification of the specimen does not occur so that obvious plateau-stress region does not appear. Thus, the deformation behavior in lotus TiAl with brittle matrix significantly differs from that in lotus metals with ductile matrix.

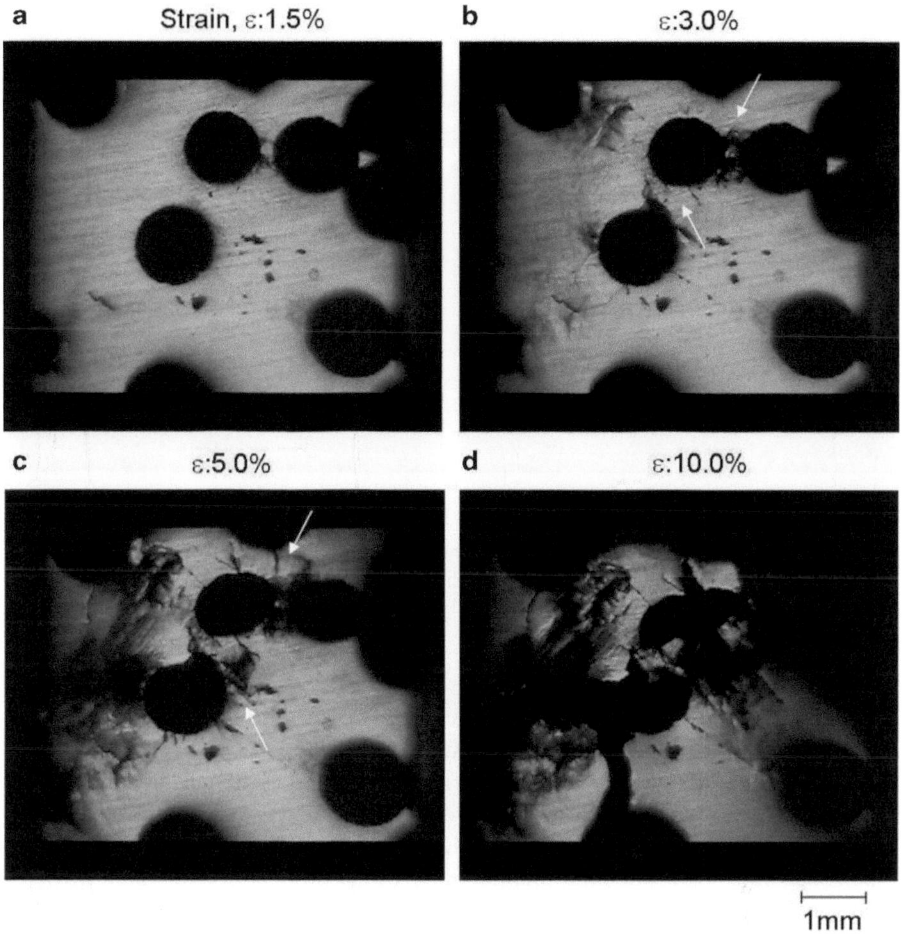


**Fig. 7.44** Optical micrographs of cross section of lotus TiAl fabricated in mixture gas of 1.0 MPa hydrogen and 1.5 MPa helium (perpendicular compression). Strain: (**a**) 1.5 %, (**b**) 3.0 %, (**c**) 5.0 %, and (**d**) 10.0 %. Porosity and average pore diameter of specimens are 27 % and 619 μm, respectively (Reprinted with permission from [44] © 2009 Elsevier B.V.)

## 7.5 Bending Strength

Bending is one of the most representative process to deform the porous structure parts and it is necessary to understand bending properties. The bending properties of lotus copper were investigated by Hyun et al. [47]. Figure 7.46 shows schematic drawings of specimens for three-point bending tests; *x*, *y*, and *z* are defined as the orientation of pore axis in the test specimens: (a) type X specimen whose pore axis is parallel to the bending direction, (b) type Y specimen, and (c) type Z specimen

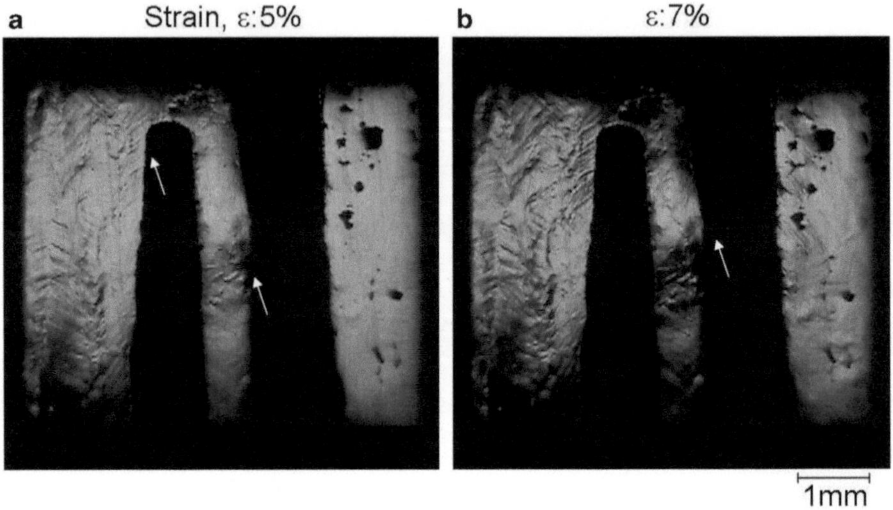


**Fig. 7.45** Optical micrographs of cross section of lotus TiAl fabricated in mixture gas of 1.0 MPa hydrogen and 1.5 MPa helium (parallel compression). Strain: (**a**) 5.0 % and (**b**) 7.0 %. Porosity and average pore diameter of specimens are 27 % and 642 μm, respectively (Reprinted with permission from [44] © 2009 Elsevier B.V.)

whose pore axes are perpendicular to the bending direction. Figure 7.47 shows the typical load–displacement curves of type X, type Y, and type Z specimens with 31.4 % porosity. The yield point and plastic deformation are observed in all specimens. The maximum load and the fracture are observed in type Y and type Z specimens, but it is not observed in the type X specimen; the load is increased continuously. Type Y and type Z specimens are fractured preferentially because the strength, which involves both tensile and compressive stresses in bending, is lower than that of type X specimen. Such result can be explained by the behavior of the tensile and compressive strength on the lotus copper; the specimen with cylindrical pores parallel to the tensile and compressive direction has higher strength than that with pores perpendicular to the direction [20, 38]. Furthermore, it is found that the absorbed energy, which is defined by the value calculated from the sectional area of the load–displacement curve of type X specimen, is higher than that of Y and Z specimens.

After the bending tests, the optical micrographs are shown in Fig. 7.48. In type X specimen no crack is observed after bending, while the crack is observed in other type of specimens: type Y and Z. Especially the crack is shown on the upper surface where the tensile stress is loaded during bending. Since it is known that the strength and elongation of the specimen with pores parallel to the tensile direction are much higher than those perpendicular to the tensile direction, it is reasonable that type X specimen can be deformed without crack. Thus, it can be explained that the bendability of type X specimen is much better than that of type Y and Z specimens.

In order to understand the bending properties of type X specimens, the bending properties of this type of specimens were measured as functions of porosity and

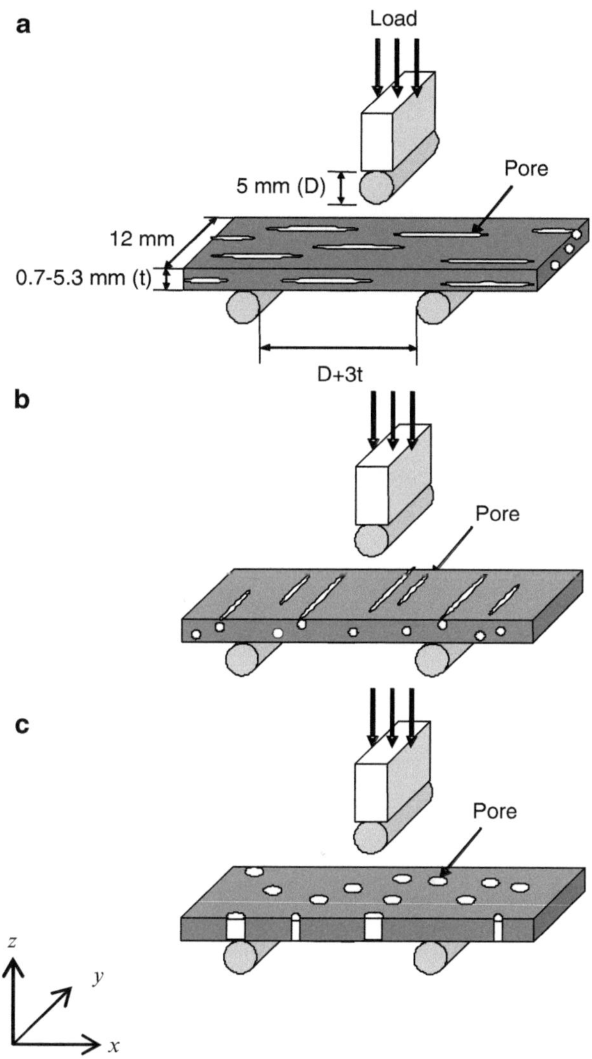


**Fig. 7.46** Schematic drawings of specimens for bending test; *x*, *y*, and *z* are defined as the orientation of pore axis in the test specimens. (**a**) Type X specimen, (**b**) type Y specimen, and (**c**) type Z specimen (Reprinted with permission from [47] © 2004 Elsevier B.V.)

thickness of the specimens. The bending load–displacement curves for type X specimens with different porosities are shown in Fig. 7.49. The slope of the curves decreases with increasing porosity. This tendency agrees with the results of tensile and compressive properties shown before [20, 38]. Some interesting features are observed; the crack propagation is ceased by the elongated pores as shown in Fig. 7.50b, while the cracks occurred locally from several parts on the surface as shown in Fig. 7.50c. Thus, it is suggested that the elongated pores prevent cracks from propagating and then the specimen after bending is not broken down perfectly by the cracks.

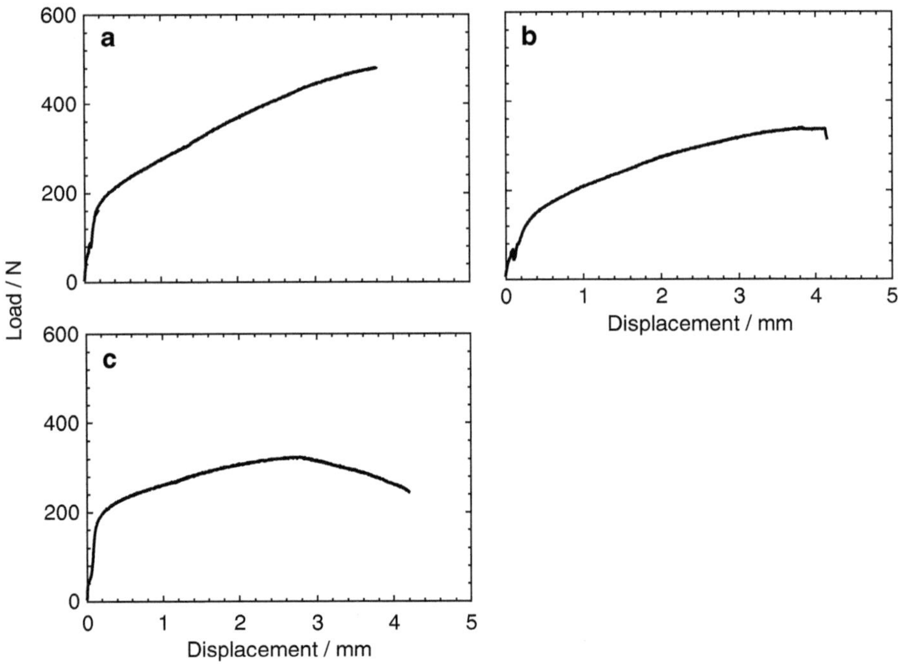


**Fig. 7.47** Bending load–displacement curves of lotus copper with 31.4 % porosity at 3.3-mm thickness. (**a**) Type X specimen, (**b**) type Y specimen, and (**c**) type Z specimen (Reprinted with permission from [47] © 2004 Elsevier B.V.)

## 7.6 Fatigue Strength

For practical use of lotus metal as structural materials, the knowledge of fatigue behavior is important because the effect of pore morphology on the fatigue behavior probably differs from that on tensile (or compressive) strength. So far the fatigue behavior of conventional porous metals with irregular pores has been investigated [32, 48–51], and the effects of porosity, the local strength of cell walls, etc., were clarified. For lotus metals, however, detailed studies have not been carried out, e.g., the effects of pore morphology on fatigue strength at finite life and fracture behavior have not been investigated. From a standpoint of practical use, the fatigue behavior of lotus metals has to be investigated in details.

Lotus copper was employed as a model of lotus metals to study the effects of pore morphology and applied-stress direction on the fatigue strength at finite life and fracture surface characteristics [52]. Dogbone-type specimens with a gauge diameter of 4 mm and gauge length of 6 mm were turned from the square rods. The angles θ between the longitudinal direction of the dogbone-type specimens and the longitudinal direction of cylindrical pores were 0, 16, 19, 21, 40, and 90°. The directions of $\theta = 0°$ and 90° are the directions parallel and perpendicular to the longitudinal direction of pores, respectively. Constant stress amplitude fatigue tests

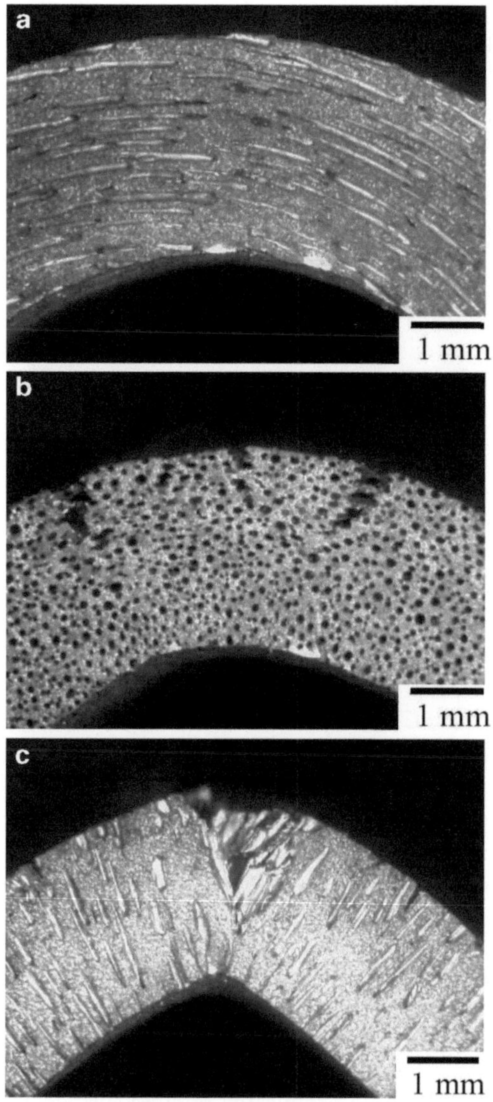


**Fig. 7.48** Optical micrographs of lotus copper with 31.4 % porosity after bending test. (**a**) Type X specimen, (**b**) type Y specimen, and (**c**) type Z specimen (Reprinted with permission from [47] © 2004 Elsevier B.V.)

were carried out with a servovalve-controlled electrohydraulic testing machine in air at room temperature. Cyclic tension–compression stress of $R = -1$ was applied to nonporous and lotus copper specimens, where $R$ is the ratio of the minimum stress to the maximum stress. The frequency of cyclic stress was 5 Hz. Stress amplitude $\sigma_a$ was calculated from an applied load and cross-sectional area including pores. Constant stress amplitude fatigue tests with various stress amplitudes $\sigma_a$ were carried out, and S–N curves which give the relation between stress amplitude (S) and the number of cycle to fatigue (N) were obtained.

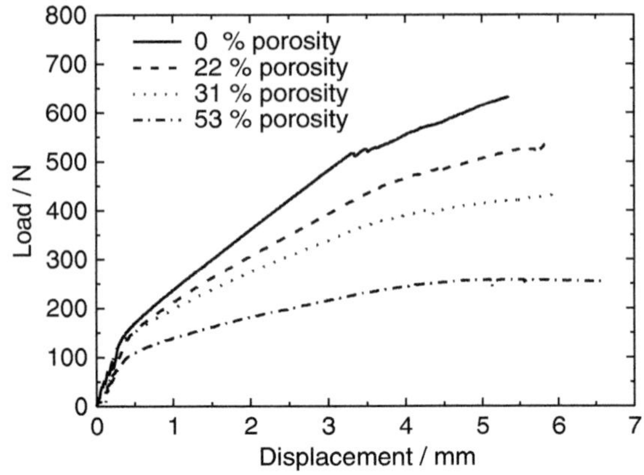


**Fig. 7.49** Bending load–displacement curves for nonporous copper and X specimens (lotus copper) of thickness of 3.3 mm with different porosities (Reprinted with permission from [47] © 2004 Elsevier B.V.)

Figure 7.51 shows the log plots of $\sigma_a$ against the number of cycles to failure $N_f$ for nonporous and lotus copper, where cyclic stress was applied in the direction (a) parallel and (b) perpendicular to the longitudinal axis of pores. The lines denote fittings of the following function to the experimental data:

$$\log\ \alpha_a = C \log N_f + D, \tag{7.17}$$

where $C$ and $D$ are fitting coefficients. For both directions the numbers of cycles to failure of lotus copper and nonporous copper decrease with increasing stress amplitude. The fatigue strength at finite life decreases in both directions with increasing porosity. Nonporous copper does not show anisotropy in the fatigue strength at finite life. On the other hand, the fatigue strength at finite life of lotus copper shows significant anisotropy; the fatigue strength in the perpendicular direction is lower than that in the parallel direction.

Figure 7.52 shows the number of cycles to failure of lotus copper with porosity of 40 %, when cyclic stress of $\sigma_a = 32$ MPa was applied in the direction of $\theta = 0$, 16, 19, 21, 40, and 90°. The number of cycles to failure decreases with increasing angle $\theta$. Thus, the fatigue life is the longest, when stress is applied in the direction parallel to the longitudinal axis of the pores ($\theta = 0°$).

Figure 7.53 shows the fracture surface of lotus copper with porosity 30 % when cyclic stress of $\sigma_a = 47$ MPa was applied in the direction parallel to the longitudinal axis of the pores, where the number of cycles to failure $N_f$ was 541,336. As shown in Fig. 7.53b, a stage I-type facet is formed in the lower right region on the entire fracture surface. This indicates that a crack initiated near this region. As shown in Fig. 7.53c, a striation of stage II is formed in the center of the entire fracture surface. The solid arrow denotes the direction in which a primary crack propagates. However, the direction of crack propagation changes when a crack reaches a pore as denoted by the broken arrow in Fig. 7.53c. A dimple pattern is not formed. Instead, the elongation of copper matrix that is analogous to a mountain ridge is formed as shown in Fig. 7.53d. The direction of the primary crack propagation indicates that

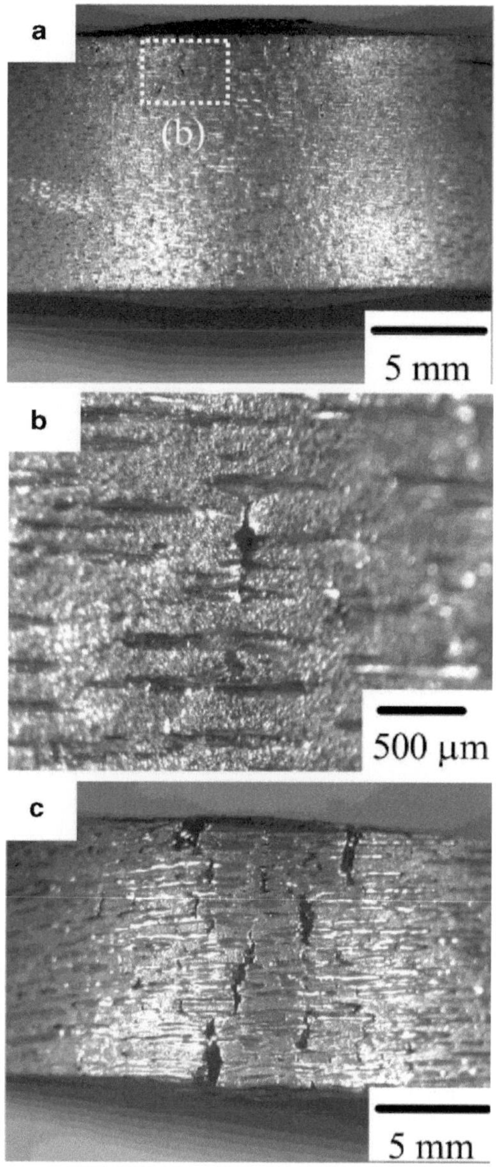


**Fig. 7.50** Optical micrographs of the X specimens appeared on the cracks on the surface after bending of lotus copper with 5.3-mm thickness. (**a**) and (**b**): the porosity is 31.4 % and (**c**) the porosity is 53.3 % (Reprinted with permission from [47] © 2004 Elsevier B.V.)

this surface is the final fracture surface of stage III. The fracture surface characteristics for the parallel loadings were independent of the number of cycles to failure $N_f$.

Figure 7.54 shows the fracture surface of lotus copper with porosity of 30 % when cyclic stress of $\sigma_a = 32$ MPa was applied in the direction perpendicular to the longitudinal axis off pores, where the number of cycles to failure $N_f$ was 24,127.

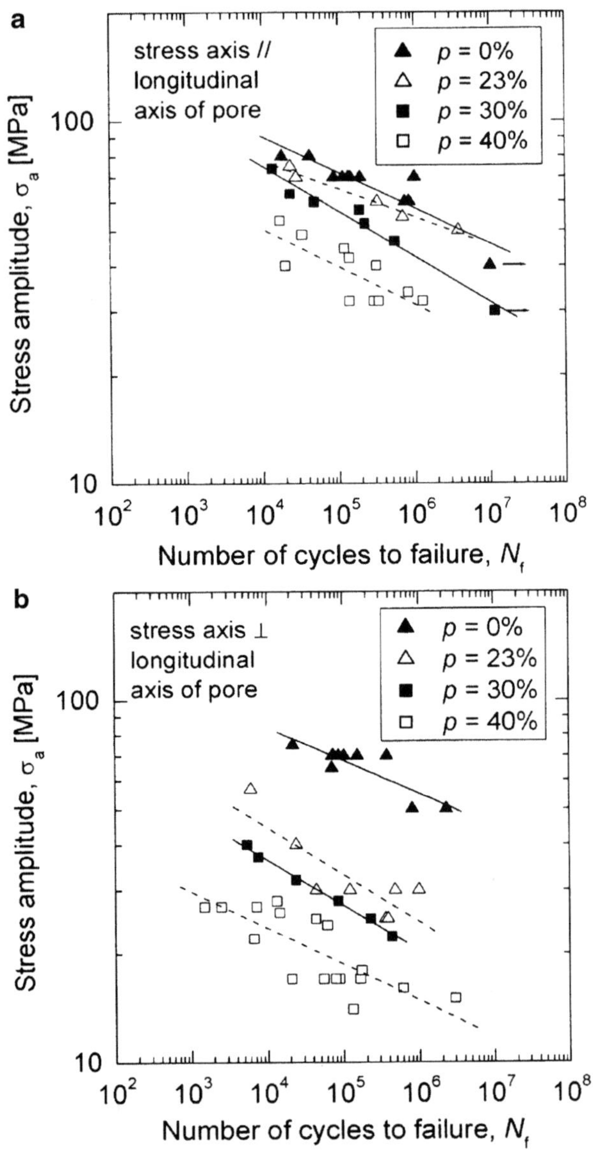


**Fig. 7.51** Log plots of stress amplitude against the number of cycles to failure for lotus copper, where cyclic stress was applied in the direction (**a**) parallel and (**b**) perpendicular to the longitudinal axis of cylindrical pores (Reprinted with permission from [52] © 2007 Materials Research Society)

Figure 7.54b shows a stage I-type facet. The regions showing this type of facet are randomly distributed on the entire fracture surface, which indicates cracks initiated in various sites. Figure 7.54c shows a striation pattern of stage II, and the regions showing this type of surface are also randomly distributed on the fracture surface. A crack does not propagate in the shortest path between pores, as shown by the arrow in Fig. 7.54c. Furthermore, a primary direction in which a crack propagates

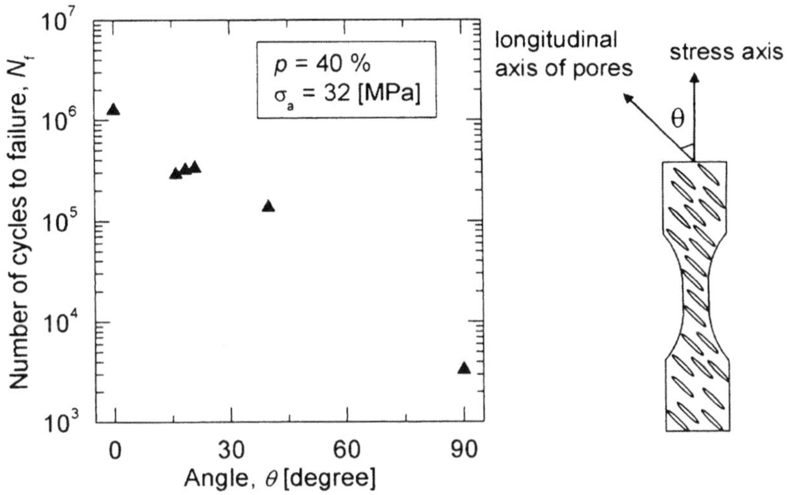


**Fig. 7.52** The number of cycles to failure of lotus copper with porosity of 40 % as a function of angle $\theta$. $\Theta$ is an angle from longitudinal axis of pore, and cyclic stress of 32 MPa was applied in the direction of $\theta$ (Reprinted with permission from [52] © 2007 Materials Research Society)

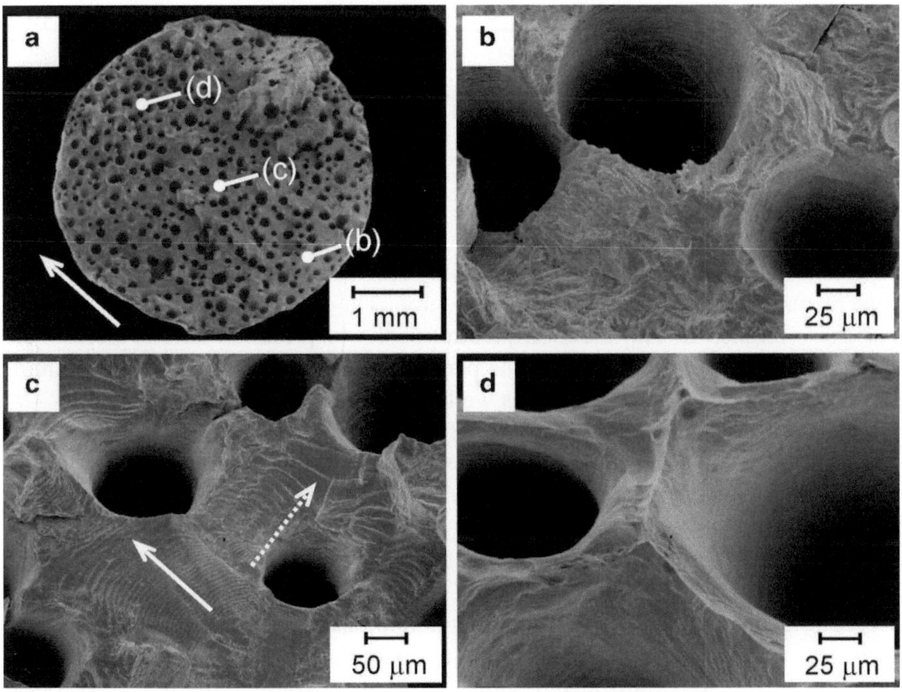


**Fig. 7.53** The fracture surface of lotus copper with porosity of 30 % when cyclic stress of 47 MPa was applied in the direction parallel to the longitudinal axis of cylindrical pores where the number of cycle to failure was 541,336. (**a**) Entire fracture surface and (**b–d**) higher magnifications of (**a**). The magnified points are denoted in (**a**) (Reprinted with permission from [52] © 2007 Materials Research Society)

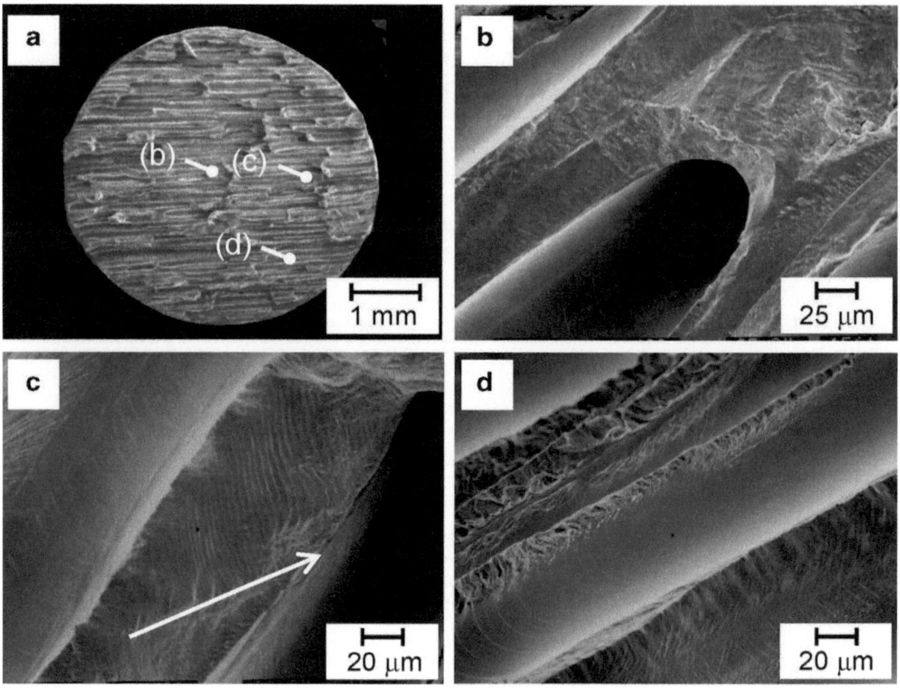


**Fig. 7.54** The fracture surface of lotus copper with porosity of 30 % when cyclic stress of 32 MPa was applied in the direction perpendicular to the longitudinal axis of cylindrical pores where the number of cycle to failure was 24,127. (**a**) Entire fracture surface and (**b–d**) higher magnifications of (**a**). The magnified points are denoted in (**a**) (Reprinted with permission from [52] © 2007 Materials Research Society)

does not exist. Figure 7.54d shows the final fracture surface of stage III, and this type of surface is formed near only this region. The ridge of a mountain-ridge-like pattern is aligned in the direction parallel to the longitudinal axis of pores. The fracture surface characteristics for the perpendicular loadings were independent of the number of cycles to failure $N_f$ as well as that for the parallel loadings.

It is well known that the fatigue strength of some metals is related to the ultimate tensile strength [53]. The ultimate tensile strength of lotus copper in the parallel and perpendicular directions can be expressed by the following simple formula [20]:

$$\sigma = \sigma_0(1 - p)^m,$$

where $\sigma_a$ is the ultimate tensile strength of nonporous copper, $m = 1$ for the parallel direction, and $m = 3$ for the perpendicular direction. The values of $m$ were determined in the previous study [20]. Figure 7.55 shows the fatigue strength at $N_f = 10^5$ of lotus copper with various porosities in the parallel and perpendicular directions as a function of the ultimate tensile strength of lotus copper with the same

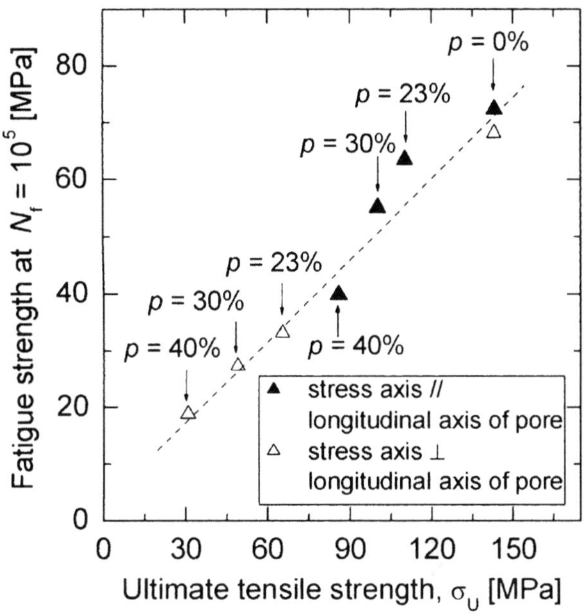


**Fig. 7.55** The fatigue strength at $N_f = 10^5$ of nonporous and lotus copper with porosity of 23, 30, and 40 % as a function of the ultimate tensile strength of lotus copper (Reprinted with permission from [52] © 2007 Materials Research Society)

porosities. The fatigue strength at $N_f = 10^5$ was estimated from the fitting curves in Fig. 7.51. The ultimate tensile strength was calculated by using Eq. (7.3) and $\sigma_0 = 143$ MPa [54]. The fatigue strength is proportional to the ultimate tensile strength. This implies that the fatigue strength follows a simple formula of $\sigma = \sigma_0\,(1-p)^m$, where $\sigma_0$ is the fatigue strength at finite life of nonporous copper. The proportional relationship between the fatigue strength and ultimate tensile strength indicates that the value of $m$ for the fatigue strength is almost the same as those of the ultimate tensile strength.

Most of the pores on the surface of specimens were closed by thin-film walls owing to machining, and some surface pores were open. In general, fatigue strength depends on defects on the surface of the specimens, because the defects become a crack-initiation site. There are many possible crack-initiation sites in lotus copper specimen, because there are many pores corresponding to the defects. Therefore, the effect of the surface open pores on the fatigue strength is probably negligible compared with that of inner pores.

# References


1. Nakajima H, Ikeda T, Hyun SK (2003) In: Banhart J, Fleck A (eds) Cellular metals: manufacture, properties, applications. MIT, Berlin, pp 191–202
2. Nakajima H, Ikeda T, Hyun SK (2004) Adv Eng Mater 6:377–384
3. Demarest HH Jr (1971) J Acoust Soc Am 49:768–775

4. Ohno I (1976) J Phys Earth 24:355–379
5. Ichitsubo T, Tane M, Ogi H, Hirao M, Ikeda T, Nakajima H (2002) Acta Mater 50:4105–4115
6. Tane M, Ichitsubo T, Hirao M, Ikeda T, Nakajima H (2004) J Appl Phys 96:3696–3701
7. Hyun SK, Ikeda T, Nakajima H (2004) Sci Tech Adv Mater 5:201–205
8. Tane M, Ichitsubo T, Nakajima H, Hyun SK, Hirao M (2004) Acta Mater 52:5195–5201
9. Phani HH (1986) Am Ceram Soc Bull 65:1584–1586
10. Kovacik J (1998) Acta Mater 46:5413–5422
11. Kovacik J, Simancik F (1998) Scr Mater 39:239–246
12. Rotter CA, Smith CS (1966) J Phys Chem Solids 27:267–276
13. Simmons G, Wang H (eds) (1971) Single crystal elastic constants and calculated aggregate properties: a handbook, 2nd edn. MIT, Cambridge
14. Ota K, Ohashi K, Nakajima H (2003) Mater Sci Eng A341:139–143
15. Yoshinari O, Kobayashi T, Nakajima H, Ide T (2012) In: Hur BY, Kim BK, Kim SE, Hyun SK (eds) Porous metals and metallic foams. GSIntervision, Seoul, pp 479–485
16. Smithells Metals Reference Book (1992) 7th edn. edited by Brandes EA, Brook GB, Butterworth- Heinemann, Oxford, pp 12–13
17. Kê TS (1947) Phys Rev 71:533–546
18. Wolla JM, Provenzano V (1995) Mater Res Soc Symp Proc 371:377–382
19. Simone AE, Gibson LJ (1996) Acta Metall 44:1437–1447
20. Hyun SK, Murakami K, Nakajima H (2001) Mater Sci Eng A 299:241–248
21. Eudier M (1962) Powder Metall 5:278–290
22. Balshin MY (1949) Doklady Akad Sci USSR 67:831–996
23. Boccaccini AR, Ondracek G, Mombello E (1995) J Mater Sci Lett 14:534–536
24. Peterson AR (1953) Stress concentration design factors. Wiley, New York, p 1
25. Dehoff RT, Gillard JP (1971) Mod Dev. In: Hausner HH (ed) Powder Metall, vol 5. Plenum, New York, p 281
26. Lund JA (1984) Int J Powder Metall Powder Tech 20:141–148
27. Cottrell AH (1952) Philos Mag 43:645–647
28. Lomer WM (1951) Philos Mag 42:1327–1331
29. Tane T, Okamoto R, Nakajima H (2010) J Mater Res 25:1975–1982
30. ASM Handbook (1987), vol 12. American Society for Metals, Materials Park
31. Agogino AM (1978) J Eng Mater Technol 100:348–355
32. Sugimura Y, Meyer J, He MY, Bart-Smith H, Grenstedt J, Evans AG (1997) Acta Mater 45:5245–5259
33. Motz C, Pippan R (2001) Acta Mater 49:2463–2470
34. Amsterdam E, de Vries JHB, De Hosson JTM, Onck PR (2008) Acta Mater 56:609–618
35. Gibson LJ, Ashby MF (1997) Cellular solids. Cambridge University Press, Cambridge
36. Ashby MF, Evans AG, Fleck NA, Gibson LJ, Hutchinson JW, Wadley HNG (2000) Metal foams: a design guide. Butterworth-Heinemann, Woburn
37. Simone AE, Gibson LJ (1997) J Mater Sci 32:451–457
38. Hyun SK, Nakajima H (2003) Mater Sci Eng A 340:258–264
39. Ide T, Tane M, Ikeda T, Hyun SK, Nakajima H (2006) J Mater Res 21:185–193
40. Qiu YP, Weng GJ (1992) J Appl Mech-Trans ASME 59:261–268
41. Deshpande VS, Fleck NA (2000) Int J Impact Eng 24:277–298
42. Kolsky H (1949) Proc Phys Soc London, Sect B 62:676–700
43. Tane M, Kawashima T, Yamada H, Horikawa K, Kobayashi H, Nakajima H (2010) J Mater Res 25:1179–1190
44. Ide T, Tane M, Nakajima H (2009) Mater Sci Eng A 508:220–225
45. Rao PP, Tangri K (1991) Mater Sci Eng A 132:49–59
46. Simone AE, Gibson LJ (1998) Acta Mater 46:3109–3123
47. Hyun SK, Nakajima H, Boyko LV, Shapovalov VI (2004) Mater Lett 58:1082–1086
48. Zhou J, Soboyejo WO (2004) Mater Sci Eng A369:23–35
49. Olurin OB, McCullough KYG, Fleck NA, Ashby MF (2001) Int J Fatigue 23:375–382

50. Harte A-M, Fleck NA, Ashby MF (1999) Acta Mater 47:2511–2524
51. Sugimura Y, Rabiei A, Evans AG, Harte AM, Fleck NA (1999) Mater Sci Eng A269:38–48
52. Seki H, Tane M, Otsuka M, Nakajima H (2007) J Mater Res 22:1331–1338
53. Suresh S (1998) Fatigue of materials, 2nd edn. Cambridge University Press, Cambridge
54. Gerber TL, Fuchs HO (1968) J Mater 3:359–374

# Chapter 8
# Various Physical and Chemical Properties of Lotus Metals

**Abstract** Lotus metals have anisotropic pore configuration. Such anisotropic pores yield anisotropic behavior of sound absorption, electrical and thermal conductivity, magnetization, and corrosion behavior, because the pore itself affects those materials characteristics. However, it does not affect thermal expansion, because only nonporous body controls thermal expansion. Thus, lotus metals exhibit unique characteristics.

**Keywords** Corrosion • Magnetization • Sound absorption • Thermal conductivity • Thermal expansion

## 8.1 Sound Absorption

Sound absorption materials with an advanced performance to noises are required for mufflers of cars, air-conditioner parts, pump chambers, elevated roads, etc. The glass wool and foamed aluminum with closed pores are used most frequently as marketed sound-absorbing materials at present. In most cases, these materials have low strength, though they have good sound absorption capacity. The development of the sound-absorbing material with comprehensive characteristics such as high strength, lightweight, and good sound absorption capacity is desirable. In order to investigate whether lotus metals exhibit significant sound absorption characteristics, Xie et al. [1–3] measured the sound absorption coefficient of lotus magnesium and copper plates by standing-wave method [4]. It is well known that the absorption coefficient depends on the angle between the material surface and the incident sound wave. In the experiments, the specimen surface was set perpendicular to the incident sound wave. Figure 8.1 shows the schematic drawing for measurements of the sound absorption coefficient. The specimen was set on the rigid wall in the sound tube. A pure sound with a single frequency was generated from the speaker installed at the other end. The standing-wave is caused by the interference between an incidence wave and a reflection wave in the tube.

H. Nakajima, *Porous Metals with Directional Pores*,
DOI 10.1007/978-4-431-54017-5_8, 

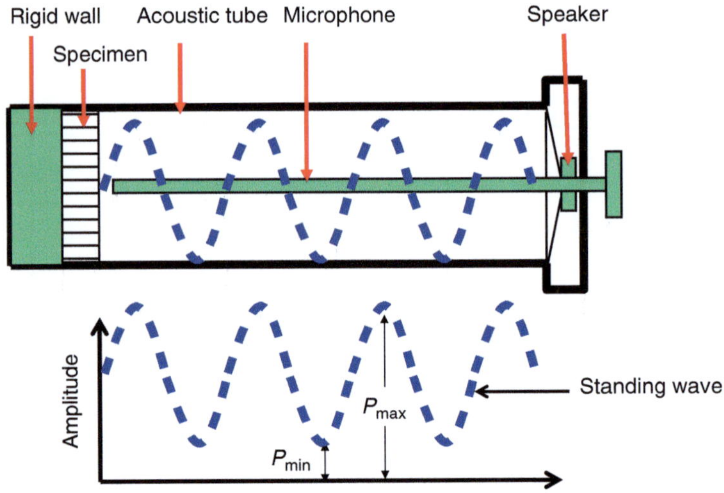


**Fig. 8.1** Schematic drawings for measurement of sound absorption coefficient by standing-wave method

The sound pressure becomes the maximum at each quarter of the wavelength. The maximum value of the sound pressure is written as $|p|_{max} = |A + B|$, where $A$ and $B$ are the amplitude of incidence wave and reflection wave, respectively. The minimum value of the sound pressure is written as $|p|_{min} = |A - B|$. The ratio between the maximum and minimum of the sound pressure, $n$, is given by

$$\frac{|p|_{max}}{|p|_{min}} = \frac{|A + B|}{|A - B|} = n. \tag{8.1}$$

The sound reflectivity of the specimen is written as

$$|r_p| = \left|\frac{B}{A}\right| = \frac{n - 1}{n + 1}. \tag{8.2}$$

The absorption coefficient, $\alpha_0$, is given by

$$\alpha_0 = 1 - |r_p|^2 = \frac{4}{n + (1/n) + 2}. \tag{8.3}$$

$|p|_{max}$ and $|p|_{min}$ are measured by moving the microphone in the tube to determine the value of $n$. Then the absorption coefficient can be calculated using Eq. (8.3). This measuring method is called a standing-wave method, which is one of the tube methods, and the details are provided in JIS A 1405–1963 standards [4].

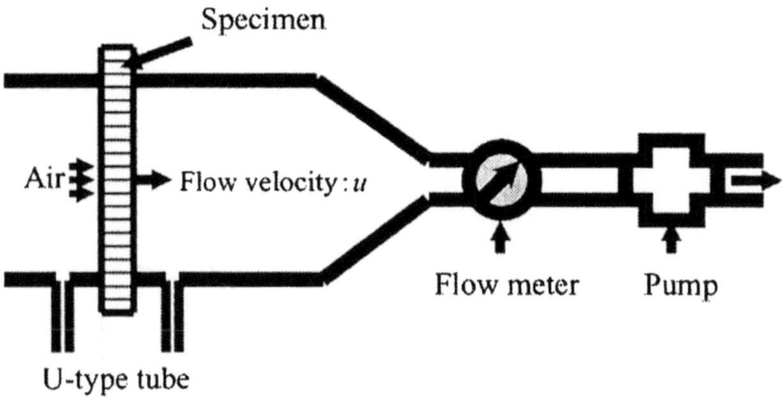


**Fig. 8.2** Schematic drawing for measurement of flow resistance (Reprinted with permission from [1] © 2004 Elsevier B.V.)

Both the flow resistance and the absorption coefficient show amount of the performance of absorbing sound in the porous material. Therefore, it is necessary to measure the flow resistance of the lotus copper. The flow resistance of the sound-absorbing material is basically the same as the ventilation resistance, used to show the ventilations such as cloth and paper. Unit area flow resistance of the porous sound-absorbing material is defined by

$$R_f = \frac{\Delta p}{u}, \tag{8.4}$$

where $u$ is the flow ratio when the constant air is passed through the vertical direction on the surface of the material and $\Delta p$ is the difference in pressure at both sides of material. A pump is operated as shown in Fig. 8.2, and the differential pressure between both sides of the specimen is measured with a U manometer. The flow ratio $u$ is written as

$$u = \frac{Q}{S}, \tag{8.5}$$

where $S$ and $Q$ are area of specimen and flow volume of air, respectively.

Zie et al. measured the absorption coefficients $\alpha_0$ for the lotus copper as a function of pore diameter under the condition that the specimen thickness and the porosity were constant [1]. As shown in Fig. 8.3, $\alpha_0$ increases with decrease in pore diameter from 660 to 460 μm in the whole frequency range up to 4 kHz. Figure 8.4 shows the porosity dependence of the absorption coefficient for constant pore diameter 380 μm and constant specimen thickness 10 mm. The absorption coefficient increases with increasing porosity from 43 to 62 %. There are some data scattering, which is attributed to coexistence of non-permeable and permeable pores. Figure 8.5 shows the dependence of the absorption coefficient on specimen

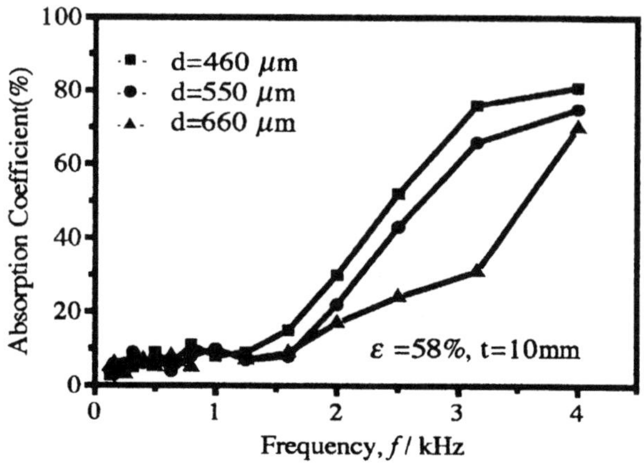


**Fig. 8.3** Effect of pore diameter on sound absorption coefficient of lotus copper (Reprinted with permission from [1] © 2004 Elsevier B.V.)

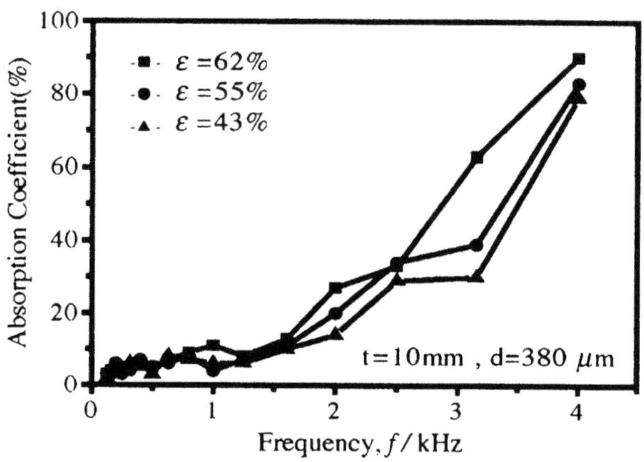


**Fig. 8.4** Effect of porosity on sound absorption coefficient of lotus copper (Reprinted with permission from [1] © 2004 Elsevier B.V.)

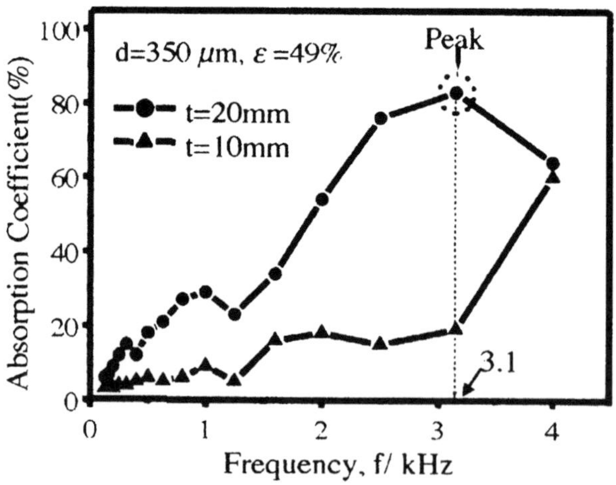


**Fig. 8.5** Effect of specimen thickness with 10 and 20 mm on sound absorption coefficient of lotus copper (Reprinted with permission from [1] © 2004 Elsevier B.V.)

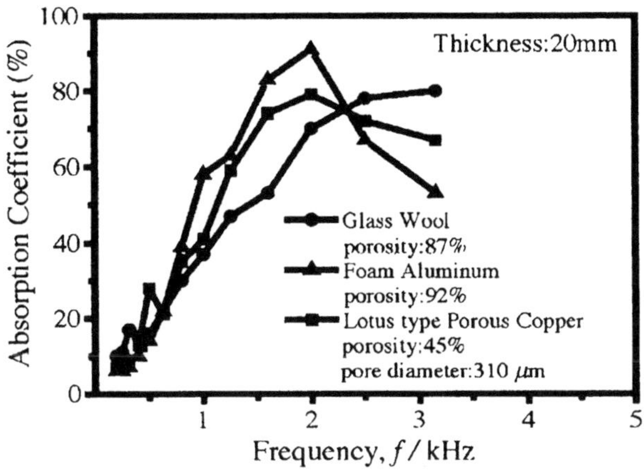


**Fig. 8.6** Comparison of absorption coefficient of various materials (Reprinted with permission from [1] © 2004 Elsevier B.V.)

thickness when the pore diameter and porosity are constant. The absorption coefficient increases with increasing thickness. Especially, the absorption coefficient increases significantly in high-frequency range. The maximum of $\alpha_0$ was observed at 3.1 kHz in the specimen of 20 mm thick, while such maximum value was not found until 4 kHz in the specimen of 10 mm thick. A similar tendency was observed for lotus magnesium [5].

The glass wool has a peculiar mechanism of absorbing sound and is used widely as the sound-absorbing material. The absorption coefficients of lotus copper, the foam aluminum, and the glass wool with the same thickness, in the same frequency region, were compared as shown in Fig. 8.6. All of them were measured by a standard-wave method. The glass wool [6] and foam aluminum [7] exhibit superior absorption capacity. The foam aluminum is composed of many independent closed pores. Continuous pores are necessary to have high-sound absorption capacity [6] so that minute cracks are introduced by rolling to connect the pores of the foam aluminum.

For mechanism of sound absorption, it is thought that the viscosity resistance of air in pores plays an important role in absorbing the sound for the porous material. The sound is absorbed by the resistance in the fiber and this space of pores, when the sound enters into the open pores in porous materials [8]. The sound is also absorbed by disturbance of the movement of air. The absorption of the sound in porous material is considered to be mainly due to the consumption of the sound energy by the viscosity and the thermal conduction when the sound is propagating into this tube. It is difficult to analyze this strictly because the pores in the porous material are arranged to have complex shape.

The lotus metal can be considered as an assembly of many parallel thin tubes. In order to simplify the analysis, first consider how a sound propagates in only one tube. When a sound propagates in a thin tube, the attenuation of a sound depends on the material of the tube. The attenuation in a smooth metal tube has been reported to be larger than the attenuation in air [9]. According to Igarasi [10], the attenuation constant $\beta$ is expressed as

$$\beta = \frac{0.0102}{cr} f^{1/2}, \tag{8.6}$$

where $c$ and $r$ are speed of sound and radius of tube, respectively. The attenuation of sound can often be disregarded when the inner diameter exceeds several centimeters because attenuation is reversely proportional to the inner diameter in Eq. (8.6). On the other hand, since the radius of pores of the lotus copper is from 200 μm to 1 mm, the attenuation increases when the sound enters lotus copper. The attenuation constants of $N$ pores are given by

$$\beta_N = \frac{\beta}{N} = \frac{0.0102}{crN} f^{1/2}. \tag{8.7}$$

The relation between the pore number and the porosity is written as

$$N = \frac{r_1^2 \varepsilon}{r^2}, \tag{8.8}$$

where $r_1$ is a radius of the specimen. From Eqs. (8.7) and (8.8), the attenuation constant in the lotus copper is given by

$$\beta_N = \frac{0.0102}{c\varepsilon r_1^2} f^{1/2}. \tag{8.9}$$

It is thought that the reason of absorbing sound is mainly due to viscous friction of air in the lotus copper. The absorption effect by only permeable pores is taken into consideration. Therefore, it is necessary to measure the porosity of only permeable pores. The thicker the specimen is, the more pores become difficult to be permeable because the length of elongated pores is limited in the lotus copper. It has been known that the pores are hardly permeable in the specimen of 20 mm or more in thickness. Therefore, some differences were seen in the absorption coefficient–frequency curve when the porosity was increased from 43 to 62 % in Fig. 8.4.

The attenuation mechanism of the lotus copper is considered to be the change of absorption sound energy into thermal energy by the viscous friction in the pores. From this viewpoint, the attenuation constant in the lotus copper is related to the radius, the porosity, the pore diameter, the thickness of the specimen, and the frequency. Figure 8.7 shows the relation of the absorption coefficient and attenuation constant by calculating from Eq. (8.9). Thus, it is found that the absorption coefficient is related to the attenuation constant of the specimen.

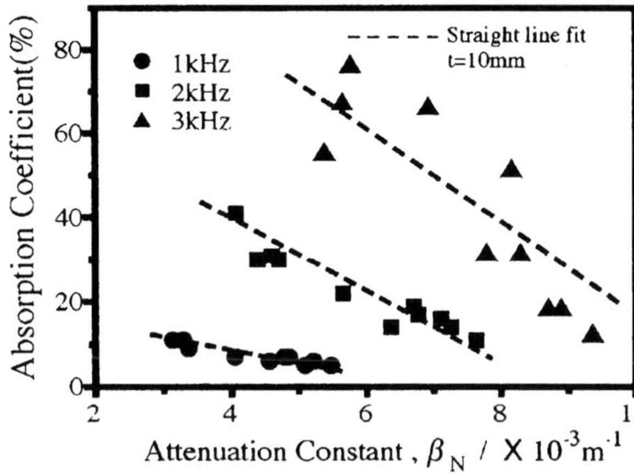


**Fig. 8.7** A relation between absorption coefficient and attenuation constant of lotus copper with 10-mm thickness (Reprinted with permission from [1] © 2004 Elsevier B.V.)

## 8.2 Thermal Conductivity

To use lotus copper effectively as heat sinks mentioned later, it is very important to know its effective thermal conductivity and consider the pore effect on the heat flow. There is much in the literatures devoted to the effective thermal conductivity of composite materials with cylindrical inclusions. Behrens [11] analytically investigated the effective thermal conductivities of composite materials under the assumption of orthorhombic symmetry and proposed a simple equation for predicting the effective thermal conductivity. Perrin et al. [12] proposed a method for predicting transport properties that included the thermal conductivity of circular cylinders in square and hexagonal arrays. Han and Cosner [13] concluded a numerical investigation of the effective thermal conductivities of composites with uniform fibers in unidirectional orientation and layered composites with fibers laid alternately along two mutually perpendicular directions. A numerical method was devoted by Sangani and Yao [14] to determine the effective thermal conductivity of a composite medium consisting of parallel circular cylinders in random arrays. They cited that the conductivity appeared to be a relatively weak function of the detailed arrangement of the cylinders. Mityushev [15] extended analysis of the resolution of the Laplace equation in composite material with a collection of non-overlapping, identical, circular disks. Moctezuma-Berthier et al. [16] showed the predominant influence of the total porosity on the thermal properties of vulgar porous media. Ogushi et al. [17] investigated the effective thermal conductivities parallel and perpendicular to the pore axis of lotus copper both experimentally and analytically. Since the thermal conductivity of the fluid in the pores is negligible in comparison to lotus copper material in the application of lotus copper to heat sinks, a very simple equation can describe the thermal conductivities of lotus copper.

### 8.2.1 Measurement of Effective Thermal Conductivity of Lotus Copper

The effective thermal conductivity of lotus copper $k_{\mathrm{eff}}$ is defined by

$$q = \frac{Q}{A} = -k_{\mathrm{eff}} \nabla T, \tag{8.10}$$

where $q$ is the heat flux from heat flow $Q$ divided by heat flowing through the cross-sectional area $A$ in lotus copper including the pores, and $T$ is the temperature in lotus copper. The tensor $k_{\mathrm{eff}}$ is orthorhombic and is expressed as

$$k_{\mathrm{eff}} = \begin{pmatrix} k_{\mathrm{eff}||} & & \\ & k_{\mathrm{eff}} & \\ & & k_{\mathrm{eff}} \end{pmatrix}. \tag{8.11}$$

The effective thermal conductivity of lotus copper is anisotropic. The parallel and perpendicular effective thermal conductivities, $k_{\mathrm{eff}//}$ and $k_{\mathrm{eff}\perp}$, of lotus copper are defined as the thermal conductivities for heat flow parallel and perpendicular to the pore axis, respectively.

Figure 8.8 presents the experimental apparatus for measuring the effective thermal conductivity. A cylindrical specimen with a diameter of 30 mm and a length of 30 mm was located between upper and lower copper rods of known thermal conductivity. The upper rod was heated by electrical heaters from the top surface, while the bottom surface of the lower rod was cooled by cooling water in order to transmit a certain amount of heat through the specimen. K-type thermocouples were located each at 5-mm spacing in the specimen and the upper and the lower rods to measure the temperature. An example of the temperature distribution of the experimental setup is shown in Fig. 8.9. As the heat flowed in one direction through the rods and specimen from top to bottom, heat flux $q$ through the specimen was obtained by the following one-dimensional equations:

$$q = \frac{q_1 + q_2}{2}, \tag{8.12}$$

$$q_1 = -k_{\mathrm{Cu}} \left( \frac{\partial T}{\partial x} \right)_1, \tag{8.13}$$

$$q_2 = -k_{\mathrm{Cu}} \left( \frac{\partial T}{\partial x} \right)_2, \tag{8.14}$$

where $q_1$ is a heat flux from the upper rod to the specimen and $q_2$ is the heat flux from the specimen to the lower rod, $k_{\mathrm{Cu}}$ is the thermal conductivity of the rod, and $x$ is the direction of heat flow from the upper to the lower rod. From

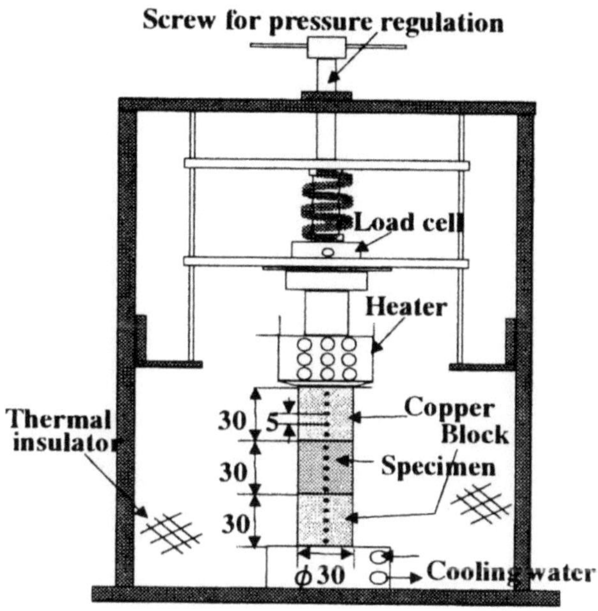


**Fig. 8.8** Experimental apparatus for measuring effective thermal conductivity (Reprinted with permission from [17] © 2004 American Institute of Physics)

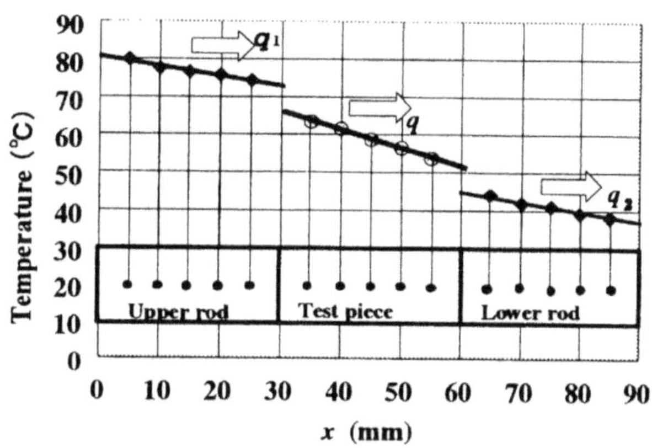


**Fig. 8.9** Temperature distribution in specimen and rods (Reprinted with permission from [17] © 2004 American Institute of Physics)

Eqs. (8.10)–(8.14), effective thermal conductivity $k_{\mathrm{eff}//}$ and $k_{\mathrm{eff}\perp}$ are obtained by the following equation:

$$k_{\mathrm{eff}||},\quad k_{\mathrm{eff}\perp} = -\frac{q_1 + q_2}{2\left(\frac{\partial T}{\partial x}\right)_{\mathrm{lotus}}}, \tag{8.15}$$

where $(\partial T/\partial x)_{\mathrm{lotus}}$ is the temperature gradient in the specimen with pores parallel or perpendicular to heat flow direction $x$.

### 8.2.2 Analysis of Effective Thermal Conductivity of Lotus Copper

Since the heat flow cross-sectional areas parallel to the pore axis in lotus copper is proportional to $(1-\varepsilon)$, the effective thermal conductivity $k_{\mathrm{eff}//}$ is expressed by the following equation:

$$\frac{k_{\mathrm{eff}||}}{k_s} = 1 - \varepsilon, \tag{8.16}$$

where $k_s$ is the thermal conductivity of nonporous copper and $\varepsilon$ is the porosity expressed by the volume ratio of pores versus the total volume of lotus copper.

Behrens [11] derived the effective thermal conductivity of composite materials with orthorhombic symmetry. By applying his equation to the thermal conductivity of lotus copper, the effective thermal conductivity perpendicular to the pores can be expressed by the following equation:

$$\frac{k_{\mathrm{eff}\perp}}{k_s} = \frac{(\beta + 1) + \varepsilon(\beta - 1)}{(\beta + 1) - \varepsilon(\beta - 1)}, \tag{8.17}$$

where $\beta(=k_p/k_s)$ is the conductivity ratio, that is, pore conductivity $k_p$ divided by material conductivity $k_s$ of lotus copper. Because the thermal conductivity of the hydrogen gas or air in the pores of lotus copper is negligible compared with that of lotus material, the effective thermal conductivity of lotus copper is derived as the following equation by setting $\beta = 0$ in the above equation:

$$\frac{k_{\mathrm{eff}\perp}}{k_s} = \frac{1 - \varepsilon}{1 + \varepsilon}. \tag{8.18}$$

Han and Cosner [13] performed a numerical study on the effective thermal conductivities of fibrous composites using a unit-cell approach under a uniform fiber diameter condition. Since the diameter of lotus copper is distributed around a certain range, numerical simulation for the thermal conductivity perpendicular to the pores under a nonuniform pore diameter condition was conducted to verify the applicability of Eq. (8.18) to lotus copper.

Figure 8.10a shows a comparison between the experimental data and the results evaluated by the analytical equation, Eq. (8.16), for thermal conductivity parallel to the pores. Experimental data for $k_{\mathrm{eff}//}$ showed good agreement with the analytical results derived from the assumption that heat flow through the cross-sectional area parallel to the pore axis is proportional to $(1-\varepsilon)$. A value of 335 W/(mK) for thermal conductivity $k_s$ of the lotus copper material was used for comparison. Figure 8.10b gives a comparison between the experimental data and Eq. (8.18), where effective thermal conductivity $k_{\mathrm{eff}\perp}$ perpendicular to the pores was lower than that of the parallel ones($k_{\mathrm{eff}//}$) and was 40 % of lotus copper material $k_s$ with a

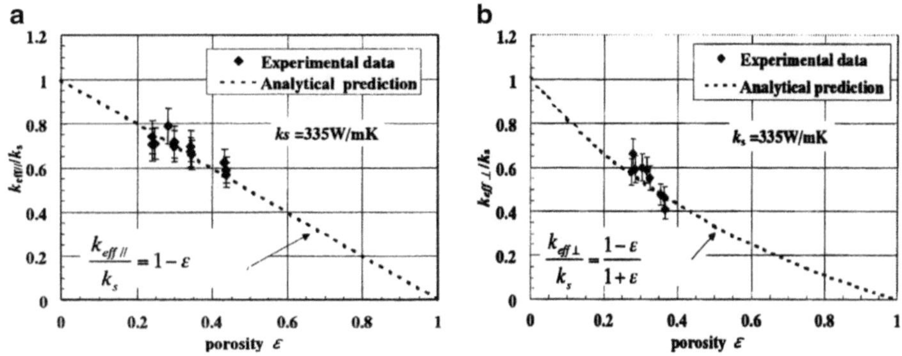


**Fig. 8.10** (**a**) Comparison of experimental data of the effective thermal conductivity of lotus copper parallel to pores with analysis data. (**b**) Comparison of experimental results of the effective thermal conductivity of lotus copper perpendicular to pores with analysis data (Reprinted with permission from [17] © 2004 American Institute of Physics)

porosity of 0.4. The analytical values evaluated by Eq. (8.18) showed good agreement with the experimental data, indicating that Eq. (8.18) can be used to predict the effective thermal conductivity perpendicular to pores of lotus copper within experimental accuracy of $\pm 10$ %. The results show that lotus copper displayed anisotropy of the effective thermal conductivity. The effective thermal conductivity $k_{\mathrm{eff}\perp}$ perpendicular to pores was lower than that of parallel to $k_{\mathrm{eff}//}$ ones.

## 8.3 Electrical Conductivity

For the permeability and large surface area, porous metals are indispensable as electrode materials of a battery. The conventional porous electrode materials are fabricated with powder metallurgy, and therefore the porous structure is complicated, i.e., the pores are nonuniform and dispersed irregularly. This complicated structure results in a degradation of the strength and permeability of the electrolyte solution. Because of this unique porous structure of the directional porosity metals, lotus metals show more superior mechanical properties than the conventional porous metals [18]. Suematsu et al. [19] succeeded to fabricate lotus nickel; this porous nickel is promising as a new type of electrode material that possesses the superior strength and permeability. For this application, the effect of pore structure and porosity on the electrical conductivity needs to be investigated. Thus, Tane et al. [20] studied the electrical conductivity of lotus-type porous nickel. The first part describes the experimental result on the conductivity parallel and perpendicular to the longitudinal pore directions measured by four-probe method. Next, a calculation method is shown for the electrical conductivity of such an anisotropic porous structure on the basis of the effective-mean-field (EMF) theory [21]. Finally the validity of the calculation by the method is discussed, comparing with the measurement data.

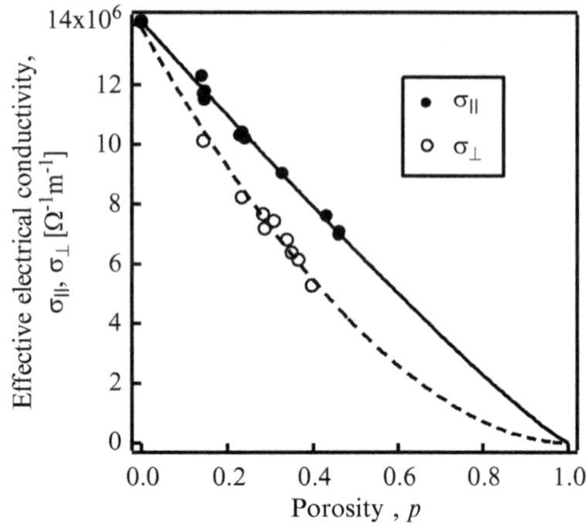


**Fig. 8.11** Porosity dependence of the electrical conductivity of lotus nickel. The *plots* indicate measured result, and the *lines* indicate fitting the equation to the measurements (Reprinted with permission from [20] © 2005 American Institute of Physics)

### *8.3.1 Measurement of Electrical Conductivity of Lotus Nickel*

Lotus nickel ingots with long straight pores aligned in one direction were fabricated through unidirectional solidification in a pressurized hydrogen or argon gas. Continuous current electrical conductivity along the longitudinal direction of the specimen was measured with the four-probe method at room temperature.

Figure 8.11 shows the porosity dependence of the normalized effective electrical conductivities $\sigma_{\|}/\sigma_0$ and $\sigma_{\perp}/\sigma_0$, where $\sigma_{\|}/\sigma_0$ and $\sigma_{\perp}/\sigma_0$ denote the normalized effective electrical conductivity parallel and perpendicular to the longitudinal pore direction, respectively. The electrical conductivity of nonporous material, $\sigma_0$, is $1.41 \times 10^7\ \Omega^{-1}\,\mathrm{m}^{-1}$. The electrical conductivity shows the anisotropy that reflects the porous structure. For the parallel to the longitudinal pore direction, the specific conductivity is almost retained. This is because the flow direction of the electrical current in the nickel matrix is almost parallel to the applied electric field. For the perpendicular direction, the electrical current needs to flow detouring around pores. Therefore, a distance that the electrical current flows increases, which results in the increase of the effective electrical resistivity (decrease of the effective electrical conductivity).

### *8.3.2 Analysis of Electrical Conductivity of Lotus Nickel*

The previous extensive researches about porous rocks elucidated that porosity dependence of the effective conductivity follows an empirical formula (called Archie's law [22]):

$$\sigma = \sigma_0(1-p)^m, \tag{8.19}$$

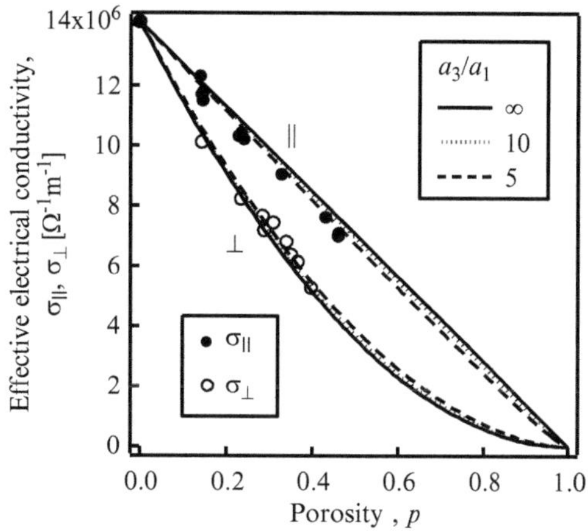


**Fig. 8.12** Porosity dependence of the effective electrical conductivity of lotus nickel parallel and perpendicular to the pore growth direction in the case of $a_3/a_1 = 5$, 10, and $\infty$, which are calculated with EMF theory. *Plots* indicate the measurement data, and *lines* indicate the calculation results with EMF theory (Reprinted with permission from [20] © 2005 American Institute of Physics)

where $\sigma$ and $\sigma_0$ are the effective electrical conductivity of porous and nonporous material, respectively, and $m$ is the coefficient determined empirically. Eq. (8.19) was fitted to the measurement data, and lines in Fig. 8.11 show the fitting curves. It is found that power-law relation holds in the anisotropic porous metals. The coefficient $m$ for parallel and perpendicular conductivity is estimated at 1.1 and 1.8, respectively.

Tane et al. [20] applied the concept of EMF theory to derive the effective electrical conductivity of composites. Composites consist of the matrix and one type of inclusions, whose volume fractions are denoted by $f_M$ and $f_I(= 1 - f_M)$, respectively. Spatial averages of an electrical current density $\overline{\mathbf{J}}$ (3×1 vector) and electric field $\overline{\mathbf{E}}$ (3×1 vector) of composites are expressed as $\overline{\mathbf{J}} = f_M \overline{\mathbf{J}}_M + f_I \overline{\mathbf{J}}_I$ and $\overline{\mathbf{E}} = f_M \overline{\mathbf{E}}_M + f_I \overline{\mathbf{E}}_I$, where $\overline{\mathbf{J}}_M = \sigma_M \overline{\mathbf{E}}_M$ and $\overline{\mathbf{J}}_I = \sigma_I \overline{\mathbf{E}}_I$ and $\sigma_M$ and $\sigma_I$ are electrical conductivity of matrix and inclusion, respectively ($\sigma$ is 3×3 matrix, and the component is 0 when $i \neq j$). The electric field $\mathbf{E}$ is defined as $\mathbf{E} = -\mathbf{grad}\phi$, where $\phi$ is the electric potential. Then, the effective electrical conductivity $\overline{\sigma}$ of the composites are defined as $\overline{\mathbf{J}} = \overline{\sigma}\overline{\mathbf{E}}$. When expressing $\overline{\mathbf{E}}_I = \mathbf{A}\overline{\mathbf{E}}_M$, the effective electrical conductivity $\overline{\sigma}$ of the composite can be written as

$$\overline{\sigma} = (f_M\sigma_M + f_I\sigma_I\mathbf{A})[f_M\mathbf{I} + f_I\mathbf{A}]^{-1}, \tag{8.20}$$

where $\mathbf{I}$ is the unit matrix (3×3 matrix). Using Eshelby's equivalent inclusion theory [23] and mean-field theory [24], $\mathbf{A}$ (3×3 matrix) can be expressed as $\mathbf{A} = [\mathbf{S}\sigma_M^{-1}(\sigma_I - \sigma_M) + \mathbf{I}]^{-1}$ in analogy with the elasticity problem [25]. $\mathbf{S}$ is the 2nd-rank Eshelby tensor.

When the present EMF theory is applied to lotus nickel, we assumed that the pore shape is ellipsoidal of $a_1 = a_2$, and the electric conductivity for the pore $\sigma_I$ is 0, where $a_1$, $a_2$, and $a_3$ denote the radii of the ellipsoidal inclusion. Figure 8.12 shows the porosity dependence of $\sigma_{\parallel}/\sigma_0$ and $\sigma_{\perp}/\sigma_0$ in the case of $a_3/a_1 = 5, 10$

and $\infty$. Measurement data are shown for comparison. Practical $a_3/a_1$ of lotus nickel is 5–10 [26], and in this range the calculations agree well with the measurements. Furthermore, the calculation can fully simulate the power-law formula. When $a_3/a_1 \to \infty$, i.e., in the case of typical lotus metal, the effective electrical conductivities parallel and perpendicular to the longitudinal pore direction are expressed with the power-law formula:

$$\left.\begin{aligned} \sigma_{\parallel} &= \sigma_0(1-p)^1, \\ \sigma_{\perp} &= \sigma_0(1-p)^2. \end{aligned}\right\} \tag{8.21}$$

The electrical conductivity of lotus nickel with cylindrical pores aligned unidirectionally was measured. It is found that lotus nickel shows the anisotropy in the electrical conductivity because of the anisotropic porous structure; the effective conductivity parallel to longitudinal pore direction decreases linearly, while that perpendicular to the pore direction decreases steeply. These porosity dependencies follow Archie's power-law formula. The electrical conductivity of lotus nickel calculated with this theory is consistent with measurement data, and the EMF theory can fully simulate Archie's power-law formula.

## 8.4 Magnetization

In the application of lotus materials to magnetic materials, lotus materials are expected to have a useful merit that the geometrical magnetic anisotropy, called shape magnetic anisotropy, can be controlled by the growth direction of pores, without any change in the outer shape of material. For example, although the perpendicular direction of platelike samples is a hard axis of magnetization, it is possible to make the perpendicular direction be an easy axis of magnetization by introducing elongated pores aligned perpendicular to the plate. With regard to magnetic properties of porous materials, the effect of pore size has been widely studied for materials possessing isotropic pores [27, 28], but lotus materials have never been investigated. Onishi et al. [29] measured magnetization curves of lotus nickel and cobalt to characterize the anisotropy of the magnetization process, and the effect of porosity on the anisotropy was analyzed in terms of the empirical Archie's law.

The saturation magnetization of the porous and nonporous specimens was measured in the magnetic fields applied parallel and perpendicular to the pore growth direction. Due to the fact that the magnitudes of saturation magnetization of normal ferromagnetic materials such as nickel are given by the atomistic origin, it does not depend on the shapes of specimens and the directions of applied magnetic field, and it was confirmed that the saturation magnetization observed for porous nickel decreased linearly with increasing porosity. The same behavior was observed for cobalt. The relationship between the saturation magnetization of porous nickel and cobalt and their porosities is summarized in Fig. 8.13. The ratios

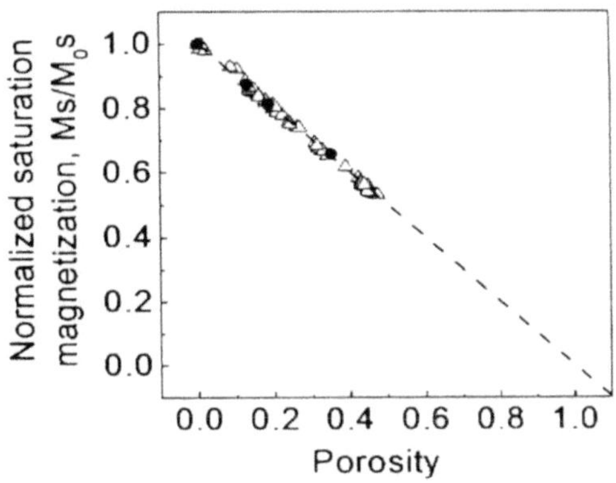


**Fig. 8.13** Porosity dependence of the normalized saturation magnetization ($M_{sat}/M_{0sat}$) of lotus nickel (*filled circles*) and cobalt (*triangles*) (Reprinted with permission from [29] © 2008 American Institute of Physics)

of the saturation magnetization of the porous materials to that of the nonporous material naturally correspond with the volume ratios of the net metal to the whole porous material ($1-p$) [30]. As a result, the saturation magnetization of porous materials is expressed by the following formula:

$$M_{sat} = M_{0sat}(1-p), \tag{8.22}$$

where $M_{sat}$ and $M_{0sat}$ denote the saturation magnetization of the porous material and that of the nonporous material, respectively. This result shows that the porosity of a porous material can be estimated easily by comparing its saturation magnetization with that of nonporous material.

Next, the magnetic field dependence of magnetization was in detail measured for the lotus nickel and cobalt. Although there is no remarkable difference in the obtained results between nickel and cobalt, effects of the pore direction are clearer in nickel. This is due to the soft magnetic properties of nickel. Figure 8.14a, b shows the normalized magnetization curves if lotus nickel, i.e., the magnetic field dependence of the ratio of the magnetization of porous specimens ($M$) to their saturation magnetization ($M_{sat}$), in the magnetic field parallel and perpendicular to the pore growth direction. It was observed that for all the cases, the $M/M_{sat}$ values were close to unity above 3 kOe, showing the nearly saturated state. When the magnetic field was applied parallel to the pore growth direction, the normalized magnetization curve of the porous specimen corresponded with that of the nonporous specimen, as shown in Fig. 8.14a. In the case of perpendicular direction, on the other hand, different slopes of the normalized magnetization curves appeared at low-magnetic fields and the $M/M_{sat}$ value decreased with the increase in the porosity as shown in Fig. 8.14b. The different slopes between the parallel and perpendicular directions were caused by geometric magnetic anisotropy of specimens containing anisotropic pores. In Fig. 8.15, $M$ was normalized against the magnetization of the nonporous specimen ($M_0$) in order to discuss the porosity dependence of the magnitude of anisotropy in terms of Archie's formula. Only in the case of the magnetization perpendicular to the pore growth direction, significant lowering of the $M/M_0$ was

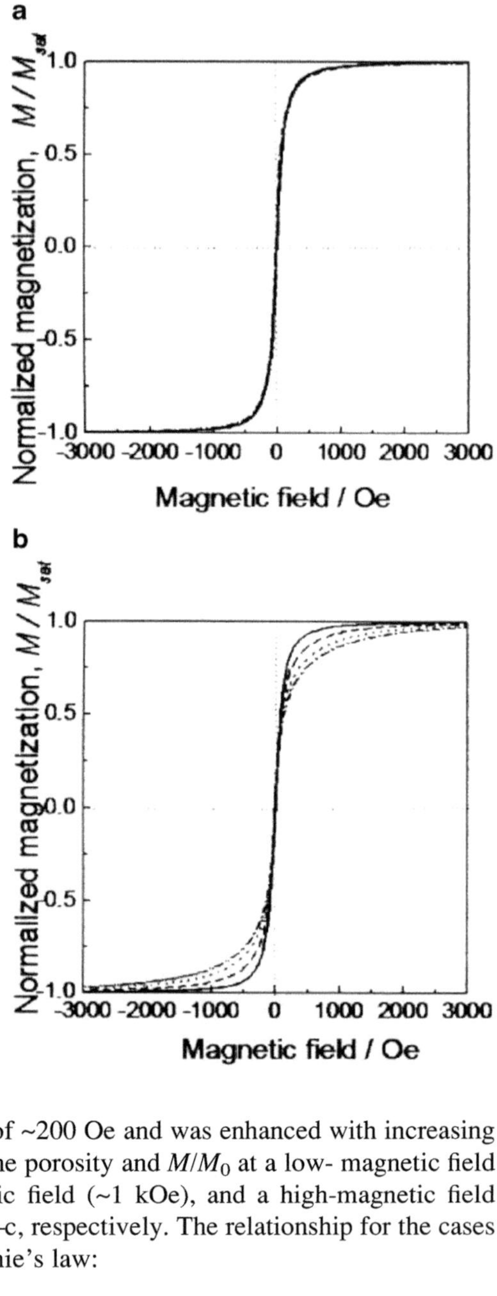


**Fig. 8.14** Magnetization curves for specimens of 0 % porosity (*solid line*), 11 % porosity (*broken line*), 19 % porosity, and 35 % porosity (*chain line*) in the low-magnetic field (**a**) parallel and (**b**) perpendicular to the pore growth direction (Reprinted with permission from [29] © 2008 American Institute of Physics)

observed, at the low-magnetic field of ~200 Oe and was enhanced with increasing porosity. The relationship between the porosity and $M/M_0$ at a low- magnetic field (~200 Oe), an intermediate magnetic field (~1 kOe), and a high-magnetic field (~10 kOe) were shown in Fig. 8.16a–c, respectively. The relationship for the cases could be fitted by the following Archie's law:

$$M = M_0(1 - p)^n, \tag{8.23}$$

where the exponent $n$ is a fitting parameter. In this equation, it is considered that difference between parallel and perpendicular directions represents the relationship

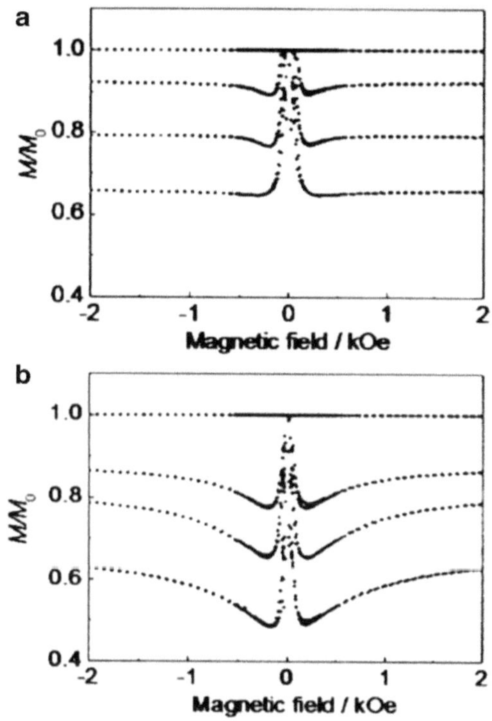


**Fig. 8.15** Normalized magnetization for specimens of (I) 0 % porosity, (II) 11 % porosity, (III) 19 % porosity, and (IV) 35 % porosity in the low-magnetic field (**a**) parallel and (**b**) perpendicular to the pore growth direction (Reprinted with permission from [29] © 2008 American Institute of Physics)

between anisotropy and porosity. The $n$ value decreased toward unity with increasing applied magnetic field. The relationship between the applied magnetic field and $n$ is shown in Fig. 8.17. The $n$ values take a maximum, approximately 1.8 (perpendicular) and approximately 1.1 (parallel) at around 200 Oe. The anisotropy of magnetization clearly appears when $n > 1.1$, i.e., magnetic field less than 2.3 kOe. The $n$ values of 1.8 and 1.1 are coincident with those reported by Tane et al. [20] for electrical conductivity of lotus nickel in the parallel and perpendicular direction, implying the existence of remarkable similarity between electrical conductivity and magnetization.

A possible mechanism of the coincidence in $n$ may come from similar spatial distribution of the electrical current and magnetic flux in the lotus nickel. When a low-magnetic field is applied, the magnetic anisotropy originates from the area of the pore walls perpendicular to the direction of applied magnetic field because magnetic poles created at the pore wall increase the total energy of the system, and therefore, this area acts as a resistance to magnetization. The anisotropic shape of pore is considered to cause the observed magnetic anisotropy, and at a low-magnetic field, the magnetic flux in lotus materials flows without crossing pore wall. Since electrical current also flows without crossing pore walls, the spatial distribution of electrical current and magnetic flux should be similar at a low-magnetic field, and similar values of $n$ are expected to appear in electrical conductivity and magnetization.

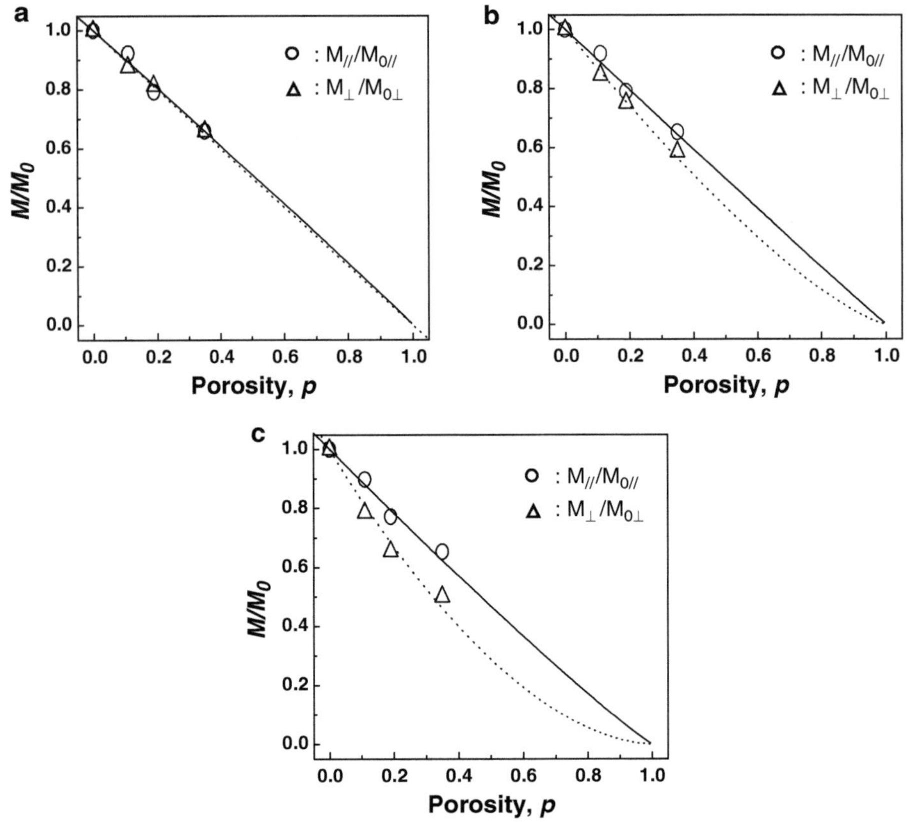


**Fig. 8.16** Porosity dependence of the normalized magnetization at the magnetic field of (**a**) 10 kOe, (**b**) 1 kOe, and (**c**) 200 Oe. The *plots* indicate the experimental data, and the *lines* indicate the fitting results of formula (8.23) to the experimental data (Reprinted with permission from [29] © 2008 American Institute of Physics)

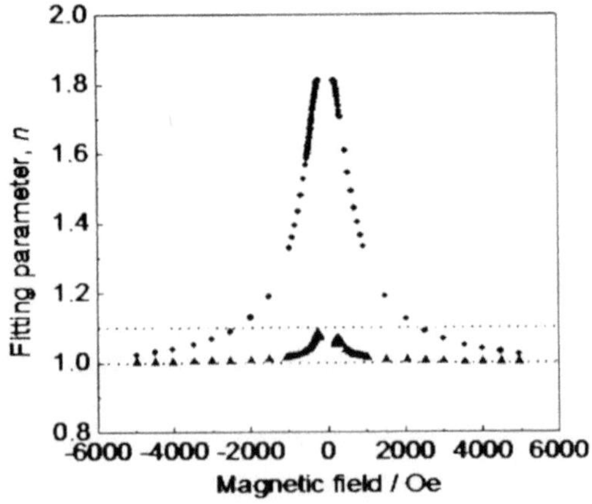


**Fig. 8.17** Dependence of the fitting parameter n on magnetic field applied to parallel (*filled triangles*) and perpendicular (*filled circles*) to the pore growth direction. The *broken lines* indicate $n = 1$ (the case in which all the pores are penetrated) and $n = 1.1$ (10 % anisotropy at the point) (Reprinted with permission from [29] © 2008 American Institute of Physics)

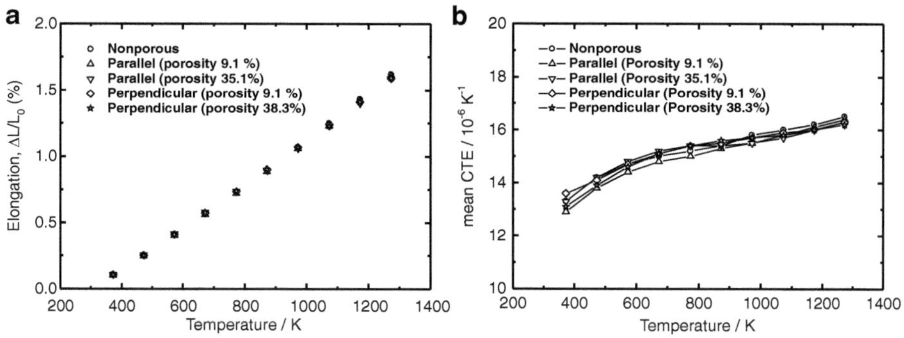


**Fig. 8.18** (**a**) Elongation of lotus nickel in the direction parallel and perpendicular to the pore axis with different porosities as a function of temperature. (**b**) Coefficient of linear thermal expansion of lotus nickel in the direction parallel and perpendicular to the pore axis with different porosities as a function of temperature (Reprinted with permission from [31] © 2006 Elsevier Ltd)

## 8.5 Thermal Expansion

The understanding of the properties of thermal expansion of lotus-type porous metals is important for the practical application such as the electrode, heat sink, and high-temperature parts. Measurements of thermal expansion were carried out by Tane et al. [31] in a thermal dilatometer. The equipment uses an alumina pushrod in contact with the specimen to record the change in linear dimension during a heating. The samples were in the shape of rods of nonporous and porous nickel of 6 mm in diameter and 15 mm in length. The temperature was increased from 294 to 1,273 K at a rate of 1 K $min^{-1}$. Measurements were performed under mechanical-pump vacuum which was sufficient to prevent from oxidation, as verified by measuring negligible mass changes after the experiments.

The thermal expansion data for nickel samples with different porosities are shown in Fig. 8.18a, where $\Delta L = L - L_0$ and $L_0$ is the original length of samples at room temperature. In general, it is convenient to express the thermal expansion behavior by the coefficient of linear thermal expansion (CTE) as follows:

$$\alpha = \frac{1}{L_0}\frac{\Delta L}{\Delta T}, \tag{8.24}$$

where α is the coefficient of thermal expansion ($K^{-1}$), $L_0$ is the original sample length (m), and $\Delta T = T - T_0$ is the temperature interval ($T_0 = 293$ K). The values of CTE α for specific temperature ranges are shown in Fig. 8.18b. These results show no significant effect of the porosity and the pore direction in the lotus-type porous nickel on the CTE.

For prediction of the CTE of metal-matrix composite, the models are derived by Kerner [32] and Schapery [33]. Kerner's model gives the CTE of composite as

$$\alpha_c = \bar{\alpha} + V_p(1 - V_p)(\alpha_p - \alpha_m)\frac{K_p - K_m}{(1 - V_p)K_m + V_pK_p + (3K_pK_m/4G_m)}, \tag{8.25}$$

where $\bar{\alpha}$ is the rule-of-mixtures value given by $V_p\alpha_p + (1 - V_p)\alpha_m$. $K$ and $G$ are the bulk and shear moduli, $V$ is the volume fraction, and α is the coefficient of thermal expansion. The subscripts c, p, and m denote composite, particle, and matrix, respectively.

The Schapery's model is written as

$$\alpha_c = \alpha_p + (\alpha_p - \alpha_m)\frac{(1/K_c) - (1/K_p)}{(1/K_m) - (1/K_p)}. \tag{8.26}$$

In the case of porous metals, the $K_p$ in Eqs. (8.25) and (8.26) is zero and then $\alpha_c = \alpha_m$. This indicates the CTE of lotus-type porous metals depends on that of the matrix, which the effect of porosity on CTE can be neglected.

## 8.6 Corrosion

It is well known that the properties of passive films on stainless steels depend on the type of solution used for the corrosion tests. Stainless steels exhibit different corrosion resistance in alkaline, neutral, and acidic environments, since the mechanisms of formation and breakdown of the passive film are different [34]. AISI 316L stainless steel has been used as a surgical implant material for many years exhibiting sufficient corrosion resistance under body fluids which contain $Cl^-$ ion as well as many other components such as water, dissolved oxygen, bacteria, cells, enzymes, and proteins. The improved corrosion resistance of type AISI 316L stainless steel in a chloride environment is attributed to the addition of molybdenum which is effective in stabilizing the passive film in the presence of chlorides [35]. Recently it has been found that nitrogen addition to austenitic stainless steel improves the pitting corrosion resistance [36–38]. In addition, nitrogen as an alloy component allows a nickel content reduction in the austenitic stainless steel, since the allergic reactions caused by this element is suppressed.

Potential applications of this new family of stainless steels include components subjected to corrosion and wear mechanisms, structural components, as well as medical and biocompatible parts, because nitrogen confers high-yield strength, ultimate tensile strength and ductility, as well as fracture toughness, low-magnetic permeability, very favorable localized corrosion resistance, and higher wear and abrasion resistance. Due to the combination of these excellent properties, currently there is also much interest in the fabrication of porous parts made of high-nitrogen Ni-free stainless steels.

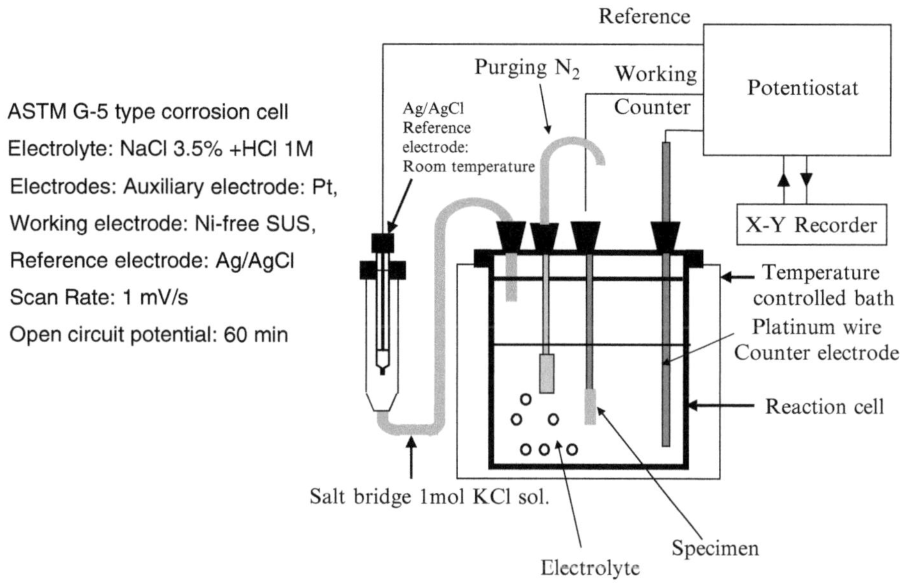


**Fig. 8.19** Experimental setup for measurement of electrochemical polarization curves

The corrosion behavior of porous materials in general differs from that of the bulk materials of the same composition. Moreover, the pore morphology, porosity, pore size, and pore surface condition also interfere in the degradation process, in view of the fact that a porous material is attacked not only on its surface but also from inside. Alvarez et al. measured the corrosion rates, corrosion potentials, and breakdown potentials in dehydrogenized porous and nonporous specimens in order to investigate the corrosion resistance of lotus high-nitrogen Ni-free stainless steels and to compare it with the nonporous material [39].

Electrochemical polarization experiments were conducted using a potentiostat unit connected to a potential and current x–y recorder. The results were finally processed in a personal computer using a data acquisition program. The potentiodynamic polarizations were performed at a scan rate of 1 mV $s^{-1}$, and the scans were started at +100 mV with respect to the open-circuit potential. The details of the experimental setup are illustrated in Fig. 8.19.

The nitrogen effect in the polarization curves was examined firstly in nonporous samples. Potentiodynamic polarization curves for the nonporous nitrided and non-nitrided alloys are shown in Fig. 8.20, where the current density is a measure of the corrosion rate. It was confirmed that the nitrogen alloying suppressed the pitting corrosion even in such a low pH and high $Cl^{-1}$ concentration solution. A more stable passivity region was established due to the nitrogen alloying. Moreover, the solution-nitrided samples exhibited lower passivation current $I_{pass}$. The breakdown potentials for all the solution-nitrided samples were found to be more noble than the non-nitrided samples, thus exhibiting improved corrosion resistance in 3.5 wt% NaCl + 1 M HCL. The high-pitting potential of high-nitrogen

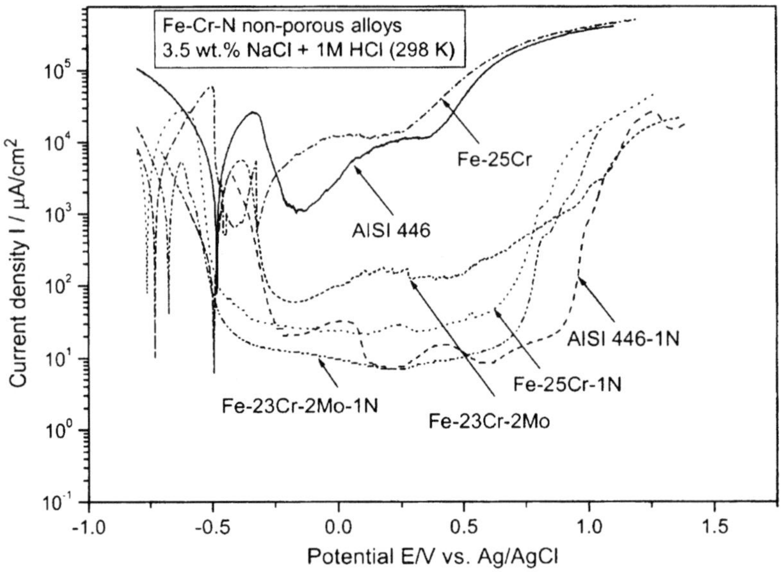


**Fig. 8.20** Nitrogen alloying effect in the potentiodynamic polarization curves of the investigated samples (Reprinted with permission from [39] © 2007 Elsevier Ltd)

steels is attributed to the amount of N dissolved within the austenite. Together with a sufficient Cr and Mo content, which bring about a passive layer (Cr) and a pre-passive film (M), respectively, nitrogen leads to an increase of the pH value within the surrounding medium by a decomposition of $NH_3$ to $NH_4{}^{+} + OH^{-}$ [40]. Indeed, Fe-23wt%Cr-2wt%Mo-1wt%N alloy resulted to have the higher corrosion potential. Inspection of the non-nitrided samples after polarization showed pitting in the surfaces of the AISI 316L and also in AISI 446 stainless steel and Fe-25wt%Cr, Fe-23wt%Cr-2tMo alloys, while in the solution-nitrided samples, the surface after the polarization looked only etched by the electrolyte.

When comparing results for the porous and nonporous alloys, in the former, a clear susceptibility to the attack of chloride ions was found. Figure 8.21 shows the polarization curves of the porous and nonporous solution nitride alloys. A clear defined passive region was not observed in the porous alloys; in contrast, the nonporous alloys exhibited a very wide passive region. The passive current was 2 orders of magnitude higher for the porous alloys than for nonporous alloys. Regarding the breakdown potential, the nonporous alloys exhibited a slightly more noble breakdown potential. The porous configuration could have introduced many small crevices which are conductive to localized corrosion, especially in the internal surface of the pores that cannot be mechanically polished. In these sites a marked increase in the acidity within a confined space could be able to accelerate localized corrosion.

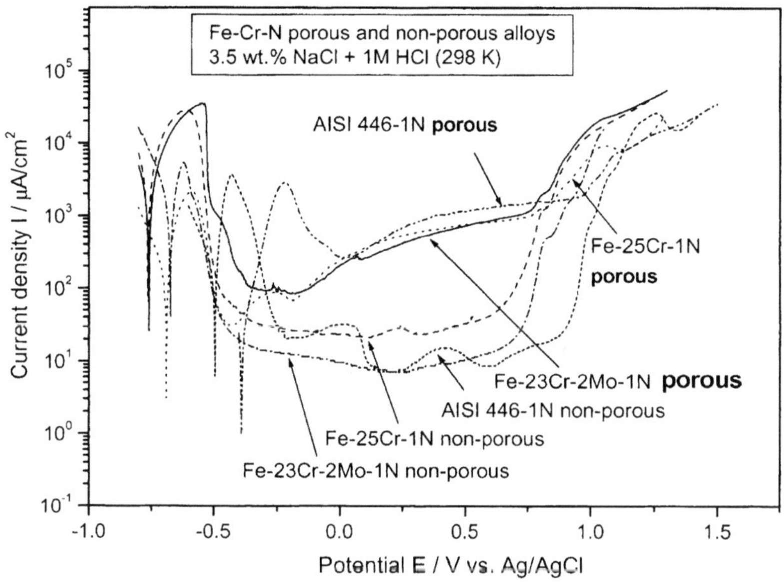


**Fig. 8.21** Potentiodynamic polarization curves of lotus Fe-25wt%Cr-1wt%N, Fe-23wt%Cr-2wt%Mo-1wt%N, and AISI446-1wt%N steels and nonporous steels in 3.5wt%NaCl + 1 M HCl solution (Reprinted with permission from [39] © 2007 Elsevier Ltd)

Figure 8.22 shows the electropolishing effect in the porous specimens. This study revealed that solution nitriding + electropolishing produces a polarization curve with a very wide passive region. The current density in this passive region is within 14–300 μA cm$^{-2}$, almost as low as the nonporous samples' current density.

Therefore, a way of increasing the corrosion resistance of lotus alloys is to electropolish the porous surface in order to remove deformed layer, improve surface roughness, and also form a durable passive film on the surface. This more uniform oxide layer resulting from electropolishing will increase corrosion resistance in stainless steels. It is worth noting that the electropolishing treatment applied in the porous alloys did not alter the breakdown potential significantly. However, the passive current densities in the active–passive transition were greatly reduced and the passivating ability, given by the wide passivation range of the polarization curves, also increased by this final surface finish. These results indicate that a higher corrosion resistance is obtained after the electropolishing surface treatment.

The results of pitting corrosion susceptibility can also be interpreted using the images of the surface microstructure after corrosion obtained by scanning electron microscopy. Selected SEM micrographs of lotus nitrided alloys after the polarization test in 3.5wt%NaCl + 1 M HCl are shown in Fig. 8.23a–d. SEM observation revealed that lotus AISI 316L stainless steel experienced severe localized

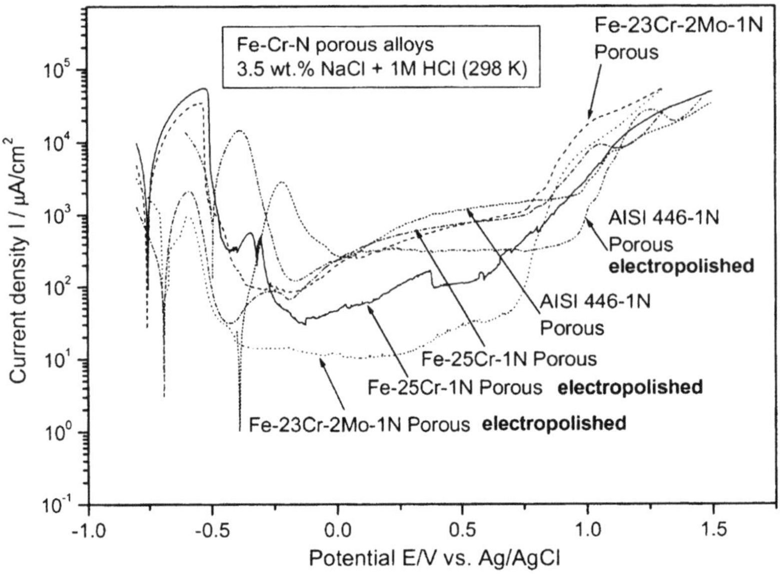


**Fig. 8.22** Potentiodynamic polarization curves of lotus Fe-25wt%Cr-1wt%N, Fe-23wt%Cr-2wt%Mo-1wt%N, and AISI446-1wt%N steels with and without electropolishing using an electrolyte composed in phosphoric acid 56 %, sulphuric acid 27 %, glycerine 7 %, and DI water at 333 K (Reprinted with permission from [39] © 2007 Elsevier Ltd)

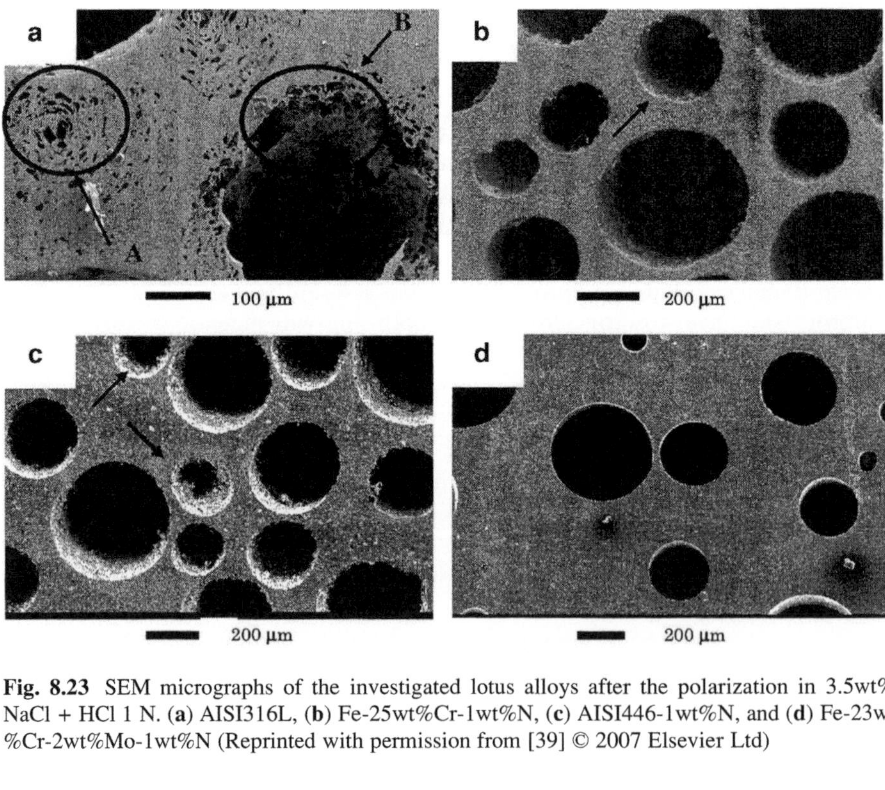


**Fig. 8.23** SEM micrographs of the investigated lotus alloys after the polarization in 3.5wt% NaCl + HCl 1 N. (**a**) AISI316L, (**b**) Fe-25wt%Cr-1wt%N, (**c**) AISI446-1wt%N, and (**d**) Fe-23wt%Cr-2wt%Mo-1wt%N (Reprinted with permission from [39] © 2007 Elsevier Ltd)

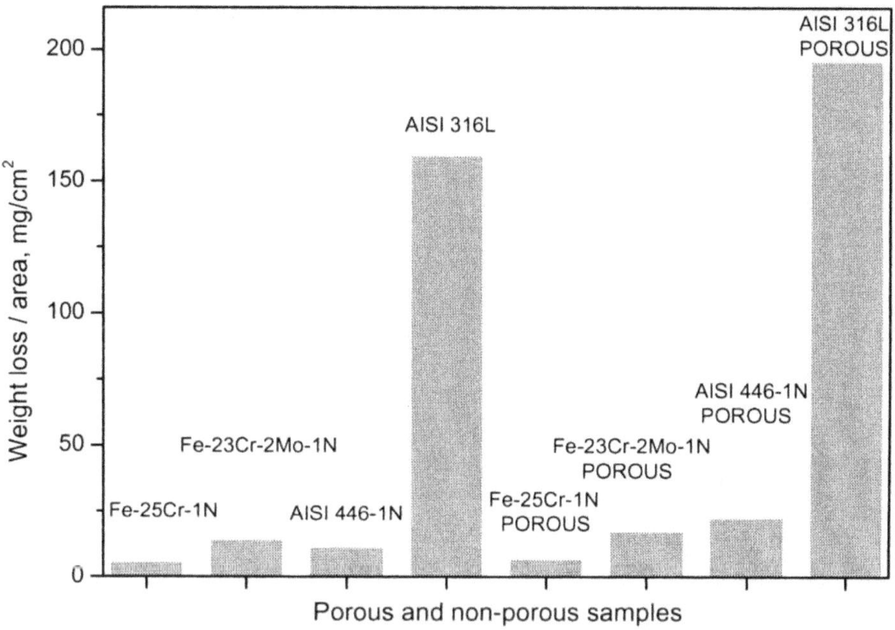


**Fig. 8.24** Weight loss in 10wt%$FeCl_3\cdot 6H_2O$ after 48 h immersion for the porous and nonporous high-nitrogen Ni-free stainless steels (Reprinted with permission from [39] © 2007 Elsevier Ltd)

corrosion. The metal base (zone A in Fig. 8.23a) as well as the internal surroundings of the pores (zone B in Fig. 8.23a) were attacked. The pits found in the metal base of AISI 316L consist of an open center, perforated middle region, and an outer ring. The center and the middle regions are typical perforated or "lacy" pit morphology in which subsurface undercutting expands the pit from the central initiation point. Contrastingly, no obvious corrosion pits were found in any of the high-nitrogen Ni-free alloys (Fig. 8.23b–d). In the case of Fe-25wt%Cr-1wt%N and AISI 446-1wt %N (Fig. 8.23b, c, respectively), it was observed that the internal wall of the pores, marked with arrows in Fig. 8.23b, c, seems to be degraded and not smooth like in the case of Fe-23wt%Cr-2wt%Mo -1wt%N Fig. 8.23d). This difference in the microstructure of the inner part of the pores in the high-nitrogen Ni-free stainless steels is probably because Fe-23wt%Cr-2wt%Mo-1wt%N alloy exhibited the lower corrosion rate under the experimental conditions. These results confirm the higher corrosion susceptibility of AISI 316L compared to high-nitrogen Ni-free stainless steels. In every high-nitrogen Ni-free alloy investigated, severe open perforated pit structure was not observed after the polarization test.

The results after immersion test in 10wt%$FeCl_3\cdot 6H_2O$ solution for 172.8 ks are shown in Fig. 8.24. It was observed that lotus AISI 316L exhibited the highest weight losses. Metallographic observations revealed that AISI 316L stainless steel coupons displayed much more pits than the rest of the solution nitride samples. Moreover, the pits developed on AISI 316L were wider and deeper with average size of 100 μm for the nonporous specimen and 150 μm for the lotus specimen. On the contrary the high-nitrogen Ni-free stainless steel specimens after 172.8 ks immersion showed a very smooth and bright surface with a small number of shallow pits with a diameter not exceeding 20 μm. The corrosion rate determined by weight loss method of the high-nitrogen Ni-free stainless steel specimens in this highly oxidizing chloride environment was less than 0.04 mm $y^{-1}$. In contrast, AISI 316L stainless steel-corroded samples showed a very high average corrosion rate. Corrosion rate of duplicate runs determined by weight loss method showed average values of 0.86 and 1.52 mm $y^{-1}$ for nonporous and lotus AISI 316L stainless steel, respectively. Therefore, it can be concluded that lotus high-nitrogen Ni-free stainless steels possess the general corrosion resistance much better than conventional austenitic stainless steels in acidic chloride environment.

The results of this investigation showed that the general corrosion resistance as well as the localized corrosion resistance of lotus high-nitrogen Ni-free stainless steels is higher than AISI 316L. From the microscopic observation and the electrochemical characterization, the alloy with best performance was lotus Fe-23wt%Cr-2wt%Mo-1wt%N since it resulted to be immune to pitting attack under these testing conditions. From the weight loss measurements, lotus Fe-25wt%Cr-1wt%N alloy tolerated very good the immersion in the severe aggressive media much better than any other alloy.

## References

1. Xie ZK, Ikeda T, Okuda Y, Nakajima H (2004) Mater Sci Eng A 386:390–395
2. Xie ZK, Ikeda T, Okuda Y, Nakajima H (2004) Jpn J Appl Phys 43:7315–7319
3. Xie ZK, Ikeda T, Okuda Y, Nakajima H (2004) Mater Sci Forum 449–452:661–664
4. Kimura S (1977) Architectural sound and anti-noise plan. Shokokusha Publishing, Tokyo, p 142 (in Japanese)
5. Xie ZK, Ikeda T, Okuda Y, Nakajima H (2003) J Jpn Inst Metals 67:708–713
6. Okuda Y (1990) Proc J Acoustic Soc Jpn 581
7. Akiyama S, Itoh M, Ishii E (1991) National Industrial Research Institute. Report of Kyushu, p 2928
8. Nisimaki S (1971) Electro-acoustics and vibration. Corona Publishing, Tokyo, Japan, p 79
9. Beranek LL (1950) Acoustic measurement. Wiley, New York
10. Igarasi J (1970) Sound and vibration. Kyoritsu Shuppan Co., Tokyo, p 37
11. Behrens E (1968) J Compos Mater 2:2–17
12. Perrins WT, McKenzie DR, McPhedran RC (1979) Proc Roy Soc London Ser 369:207
13. Han LS, Cosner AA (1981) J Heat Transfer 103:387–392
14. Sangani AS, Yao C (1988) Phys Fluids 31:2435–2444
15. Mityushev V (1999) Proc Roy Soc London A 455:2513–2528
16. Moctezuma-Berthier A, Vizika O, Adler PM (2002) Trans Porous Media 49:331–332
17. Ogushi T, Chiba H, Nakajima H, Ikeda T (2004) J Appl Phys 95:5843–5847
18. Hyun SK, Murakami K, Nakajima H (2001) Mater Sci Eng A 299:241–248
19. Suematsu Y, Hyun SK, Nakajima H (2004) J Jpn Inst Metals 68:257–261
20. Tane M, Hyun SK, Nakajima H (2005) J Appl Phys 97:103701-1–103701-4
21. Tane M, Ichitsubo T (2004) Appl Phys Lett 85:197–199
22. Archie GE (1942) Trans AIME 146:54–63

23. Eshelby JD (1957) Proc Roy Soc London A 241:376–396
24. Mori T, Tanaka K (1973) Acta Metall 21:571–574
25. Hatta H, Taya M (1986) J Eng Mater 24:1159–1172
26. Ohnishi Y, Hyun SK, Nakajima H (2006) In: Nakajima H, Kanetake N (eds) Porous metals and metal foaming technology. Jpn Inst Metals, Sendai, pp 423–426
27. Noudem JG, Reddy ES, Schmitz GJ (2003) Physica C 390:286–290
28. Castano FJ, Nielsch K, Ross CA, Robinson JWA, Krishnan R (2004) Appl Phys Lett 85:2873–2875
29. Onishi H, Hyun SK, Nakajima H, Mitani S, Takanashi K, Yakushiji K (2008) J Appl Phys 103:093539
30. Igarashi H, Okazaki K (1976) J Am Ceram Soc 59:371–372
31. Tane M, Hyun SK, Nakajima H (2006) Scr Mater 54:545–552
32. Kerner EH (1956) Proc Phys Soc London B 69:808–813
33. Schapery RA (1968) J Comp Mater 2:380–404
34. Dillon CP (1995) Corrosion resistance of stainless steels. Marcel Dekker, New York, p 258
35. Seah KHW (1993) Corros Sci 34:1841–1851
36. Fossati A, Borgioli F, Galvaneto E, Bacci T (2006) Corros Sci 48:1513–1527
37. Di Schino A, Barteri M, Kenny JM (2003) J Mater Sci 38:4725–4733
38. Rechsteiner A, Speidel M (1997) Proceedings of 1st European stainless steel conference, vol 2, Florence, p107
39. Alvarez K, Hyun SK, Tsuchiya H, Fujimoto S, Nakajima H (2008) Corros Sci 50:183–193
40. Ernest P, Laycock NJ, Moayed MH, Newman RC (1997) Corros Sci 39:1133–1136

# Chapter 9
# Processing of Lotus Metals

**Abstract** In order to use lotus metals for various industrial products, processing is indispensable. Here, two important topics are chosen: joining technique and severe deformation without shape change of metals.

**Keywords** Equal-channel angular extrusion process • Laser beam • Vickers hardness • Welding

## 9.1 Weldability

### *9.1.1 Weld Fusion Property of Lotus Copper*

For industrial use of the porous metals as various parts, reliable joining techniques such as welding is required as well as processing techniques. There have been some reports on laser and arc welding of foamed aluminum manufactured by using foaming agent, $TiH_2$, or $CaCO_3$ [1–3]. However, weldability of porous metals with controlled pore direction such as lotus metals has not been investigated, although the welding is necessary for fabricating engineering materials. Murakami et al. [4] investigated the weldability of lotus copper, and in their study the laser beam with narrow weld bead width and heat-affected zone was applied to the welding of lotus metal. Figure 9.1 shows schematic views of specimens using laser welding. The laser beam was delivered using a 600-μm-diameter optical fiber. The focal length for Nd:yttrium–aluminum–garnet (Nd:YAG) laser with a 1,064-nm wavelength was 100 mm. The diameter of the laser beam at the focus point was 300 μm. The laser beam was irradiated at an 80° forward angle relative to a workpiece surface damage to the optics by a reflection of laser beam. Argon was used as a shielding gas with a flow rate of $5.0 \times 10^4\ m^3\ s^{-1}$. Figure 9.2 shows a relation between the welding speed and the penetration depth of the weld bead in depth at the thickness of 5 mm from the work piece surface when the laser beam power of 3.2 kW was irradiated on both types of samples (a) and (b) shown in Fig. 9.1. The penetration depth in both directions


DOI 10.1007/978-4-431-54017-5_9, 

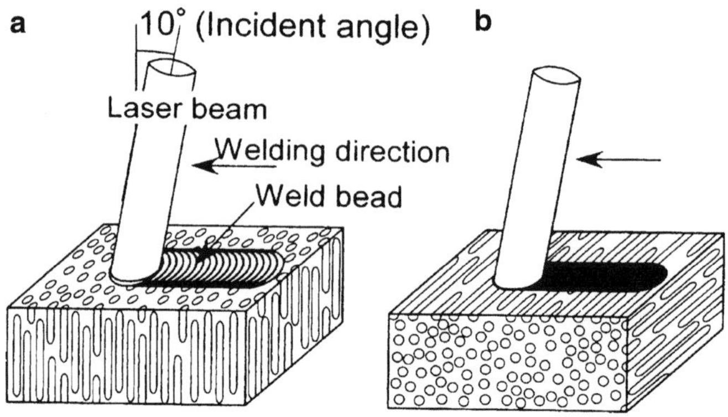


**Fig. 9.1** Schematic views of specimens during laser welding in the lotus copper with the pore growth (**a**) perpendicular and (**b**) parallel to the work piece surface (Reprinted with permission from [4] © 2003 Elsevier Science B.V.)

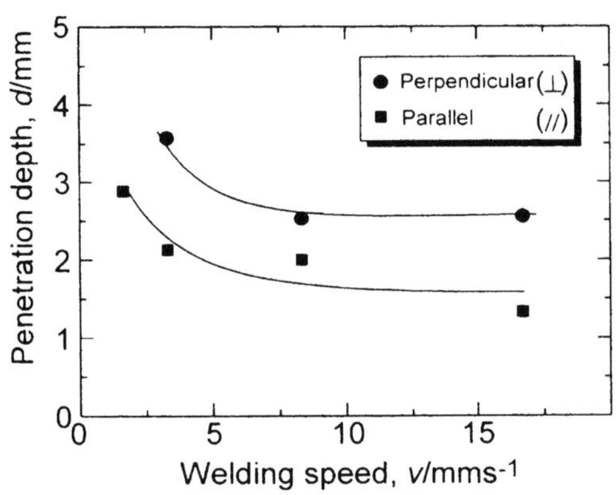


**Fig. 9.2** Relation between welding speed and penetration depth of the weld bead in depth at the thickness of 5 mm from the work piece surface when the laser beam power of 3.2 kW was irradiated on both types of sample (a) and (b) shown in Fig. 9.1 (Reprinted with permission from [4] © 2003 Elsevier Science B.V.)

decreases with increasing welding speed. The penetration depth in the parallel direction (filled square) is always shallower than that in the perpendicular direction (filled circle). Figure 9.3 shows the top view of the specimens after laser welding in the thickness of 5 mm at the constant laser beam power of 3.2 kW in the welding speed range of 1.67–16.7 mm $s^{-1}$ in (a)–(c) perpendicular and (d)–(g) parallel pore directions. In the perpendicular direction, only the groove was observed in the condition of 16.7 mm $s^{-1}$ as shown in Fig. 9.3a. The weld beads were found below the welding speed of 8.33 mm $s^{-1}$ as shown in Fig. 9.3b, c, while a smooth weld bead was observed at the welding speed of 3.33 mm $s^{-1}$. In contrast, in the parallel direction, no weld bead was made in the welding speed range from 3.33 to 16.7 mm $s^{-1}$, and a smooth weld bead on all specimens was not observed. At the

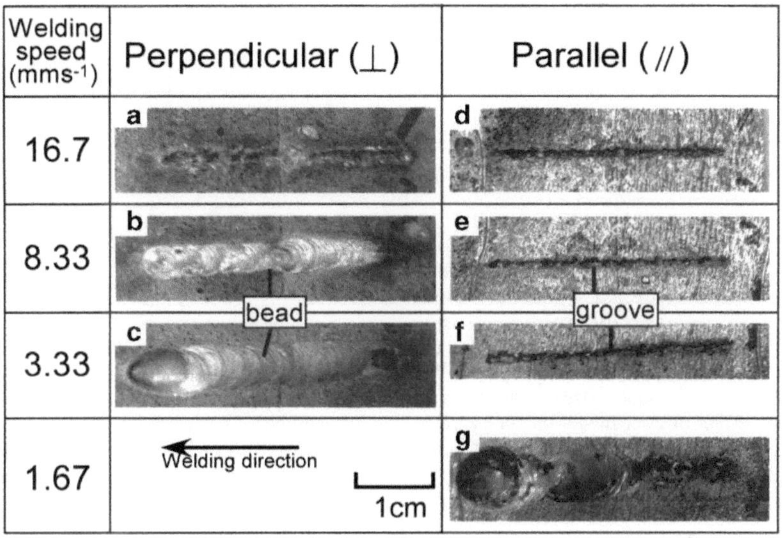


**Fig. 9.3** Top views of laser welded specimen at different welding speeds with 3.2-kW constant laser power in 5-mm thickness in (**a–c**) perpendicular and (**d–g**) parallel to the pore growth directions. Welding speed: (**a**) and (**b**) 16.7 mm $s^{-1}$, (**b**) and (**e**) 8.33 mm $s^{-1}$, (**c**) and (**f**) 3.33 mm $s^{-1}$, and (**g**) 1.67 mm $s^{-1}$ (Reprinted with permission from [4] © 2003 Elsevier Science B.V.)

welding speed of 1.67 mm $s^{-1}$, the weld bead was formed only at the end of weld, though nonuniform. These indicate that the formation of the weld bead has strong anisotropy of pore direction in relation to the irradiation direction of laser beam.

The effect of the laser power on the penetration depth in the thickness of 4 mm for two pore directions at the constant welding speed of 3.33 mm $s^{-1}$ was shown in Fig. 9.4. The penetration depth of the weld bead perpendicular to the pore growth direction shows a sharp rise at the laser beam power of 2.4 kW. On the other hand, the penetration depth in the parallel to the pore growth direction increases gradually in comparison with that in the perpendicular. These facts indicate that the keyhole formed at the laser-irradiated point perpendicular to the pore growth direction may be attributed to the metal vaporization [5] more than the laser beam power of 2.4 kW, but not in the case parallel to the pore direction. Figure 9.5 shows the cross sections of laser welding direction in the thickness of 4 mm in the welding speed of 3.33 mm $s^{-1}$ at the laser beam power of 3.2, 2.5, and 2.0 kW (a)–(c) perpendicular and (d)–(f) parallel to the pore growth directions. As shown in Fig. 9.5a, the weld part at the laser beam power of 3.2 kW in the perpendicular case was completely melted through the bottom; shortly, the fully penetrated weld bead was formed. The shape of the fusion boundary is nearly straight. The straight line of the fusion boundary is nearly parallel to the direction of the original pores. Some large and many small pores in the weld metal were observed in the weld bead. The weld metal at the laser beam power of 2.5 kW in the perpendicular shown in Fig. 9.5b did not

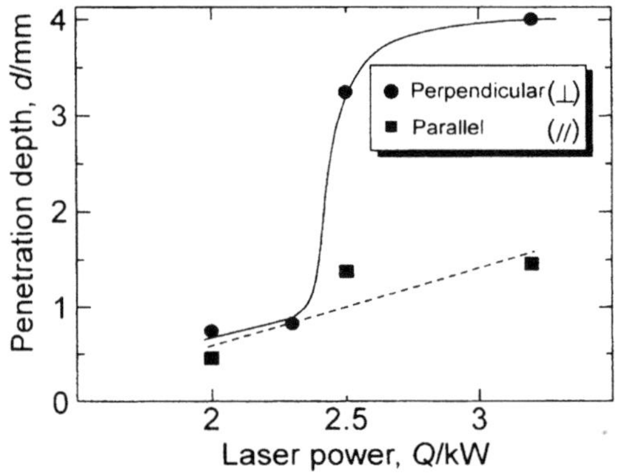


**Fig. 9.4** Effect of laser power on penetration depth of weld bead for perpendicular and parallel to the pore growth direction in depth of 4 mm thick from the work piece surface at 3.33 mm $s^{-1}$ welding speed (Reprinted with permission from [4] © 2003 Elsevier Science B.V.)

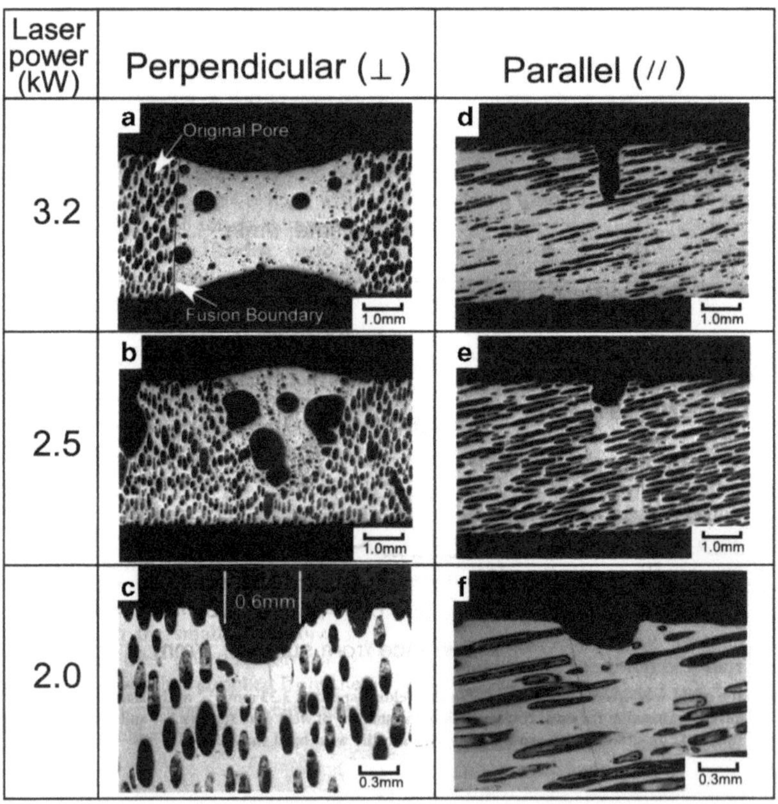


**Fig. 9.5** Cross sections of laser weld perpendicular to the welding direction in 4-mm thickness in the welding speed of 3.33 mm $s^{-1}$ at the laser beam power of 3.2, 2.5, and 2.0 kW (**a–c**) perpendicular and (**d–f**) parallel to the pore growth directions, which are slightly inclined to the work piece surface with about 10° (Reprinted with permission from [4] © 2003 Elsevier Science B.V.)

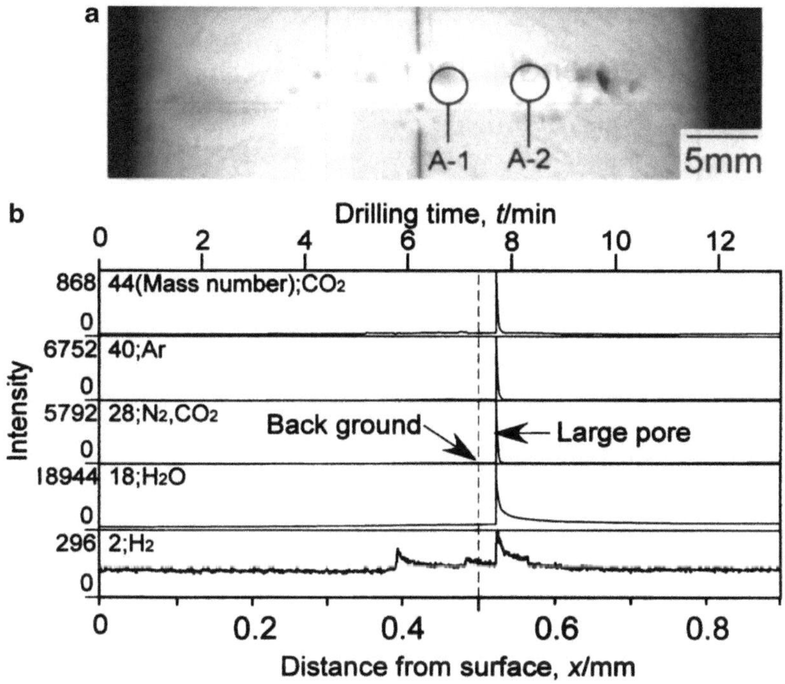


**Fig. 9.6** (**a**) X-ray radiograph of the workpiece after welding with the welding speed of 3.33 mm $s^{-1}$ at the laser beam power of 3.2-kW depth of 4 mm thick. The drilling points determined were in the *open circle* including the shadow of weld pore as A-1 and A-2. (**b**) Dependence of mass spectral intensity of gas detected from A-2 on drilling time and the distance from the weld bead surface (Reprinted with permission from [4] ©2003 Elsevier Science B.V.)

melt through the opposite side. Some larger pores evolved by the weld were observed in the weld metal. At the laser power of 2.0 kW shown in Fig. 9.5c, the weld metal was not observed, and the diameters of the grooves formed instead of weld bead were about 0.6 mm. In the parallel direction case, the shape of groove was similar to Fig. 9.5c, though the groove depth increased with increasing laser power. Obviously, the shape of the weld bead and the groove exhibits anisotropy with respect to the pore growth direction. Thus, the weldability of lotus copper in the perpendicular direction is more significant than that in the parallel direction.

X-ray radiograph of the work piece after welding with the welding speed of 3.33 mm $s^{-1}$ at the laser beam power of 3.2 kW taken in depth of 4 mm thick is shown in Fig. 9.6a. The weld bead after welding was drilled in 3-mm diameter in the vacuum chamber, and the residual gas in the pores was analyzed using a mass spectrometer. The drilling points were determined in the open circle including the shadow of the weld pores such as A-1 and A-2. Figure 9.6b shows the drilling time and the distance from the weld bead surface dependence of mass spectral intensity of residual gas detected from the pores in weld metal (Fig. 9.5a) with the welding speed of 3.33 mm $s^{-1}$ at the laser beam power of 3.2 kW in depth of 4 mm thick.

When the drilling depth increases to more than 0.5 mm, various kinds of gases were clearly detected at the same time with high intensity, which were evolved from the large weld pore at this point. These gases are identified as $H_2$, $H_2O$, $N_2$, Ar, and $CO_2$ by the mass spectrometer. No obvious change in the intensity was detected during drilling for other gases up to 100 mass numbers. The gas concentration was determined using a net intensity, which eliminated the difference between the peak intensity and the background intensity in depth of 0.5 mm from the surface. Some small hydrogen peaks with comparably high intensity were detected in depth of 0.4 and 0.48 mm. These may correspond to small weld pores which were formed during welding. The major component of A-1 was argon comprised beyond 73 vol.%, and other components were 14 vol.% nitrogen and water. The major component of A-2 was water which comprised about 60–70 vol.%, and other components were about 20 vol.% argon, several to 10 vol.% nitrogen, and several volume percent hydrogen.

It is generally considered that the pores in the laser weld bead of dense materials are formed by the shielding gas trapped in the weld pool due to keyhole instability by vaporized molten metal [6]. Therefore, it is likely that argon of the shielding gas was detected as a main component. The detected nitrogen may result in the gas from the air which is trapped together with the shielding gas. In their study, however, the presence of a large amount of water and small amount of hydrogen is not explainable in terms of only keyhole instability. The reason why those components are present in the pores is that high-pressure hydrogen gas remains in the closed original pores in the lotus copper formed during the unidirectional solidification under high pressure. It is considered that at remelting of workpiece by laser welding, the remaining hydrogen gas in the original pore was released into the molten weld metal and then coalesced into the large pore which is generated by keyhole instability. Subsequently the hydrogen gas reacted with oxygen containing in air trapped with the shielding gas, and then water vapor was evolved. Besides, there is a small amount of hydrogen gas that does not react with oxygen in some pores. Nitrogen gas in air was also trapped but remained in the pores because of no reaction with molten copper. Similar reason was supposed in the case of $CO_2$ gas formation. Therefore, the large pores in the weld metal are comprised of the shielding gas, the vaporized copper, the air trapped with the shielding gas, and the hydrogen gas. On the other hand, the small gas pores are composed of the supersaturated hydrogen supplied from the original closed pores in lotus copper.

As mentioned above, significant anisotropy in the morphology of laser weld metal was observed, depending on the pore growth direction. The laser beam mostly reflects on the copper surface due to the high reflectivity of copper [7, 8]. A part of the laser beam penetrates deeply into the pores in the lotus copper with cylindrical pores perpendicular to the welding direction, since the direction of pore growth is nearly identical to the incident laser beam direction; the multiple reflections of the laser beam at the pore inner wall increases the amount of heat input absorbed to the specimen. In addition, two-dimensional thermal conductivity on the parallel plane to the specimen surface is low, which enhances the melting of the specimen. The amount of heat input for the lotus copper with cylindrical pores parallel to the welding direction is smaller than that of the perpendicular, because of no multiple reflection and high-thermal conductivity. Moreover, the molten copper in the weld pool is blown out more significantly by high-pressure hydrogen gas released from the original closed pores.

In the laser beam welding of lotus copper plate, the pore growth direction should be perpendicular to the work piece surface. When the lotus copper plate with the original pores parallel to the work piece is used, the surface treatment should be required in order to increase the laser beam absorption.

### *9.1.2 Weld Fusion Property and Joint of Lotus Magnesium*

Figure 9.7 shows the bead appearance and macrostructures in the cross section of the welded specimen with pores perpendicular to the specimen surface at nominal laser power of 0.8, 1.0, and 1.2 kW and spot diameters of 0.3, 0.45, and 0.6 mm [9]. The fusion area of the specimen at the laser power of 0.8 kW and the spot diameter of 0.6 mm did not penetrate to the bottom although complete penetration of the fusion area was achieved for the spot diameter of 0.45 mm as well as for increased laser powers. This is because energy density in the former case is lower than that in the latter case. In the weld metal, some blowholes, which form due to the remnant hydrogen gas in the original closed pores, were observed at the laser power of 1.2 kW and the spot diameter of 0.45 mm, as shown in Fig. 9.7.

Macrostructures of the welded joint with pores parallel to the specimen surface at different nominal laser powers are shown in Fig. 9.8. The weld at the laser powers of 1.5 and 2.0 kW completely melted the bottom, while the weld at the laser power of 1.0 kW did not melt the bottom completely. The blowholes, which are indicated by black arrows in Fig. 9.8, were observed for all the conditions. There were more blowholes in the weld metal of the specimen with pores parallel to the specimen surface (parallel case) than in the weld metal of the specimen with pores perpendicular to the specimen surface (perpendicular case). This was because the blowholes were caused by the remnant hydrogen gas in the original closed pores, and the number of original closed pores for the specimen in the parallel case was much more numerous than that for the specimen in the perpendicular case.

By comparing Fig. 9.7, which represents the laser power of 1.0 kW and the spot diameter of 0.6 mm, with Fig. 9.8a, it was observed that the profiles of the weld metal in the perpendicular and parallel cases were different, although the welding condition was identical. The depth of penetration in the former case is greater than that in the latter case. In the previous report [4], the melting property of lotus copper differed greatly according to the pregrowth direction. The weld bead in the parallel case was not formed in most welding conditions, and only a groove was observed. The difference in the melting property was attributed to the high reflectivity of copper for a laser beam. On the other hand, the reflectivity of magnesium is much lower than that of copper. Therefore, it is expected that the melting property of lotus magnesium is mainly controlled by the thermal conductivity.

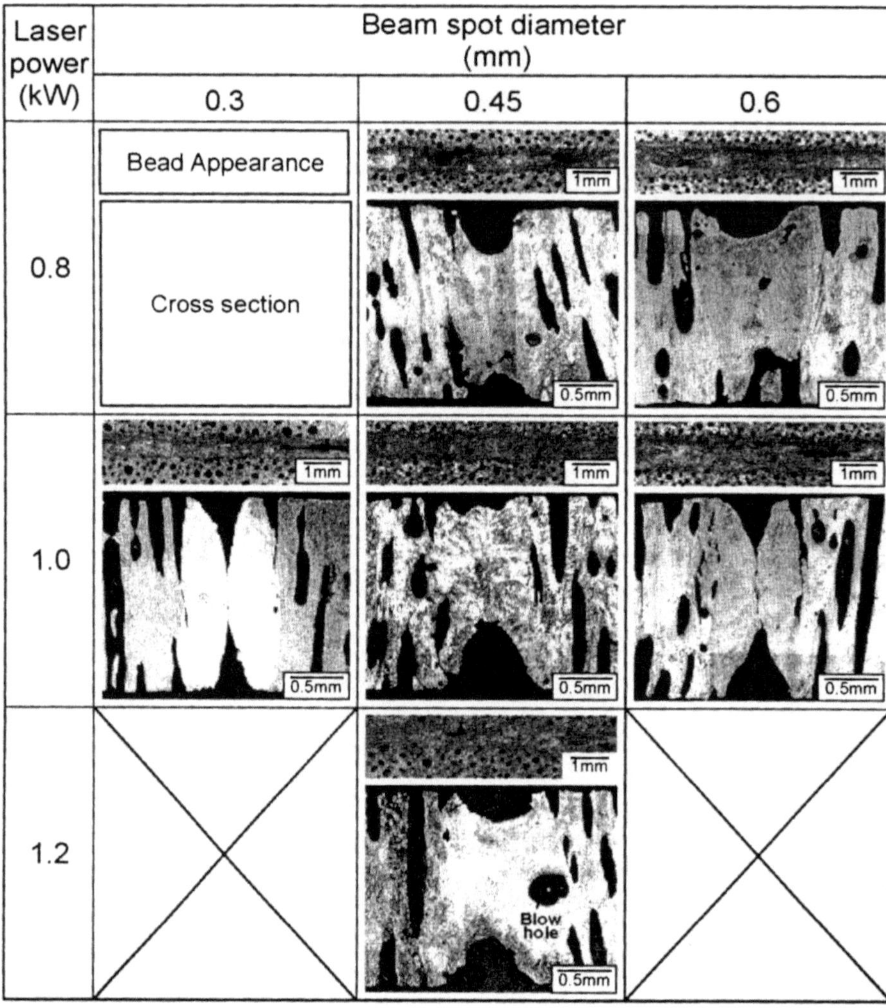


**Fig. 9.7** Bead appearance and macrostructure of laser welded specimen with the pores perpendicular to the specimen surface at the laser powers of 0.8, 1.0, and 1.2 kW and the spot diameters of 0.3, 0.45, and 0.6 mm (Reprinted with permission from [9] ©2006 Elsevier B.V.)

The effect of pore growth direction on the joint strength of lotus magnesium was investigated. Figure 9.9 shows the geometry of tensile specimens of the base metal and the welded joint. A tensile test specimen with a gage section of 1.8 mm $\times$ 5 mm was cut out of the specimen before the welding for the base metal and after the welding for the welded joint using the spark erosion wire-cutting machine. The tensile test was performed on the specimen in a tensile test machine at a rate of $1.67 \times 10^{-2}$ mm s$^{-1}$ at room temperature. The tensile strengths of the weld beads and the base metals in the parallel and perpendicular cases are shown in Fig. 9.10

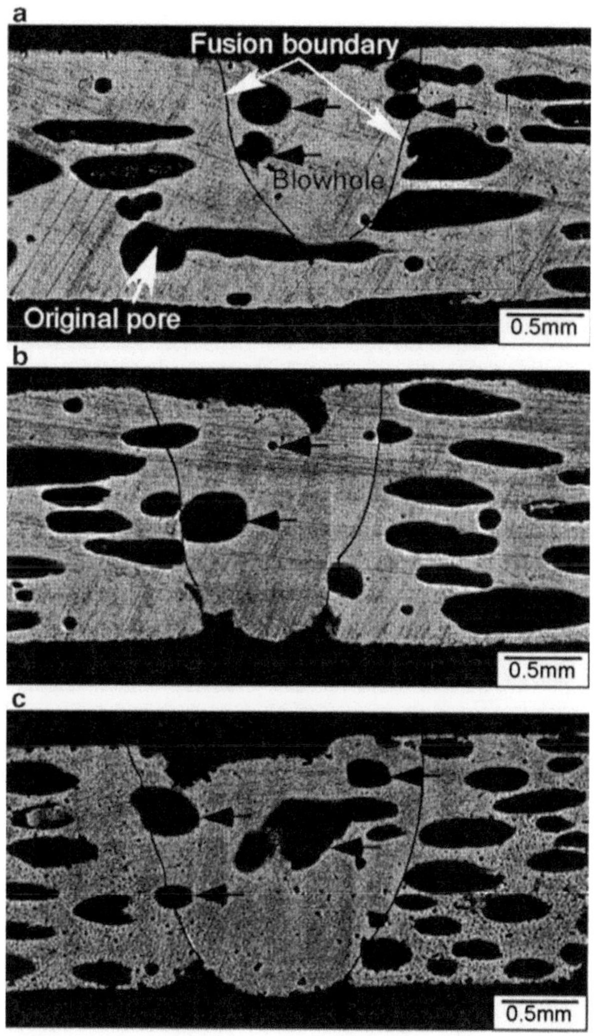


**Fig. 9.8** Macrostructure of welded specimen with the pores parallel to the specimen surface at the spot diameter of 0.6 mm and the laser powers of (**a**) 1.0 kW, (**b**) 1.5 kW, and (**c**) 2.0 kW (Reprinted with permission from [9] © 2006 Elsevier B.V.)

together with the cross-sectional view of the fractured joints. The open and closed circles in Fig. 9.10 denote the samples fractured in the base metal or in the weld bead, respectively. The average tensile strengths of the base metal in the parallel and perpendicular cases were approximately 55 and 30 MPa, respectively. The average tensile strength of the weld bead with the pores perpendicular to the specimen surface was 29 MPa, which was similar to that of the base metal because the joint was fractured in the base metal as shown in Fig. 9.10b. On the other hand, the weld bead in the parallel case was fractured at the joint interface between the weld metal and the base metal as shown in Fig. 9.10a. The average joint strength in the parallel case was 41 MP, which is lower than that of the base metal for the same case.

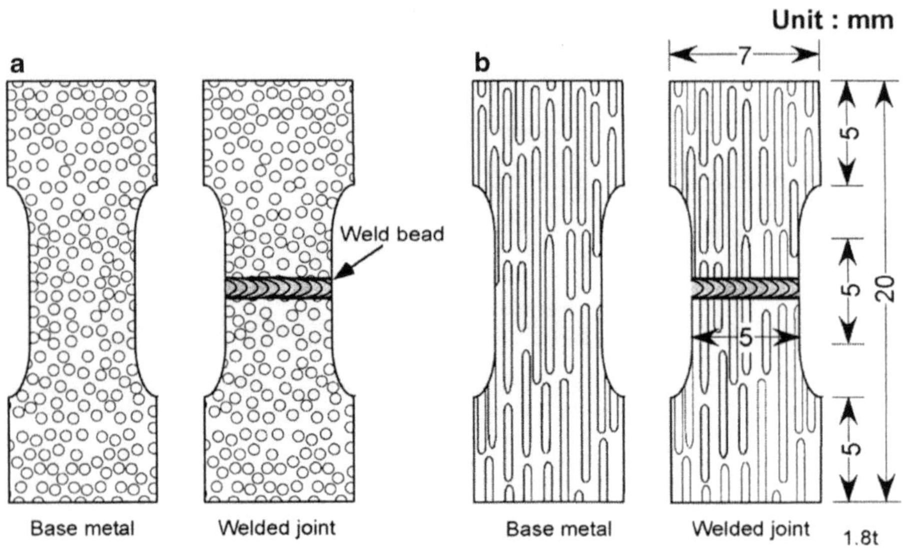


**Fig. 9.9** Geometry in the tensile test specimens of base metal and welded joint with the pores (**a**) perpendicular and (**b**) parallel to the specimen surface (Reprinted with permission from [9] © 2006 Elsevier B.V.)

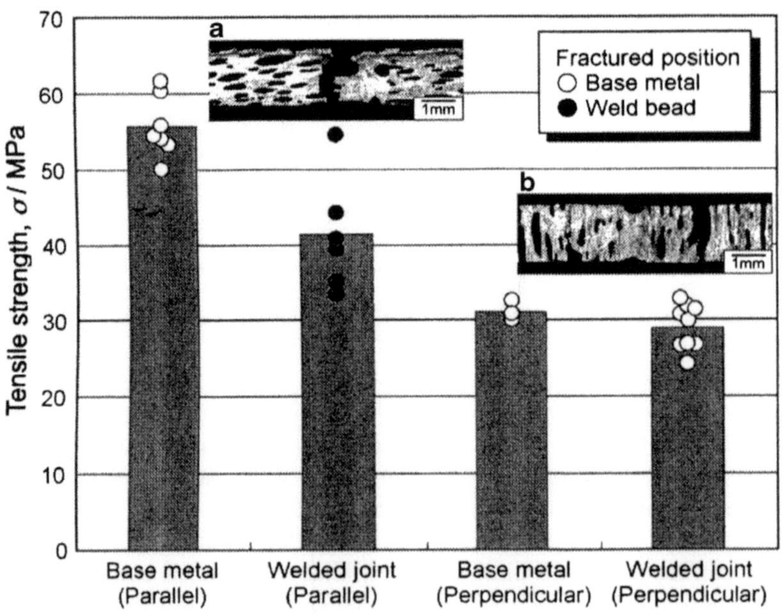


**Fig. 9.10** Tensile strength of the base metal and welded joint with the pores perpendicular and parallel to the specimen surface together with cross section of fractured joint. (**a**) Fractured in the weld bead and (**b**) fractured in the base metal (Reprinted with permission from [9] © 2006 Elsevier B.V.)

Ikeda et al. [10, 11] reported the tensile strength of nonporous magnesium produced by unidirectional solidification. The tensile strength parallel to the solidification direction was larger than that perpendicular to the solidification direction. The highly preferred orientations of the nonporous and lotus magnesium fabricated by unidirectional solidification were $\langle 0002 \rangle$ and $\langle 11\bar{2}0 \rangle$ [10, 11]. The slip system of magnesium at room temperature is only (0002) $\langle 11\bar{2}0 \rangle$ [12]. When a load perpendicular to the pore growth direction is applied to porous magnesium with a $\langle 11\bar{2}0 \rangle$ orientation, the (0002) $\langle 11\bar{2}0 \rangle$ slip occurs because the normal direction of the (0002) basal plane is perpendicular to the solidification direction. On the other hand, when a load parallel to the growth direction is applied to porous magnesium with $\langle 0002 \rangle$ and $\langle 11\bar{2}0 \rangle$ orientations, the Schmid factor is close to 0. Therefore, the anisotropic tensile strength of the lotus magnesium is controlled not only by the pore growth direction but also by the crystallographic direction.

## 9.2 Equal-Channel Angular Extrusion Process

Suzuki et al. investigated the possibility to improve the mechanical properties of the metallic part of porous metals by plastic deformation [13]. As a kind of porous metals, lotus metals were selected, which are expected to be suitable for improvement of metallic parts by plastic deformation. Since the lotus metals have more metallic parts than metallic foams with up to 90 % of porosity, it is expected that the pores do not disappear and the effect of plastic deformation is obvious. Even if the pores are compressed, metallic part work hardens so that specific strength may increase. The densified materials are expected still to have micro- or nano-pores, which induce novel functions of materials. As a plastic deformation process the equal-channel angular extrusion (ECAE) [14] was adopted as shown in Fig. 9.11. In this process a sample rod is inserted in the entry channel and extruded to the exit channel through a die with a channel angle $\Phi$. At the corner between the two channels, the sample is sheared with severe plastic deformation. Because of several advantages, nowadays this process has been investigated actively [15]. First, as the shape and the diameter of the sample rod after the extrusion are the same as that before the extrusion, it is thus possible to extrude the same sample many times through the same die. Secondly, as various processing routes can be selected arbitrarily, it is possible to vary the deformation pattern. Thirdly, lower hydrostatic stress than forging or extrusion does not cause significant densification.

Until now there is no report about ECAE of porous metals. Severe plastic deformation of porous materials has several special characteristics compared with that of nonporous materials because of the densification process and thus requires newly established plastic deformation models. Here the fundamental plastic deformation behavior of lotus copper during ECAE process is described.

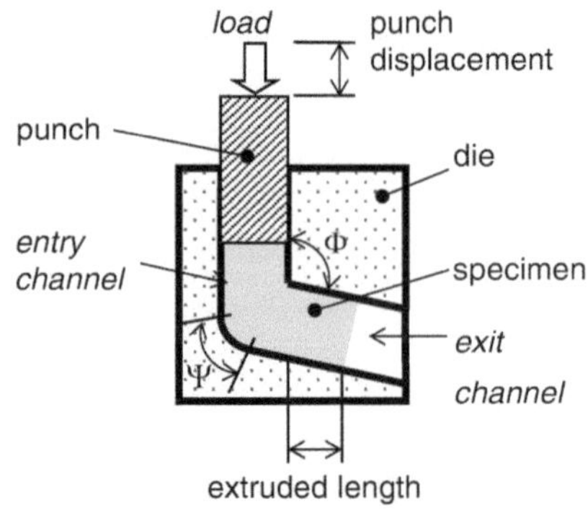


**Fig. 9.11** Schematic illustration of ECAE process (Reprinted with permission from [13] © 2008 Elsevier B.V.)

Figure 9.12 shows the cross section of lotus copper which was extruded halfway of the first pass through the die with a channel angle of 90°. In the entry channel part (Fig. 9.12b), some in polygonal shapes still remain, although a lot of pores are seen to be compressed. The pore morphologies are similar to those of specimens deformed uniaxially by compression tests [16]. In the exit channel close to the corner, almost no pores in the original size are observed but many streaks inclined toward the exit channel axis (Fig. 9.12c, d). Many pores remain in the leading end of the specimen, which is the initial unsteady part. Figure 9.13 shows the cross section of the completely extruded sample.

The volume change $\Delta V$ during the process can be calculated by

$$\Delta V(\%) = \frac{V_0 - V_1}{V_0} \times 100 = \frac{p_0 - p_1}{100 - p_1} \times 100, \tag{9.1}$$

where $V$ and $p$ are the volume and the porosity, respectively. The suffixes 0 and 1 mean the states before and after the deformation, respectively. The porosity of the sample cut from the part in the entry channel (Fig. 9.12b) was 17 %, and that in the exit channel, shown within the dotted line in Fig. 9.13 was 6 %. The volume change in the entry channel was calculated to be 40 % with $p_0 = 50$ % and $p_1 = 17$ %, and that in the exit channel was calculated to be 12 % with $p_0 = 17$ % and $p_1 = 6$ %. As shown in Fig. 9.13, the pores after the shear deformation are inclined from the central axis. This inclination can be explained by using a geometrical model shown in Fig. 9.14. One assumes a square region "klmn" on a cross section of a sample rod (Fig. 9.14a); the square turns and deforms into a parallelogram (Fig. 9.14b) shape by the shear stress at the corner. In the case of nonporous materials, the line "lm" inclines theoretically 26.5° to the exit channel axis [14].

In the case of lotus metals which are densified during the shear deformation, the area "lmn" should be decreased. As the volume reduction by shearing at the corner was 12 %, the area $S_0$ of the square "klmn" should be reduced. Therefore, the side "nm" reduces from 1 to 0.88 as shown in Fig. 9.14b, and the inclination θ of "lm" can be estimated as 28°. The observed inclination of pores agrees well with the estimation. From these results, the deformation behavior during the extrusion through ECAE process can be schematically illustrated in Fig. 9.15.

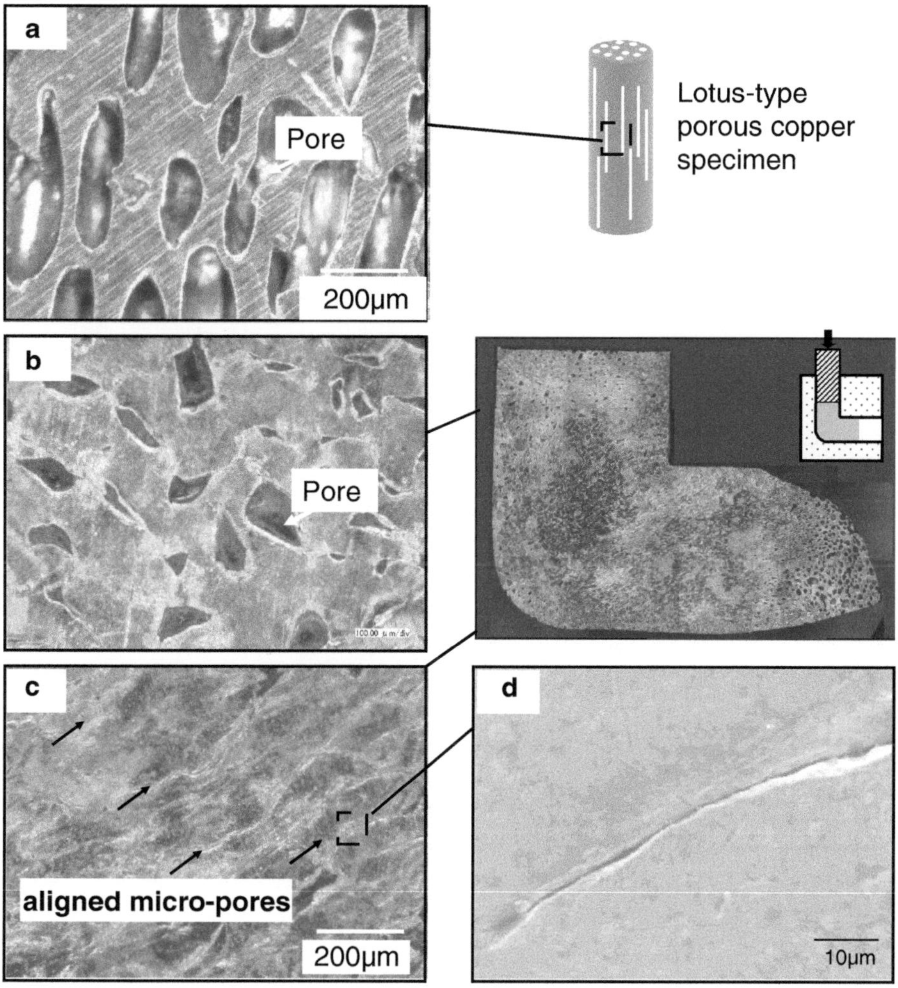


**Fig. 9.12** Cross sections of a lotus copper specimen before and after extrusion approximately 36 mm in a single pass through an ECAE die with a channel angle of 90°. (**a**) Before extrusion. (**b**) Compressed part in the entry channel. (**c**) Shear-deformed part in the exit channel close to the corner. The *arrows* indicate micropores. (**d**) Micropore magnified in the window in (**c**) (Reprinted with permission from [13] © 2008 Elsevier B.V.)

The extruded samples after the second pass are shown in Fig. 9.16a, b. On the cross section of the extruded sample via route A, these pores aligned in one direction are observed. Even after the second pass, the walls of the pores seem not to be bonded, since hydrogen captured in the pores prevented complete compression pores. On the other hand, the pores were opened again in the case of route C where the sample was rotated axially 180° from the first pass. This deformation behavior can also be explained by using the deformation model in Fig. 9.15. It may be possible to change the pore morphology by ECAE process.

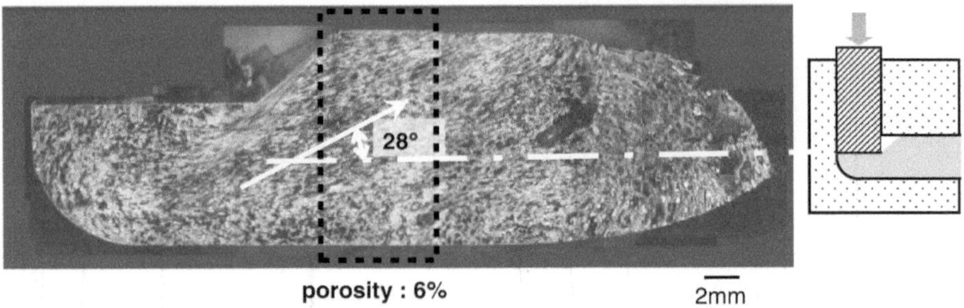


**Fig. 9.13** Cross section of a lotus copper specimen after completely extruded through an ECAE die with a channel angle of 90° in a single pass. The porous metal within the *square* drawn with a *dotted line* has a porosity of 6 % (Reprinted with permission from [13] © 2008 Elsevier B.V.)

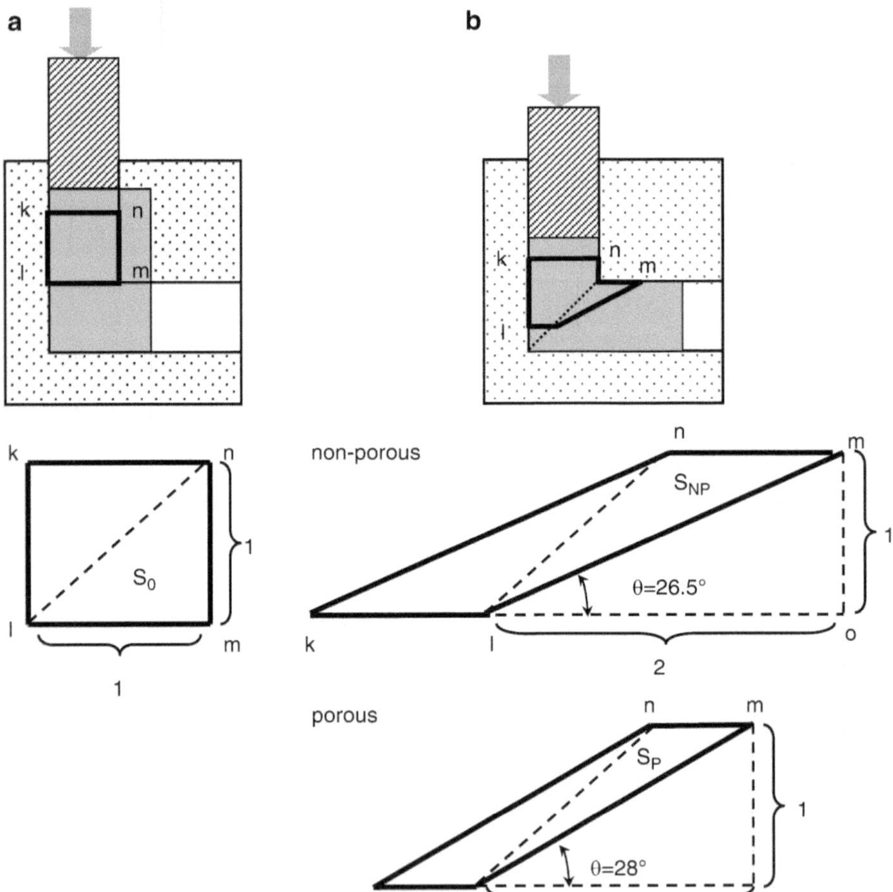


**Fig. 9.14** Geometrical model for the deformation behavior of sample rod through ECAE process with a channel angle of 90°. A square "klmn" on the cross section shown in (**a**) turns and deforms into a parallelogram shown in (**b**). The area $S_0$ of the square "klmn" in the entry channel is the same as the area SNP of "klmn" in the exit channel. $S_p$ is 12 % smaller than $S_0$ (Reprinted with permission from [13] © 2008 Elsevier B.V.)

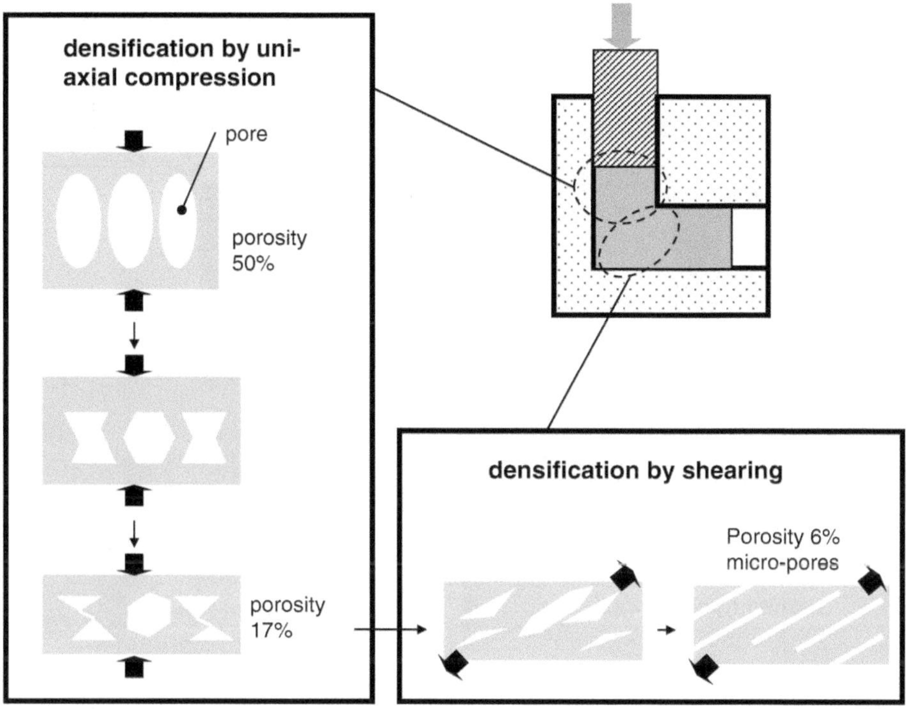


**Fig. 9.15** Schematic illustration of deformation of lotus metals through ECAE process with a channel angle of 90°. It demonstrates the case of porosity of 50 % before the extrusion (Reprinted with permission from [13] © 2008 Elsevier B.V.)

Figure 9.17 shows the diagrams of the punch load–displacement and extruded length. In the beginning of the loading, although the load increases with increasing punch displacement, the extruded length does not increase. In this stage only densification takes place in the entry channel by uniaxial compression. Then at about 20 mm of the punch displacement, the extrusion began and the increment of the extruded length is close to that of the punch displacement. As the initial length of the sample rod is 50 mm, the volume reduction is 40 % when the punch displacement is 20 mm. The value of the volume reduction estimated from the punch displacement agrees well with that estimated from the porosity change. After this stage densification in the entry channel does not take place, but the extrusion is done by shear deformation.

The mechanism of the onset of the extrusion is shown in Fig. 9.18, assuming no friction between the rod and the die. Although the sample is loaded by the punch, the normal force $F_3$ of the shear flow stress reacts against the punch load $F_1$. At the beginning of the pressing, the porosity is as high as 50 % that the compressive flow stress $\sigma$ of the entry channel by uniaxial compression is smaller than the stress caused by the punch load $F_1$, and successively the sample is compressed and densified. During this process, the compressive flow stress $\sigma$ increases with a

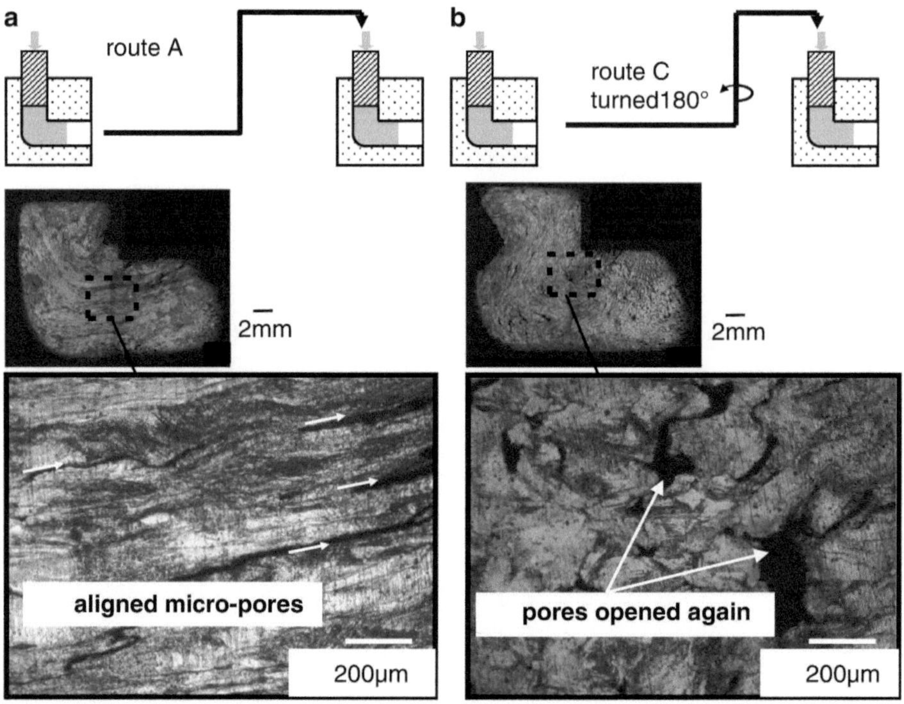


**Fig. 9.16** Cross sections of lotus copper specimens after extrusion approximately 30 mm in a second pass through an ECAE die with a channel angle of 90°. (**a**) Route A where the direction of the sample of the second pass is the same as that of the first pass. (**b**) Route C where the direction of the sample of the second pass is 180° turned axially from the first pass (Reprinted with permission from [13] © 2008 Elsevier B.V.)

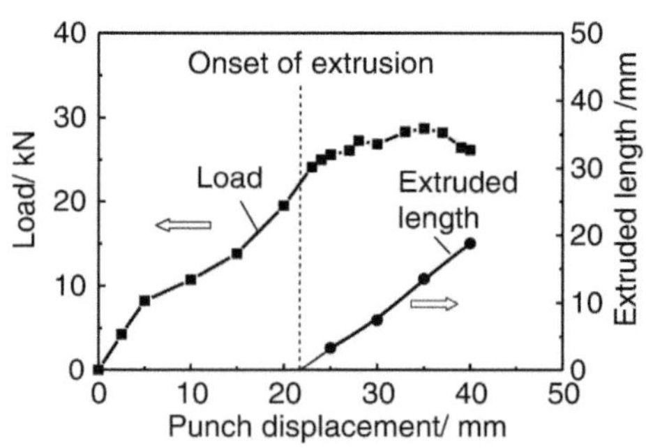


**Fig. 9.17** Change of the punch load and the extrusion length during the first pass through an ECAE die with a channel angle of 90°. The onset of the extrusion was the place where the increase of the extruded length began (Reprinted with permission from [13] © 2008 Elsevier B.V.)

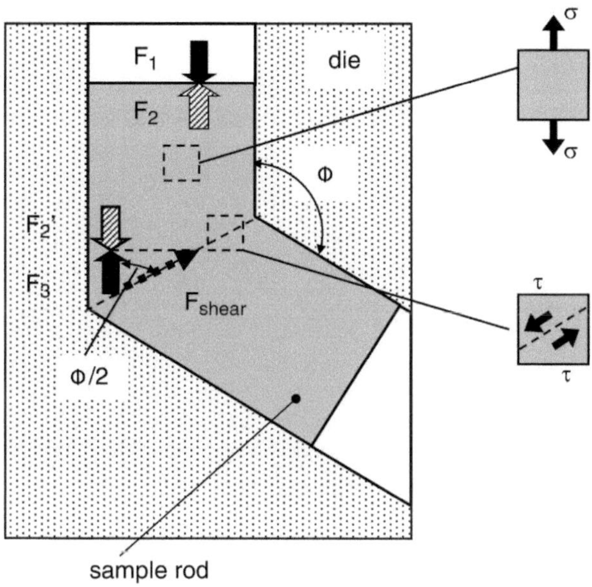


**Fig. 9.18** Mechanical balance model for ECAE process (Reprinted with permission from [13] © 2008 Elsevier B.V.)

decrease of the porosity. It is considered that the force $F_2$ caused by the compressive flow stress σ is balanced with the force $F_3$ caused by the shear flow stress $\tau$ at 20 mm of the punch displacement.

At the beginning of pressing of lotus copper, the load increases slower than that of nonporous metals. At the punch displacement of about 30 mm, the punch load achieved a plateau value. From these results, the criteria for the onset of the extrusion are expressed by the balance of the forces:

$$F_1 = F_2 > F_3.$$

Thus,

$$F_1 = A\sigma > A\tau \cot(\Phi/2), \tag{9.2}$$

where A is the area of the cross section perpendicular to the axis of the entrance channel. Thus, the criteria can be expressed as

$$\frac{\sigma}{\tau} > \cot\left(\frac{\Phi}{2}\right). \tag{9.3}$$

Or $\Phi = 90^\circ$, the criteria is expressed as $\sigma/\tau > 1$. Although the shear stress $\tau$ of lotus copper has not been measured exactly, the expression indicates the tendency that the criteria can be satisfied easily when the channel angle $\Phi$ is large.

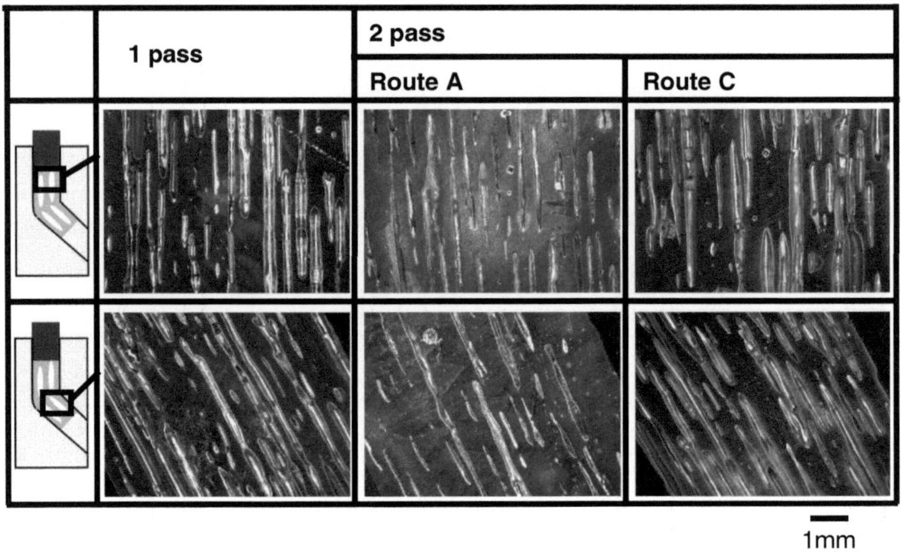


**Fig. 9.19** Cross sections of lotus copper specimens after extrusion in the first and the second pass through an ECAE die with a channel angle of 150° (Reprinted with permission from [13] © 2008 Elsevier B.V.)

If the channel angle of 150° is chosen, an extrusion without densification can be expected. Figure 9.19 shows the cross sections parallel to the sample axis of the extruded sample obtained via 1 pass, route A (2 passes) and route C (2 passes). These photos show that the pore morphology is not changed after 1 pass, and even after 2 passes. Thus, it is concluded that it is possible to extrude without densification under certain suitable conditions.

Figure 9.20a shows the change of the hardness in the case of the channel angle of 90°. During the first pass, the hardness increases in the entry channel because of the densification caused by uniaxial compression. Then, the shear deformation at the corner increases the value of HV. Furthermore, the value of HV increases through both the entry and exit channel of the second pass. There is not so significant difference between route A and C of the second pass. Figure 9.20b shows the HV in the case of the channel angle of 150°. In the entry channel of the first pass, the value of HV is not so different from that before the extrusion. Then in the end channel, HV increases. During the second pass, HV also does not increase in the entry channel and then increases in the exit channel. Compared with the sample extruded through the die with the channel angle of 90°, the increment of the HV is not so large. It is apparent that ECAE process can give work hardening to the sample in the case of channel angle of 150° maintaining the pores. The HV increases with increase of the pass number. Therefore, it is also expected that harder metal with pores after multi-pass ECAE can be obtained through the die with the channel angle of 150°.

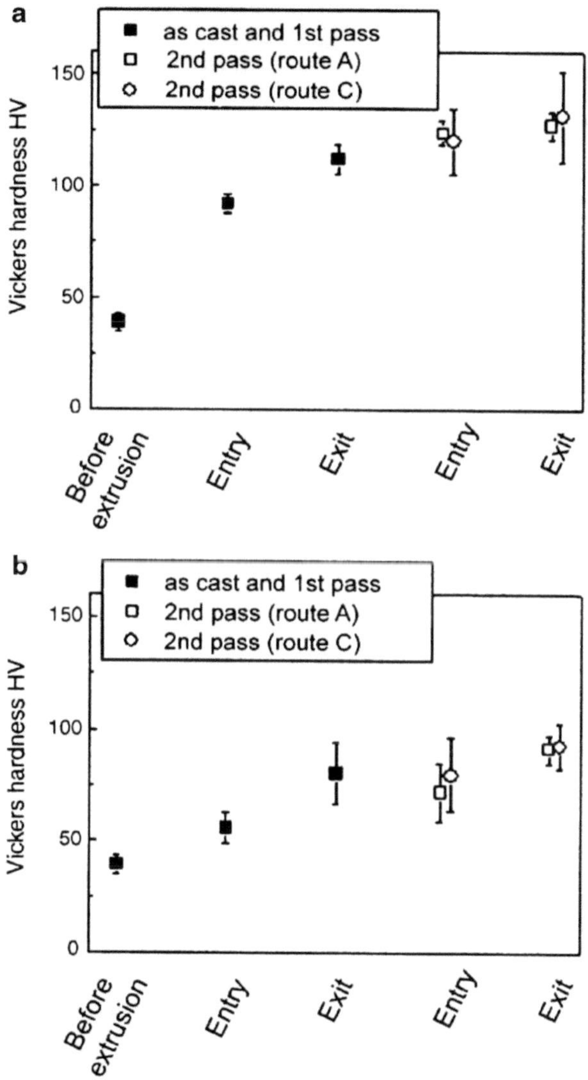


**Fig. 9.20** Change of Vickers hardness during the first pass through an ECAE. (**a**) Channel angle of 90° and (**b**) channel angle of 150° (Reprinted with permission from [13] © 2008 Elsevier B.V.)

# References

1. Pogibenko AG, Konkevich VY, Arkuzova LA, Ryazantsev VI (2001) Weld Int 156:312–316
2. Bollinghaus T, Bleck W (2001) Cellular metals and metal foaming technology. MIT, Bremen, pp 495–500
3. Haferkamp H, Ostendorf A, Goede M, Bunte J (2001) Cellular metals and metal foaming technology. MIT, Bremen, pp 479–484
4. Murakami T, Nakata K, Ikeda T, Nakajima H, Ushio M (2003) Mater Sci Eng A357:134–140
5. Yamaoka H, Yuki M, Tsuchiya K (2000) Q J Jpn Weld Soc 18:422–430
6. Seto N, Katayama S, Matsunawa A (2000) Q J Jpn Weld Soc 18:243–255

7. Brandes EA (ed) (1983) Metal reference book, 6th edn. Butterworths, London
8. Watanabe H, Susa M, Nagata K (1997) Metall Mater Trans A 28:2507–2513
9. Murakami T, Tsumura T, Ikeda T, Nakajima H, Nakata K (2007) Mater Sci Eng A 456:278–285
10. Ikeda T, Nakajima H (2002) J Jpn Foundry Eng Soc 74:812–816
11. Ikeda T, Hoshiyama H, Nakajima H (2004) J Jpn Light Metals 54:388–393
12. Asada H, Yoshinaga H (1959) J Jpn Inst Metals 23:67–71
13. Suzuki S, Utsunomiya H, Nakajima H (2008) Mater Sci Eng A 490:465–470
14. Segal VM, Rezinikov VI, Drobyshevkiy AE, Kopylov VI (1981) Russ Metall 1:99–105
15. Valiev RZ, Langdon TG (2006) Prog Mater Sci 51:881–981
16. Hyun SK, Nakajima H (2003) Mater Sci Eng A 340:258–264

# Chapter 10
# Various Applications of Lotus Metals

**Abstract** In this chapter, four examples of application of lotus metals are described. For heat sinks, penetrable pores are utilized to cool electronic devices. For artificial teeth, open pores are effective to let bone tissue intrude for fixation of the teeth into alveolar bones. For use of vibration–damping materials and golf putter, closed pores serve to damping effect. Thus, anisotropic pores in lotus metals contribute to such functions of products.

**Keywords** Biocompatibility • Golf putter • Heat sink • Implant • Vibration–damping

## 10.1 Nomenclature

$A_b$: baseplate area of heat sink [$m^2$]
$A_w$: cross-sectional area [$m^2$]
$D_e$: hydraulic diameter [m]
$dp_{mean}$: average diameter of pores [m]
$f_g$: fin gap of groove fins [m]
$f_t$ : thickness of groove fins [m]
$h_b$: heat transfer coefficient based on $A_b$ [W/($m^2$·K)]
$h_{dp}$: heat transfer coefficient on surface of pores [W/($m^2$·K)]
$H_f$: fin height [m]
$h_{fin}$: heat transfer coefficient on surface of groove fins [W/($m^2$·K)]
$h_m$: average heat transfer coefficient [W/($m^2$·K)]
$k$ : thermal conductivity [W/(mK)]
$L$: thickness of lotus type porous copper fins [m]
$L_g$: length of groove fins [m]
$N_t$: total number of pores
$p$: porosity
$Pr$: Prandtl number of fluid

H. Nakajima, *Porous Metals with Directional Pores*,
DOI 10.1007/978-4-431-54017-5_10, 

$Q$: heat flow [W]
$Re_{dp}$: Reynolds number defined by $dp$
$Re_{D_e}$: Reynolds number defined by $D_e$
$R_{fi}$ : thermal resistance of heat sink [K/W]
$R_a$ : thermal resistance of fluid [K/W]
$S_f$: total surface of pores [$m^2$]
$T_b$: temperature of baseplate [K]
$T_i$: inlet temperature of fluid [K]
$T_o$: outlet temperature of fluid [K]
$U_t$: total fluid rate [$m^3/s$]
$u_{dpmean}$: velocity of fluid through pore[m/s]
$W$: width of heat sink [m]
$\triangle P$: pressure drop [Pa]
$\eta$: fin efficiency
$a$: air
$f$: cooling water

## 10.2 Heat Sink

In recent years, heat dissipation rates in power devices and laser diodes have been increasing to more than 100 W/cm$^2$ and in high-frequency electronic device, it is increasing to more than 1,000 W/cm$^2$ under the trend of miniaturization and growing capacity. Figure 10.1 shows typical cooling loads and cooling technologies compiled by Nishio [1]. When the heat flux increases, the temperature increases, and a heat flux of 100 W/cm$^2$ is as high as an equivalent heat flux from nuclear blast at 2,000 K. However, a power device or a laser diode should be kept at normal temperature level in spite of a large-heat flux of more than 100 W/cm$^2$. Therefore, novel heat sinks with high-heat transfer performance are required to cool these devices. Among various types of heat sinks, heat sinks utilizing microchannels with channel diameters of several tens of microns are expected to have excellent cooling performance because a higher heat transfer capacity is obtained with smaller channel diameters. Wei and Joshi investigated three-dimensional stacked microchannel heat sinks to enhance the cooling performance of heat sinks with microchannels [2]. Porous material with open pores is preferable for three-dimensional microchannels because of the higher surface area per unit volume and lower product cost. Bastawros and Evans investigated cellular metals as a heat transfer medium [3]. Various porous materials such as sintered porous metals, cellular metals, and fibrous composites have been investigated for heat sink applications [4, 5]. However, heat sinks using porous materials were clarified to have a high-pressure drop because the cooling fluid flow through the pores of the porous materials is complex [6].

Among the described porous materials, a lotus metal with straight pores is preferable for heat sinks due to the small pressure drop of the cooling fluid flowing

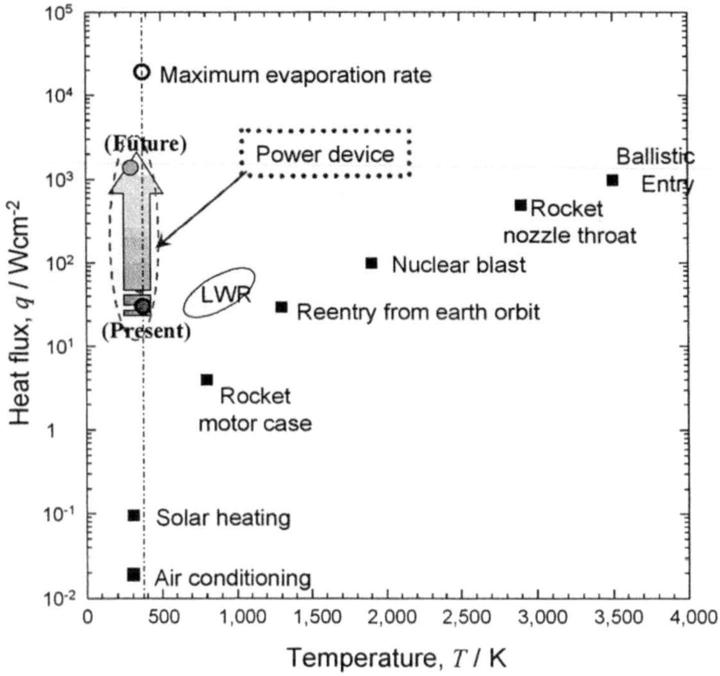


**Fig. 10.1** Typical cooling loads and cooling technology [1]

through the pores. The following are the main features of lotus metals: (1) the pores are straight, (2) the pore size and porosity are controllable, and (3) porous metals can be produced with pores as small as a hundred microns in diameter. To effectively use lotus copper as a heat sink, its effective thermal conductivity must be known, by which the effect of directional pores on the heat flow must be considered. One defined perpendicular effective thermal conductivity $k_{\mathrm{eff}\perp}$ of lotus copper as the thermal conductivity for heat flow perpendicular to the pore axis is shown in Fig. 10.2.

To predict the heat transfer capacity of a lotus copper heat sink, the correlation between heat conduction in lotus metal and heat transfer to the fluid in the pores must be considered. This correlation is generally expressed by fin efficiency. As a result of numerical analysis, the fin efficiency of the lotus copper fin was verified to be predictable by a simple straight fin model using effective thermal conductivity $k_{\mathrm{eff}\perp}$ and the surface area ratio between the surface area of lotus copper fin and that of the straight fin [7]. Chiba et al. investigated the model for predicting the heat transfer capacity of heat sink using the lotus copper. This model takes into consideration the heat conduction in lotus copper and the heat transfer to the fluid in the pores. Finally, cooling performance of lotus copper heat sink designed by using the model for air cooling and water cooling is investigated [8].

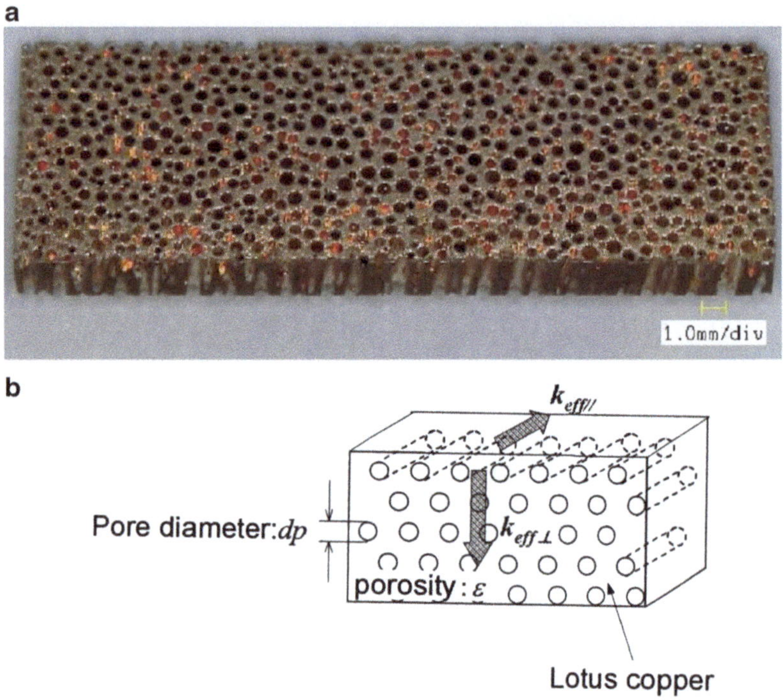


**Fig. 10.2** Lotus copper with anisotropic pores: (**a**) outer view of lotus copper and (**b**) definition of thermal conductivity (Reprinted with permission from [8] © 2011 American Society of Mechanical Engineering)

## *10.2.1 Heat Sink for Air Cooling*

Generally, a groove fin is used for cooling the power devices for air cooling or water cooling. The groove fin composes of a lot of boards connected vertically on the base. The heat transfer capacities of two types of heat sinks: (1) conventional groove fins and (2) lotus copper heat sinks, were examined. The configuration and specifications of the conventional grove fins are shown in Fig. 10.3. The conventional groove fins made of copper have a 3-mm fin gap size $f_g$, a 1-mm fin thickness $f_t$, and a 20-mm fin height $H_f$. The heat transfer capacity of the conventional groove fins is only derived from calculations because many experimental formulas of the fins have been established so far [9, 10]. Figure 10.4 shows the configuration of the heat sink using three lotus copper fins. $W$ is a width. $L_g$ is a length. Because the heat sink has 3 lotus copper fins, base area $A_b$ of the heat sink is expressed by $3WL$.

Figure 10.5 illustrates the experimental apparatus for measuring the heat transfer capacity of heat sinks. Cooling air was blown by a blower into the test duct in which the heat sink is located. The heat sink consists of fins brazed on one side of a copper

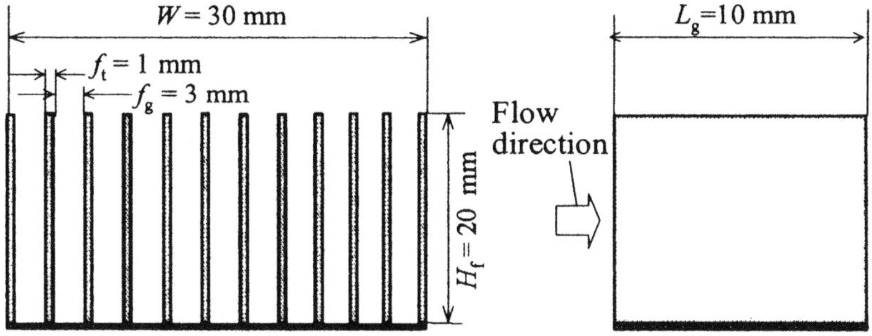


**Fig. 10.3** Configuration of conventional groove fins (Reprinted with permission from [8] © 2011 American Society of Mechanical Engineering)

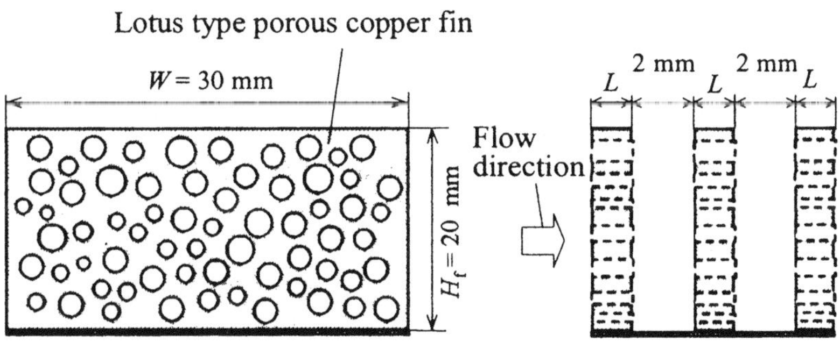


**Fig. 10.4** Configuration of lotus copper heat sink (Reprinted with permission from [8] © 2011 American Society of Mechanical Engineering)

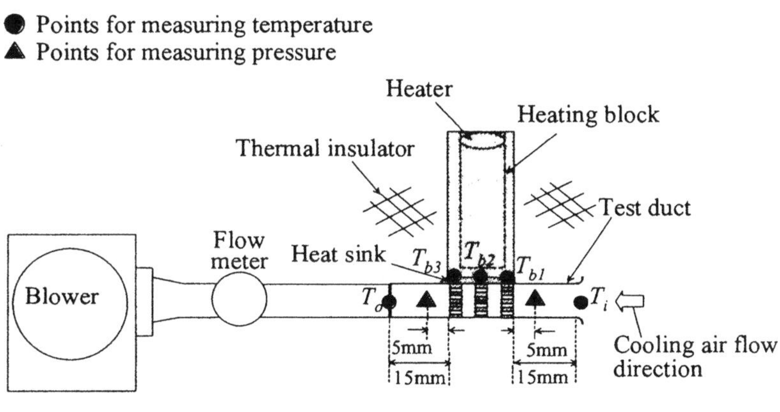


**Fig. 10.5** Experimental apparatus for measuring heat transfer capacity (Reprinted with permission from [8] © 2011 American Society of Mechanical Engineering)

baseplate and the heating block with a heater soldered to the other side of the baseplate. The inlet temperature of cooling air $T_i$, the temperature of copper baseplates $T_{b1}$, $T_{b2}$, and $T_{b3}$, and the outlet temperature of the cooling air $T_0$ are measured by K-type thermocouples.

The heat transfer capacity by heat transfer coefficient $h_b$ based on baseplate area $A_b$ of the heat sink is as follows:

$$h_b = \frac{Q}{A_b(T_b - T_i)}, \tag{10.1}$$

$$Q = \rho \cdot C_p \cdot U_t \cdot (T_0 - T_i), \tag{10.2}$$

where $Q$ is the heat input to the heat sink, $\rho$ and $C_p$ denote the density and the specific heat of the fluid, respectively. $U_t$ is the total flow rate of the air through the test duct. $T_b( =(T_{b1} + T_{b2} + T_{b3})/3)$ is the mean temperature of the copper baseplate. The pressure drop between the inlet side and the outlet side of the heat sinks was measured within an experimental accuracy of ±5 % by a pressure sensor.

Heat transfer capacity of heat sink was evaluated by using the heat transfer coefficient $h_b$ based on baseplate surface area $A_b$. $h_b$ is expressed by Eq. (10.3):

$$h_b = \frac{1}{R_{fi} \cdot A_b}, \tag{10.3}$$

where $R_{fi}$ is the thermal resistance of the heat sink between the temperature of the baseplate and the fluid's inlet temperature. $R_{fi}$ is expressed by

$$R_{fi} = \frac{T_b - T_i}{Q} = \frac{R_a}{1 - \exp\left(-\frac{R_a}{R_f}\right)}, \tag{10.4}$$

where $R_a$ is the thermal resistance between the inlet and outlet of the fluid and $R_f$ is the thermal resistance based on the logarithmic mean temperature difference between the base of the heat sink and the fluid. These thermal resistances are expressed by

$$R_f = \frac{\Delta T_m}{Q} = \frac{1}{h_{dp_{mean}} \cdot S_f \cdot \eta}, \tag{10.5}$$

$$\Delta T_m = \frac{T_0 - T_i}{\ln \frac{T_b - T_i}{T_b - T_0}}, \tag{10.6}$$

$$R_a = \frac{T_0 - T_i}{Q} = \frac{1}{\rho \cdot C_p \cdot U_t}, \tag{10.7}$$

where $h_{\mathrm{dp}_{\mathrm{mean}}}$ is the heat transfer coefficient on surface of pores $S_f$ in the lotus copper fin. As the flow characteristics through the pores in the lotus copper fins is considered to be similar in a circle pipe, heat transfer coefficient $h_{\mathrm{dp}_{\mathrm{mean}}}$ is expressed by the following correlations for the circle pipe under laminar flow regime ($Re_p < 3{,}000$) [9]:

$$Nu_p = \frac{h_{\mathrm{dp}_{\mathrm{mean}}} \cdot \mathrm{dp}_{\mathrm{mean}}}{k_{\mathrm{a}}} = 5.364\left(1 + \{(220/\pi)X^+\}^{-10/9}\right)^{3/10} \times \left\{1 + \left(\frac{\pi/(115.2X^+)}{\left[1 + (Pr_a/0.0207)^{2/3}\right]^{1/2}\left[1 + \{(220/\pi)X^+\}^{-10/9}\right]^{3/5}}\right)^{5/3}\right\}^{3/10} - 1, \tag{10.8}$$

$$X^+ = \frac{L/\mathrm{dp}_{\mathrm{mean}}}{Re_p \cdot Pr_a}, \tag{10.9}$$

$$Re_p = \frac{u_{\mathrm{dp}_{\mathrm{mean}}} \cdot \mathrm{dp}_{\mathrm{mean}}}{\upsilon_a}, \tag{10.10}$$

$$u_{\mathrm{dp}_{\mathrm{mean}}} = \frac{U_{\mathrm{t}}}{p \cdot A_w}, \tag{10.11}$$

where $Nu_p$ and $Re_p$ are Nusselt number and Reynolds number which are defined by average diameter for pore $\mathrm{dp}_{\mathrm{mean}}$, and $\nu_{\mathrm{a}}$ and $k_{\mathrm{a}}$ are dynamic viscosity and thermal conductivity of the air, respectively. $u_{\mathrm{dp}_{\mathrm{mean}}}$ is the velocity of air through the pore. $\eta$ is the fin efficiency of the lotus copper fin, which is calculated from the model by which lotus copper fin was assumed to be a straight fin model as follows:

$$\eta = \frac{\tanh(m \cdot H_{\mathrm{f}})}{m \cdot H_{\mathrm{f}}}, \tag{10.12}$$

$$m = \sqrt{\frac{h_{\mathrm{dp}_{\mathrm{mean}}} \cdot \xi \cdot 2(W + L)}{k_{\mathrm{eff}\perp} \cdot (W \cdot L)}}. \tag{10.13}$$

Pressure drop of lotus copper fin, $\Delta P$, is expressed by the following correlation for the circle pipe under laminar flow regime:

$$\Delta P_L = \left(\frac{64}{Re_p} \cdot \frac{L}{\mathrm{dp}_{\mathrm{mean}}} + \zeta\right) \cdot \frac{1}{2}\rho \cdot u_{\mathrm{dp}_{\mathrm{mean}}}{}^2, \tag{10.14}$$

where $\zeta$ is the total of the pressure loss coefficient generated when the fluid flows into and out of the pores, where $\zeta = 1.4$ was assumed.

Because the conventional groove fins have often been researched so far, empirical formula for predicting their heat transfer capacity and pressure drop are established [10]. Based on the baseplate surface area, the heat transfer coefficient $h_b$ is predicted by Eq. (10.15)

$$h_b = \frac{h_f \cdot A_f \cdot \eta}{A_b}, \tag{10.15}$$

where $h_f$ is the heat transfer coefficient on fin surface $A_f$ of the groove fins and is predicted by Eq. (10.16)

$$Nu_{f_g} = \frac{Nu_1 - Nu_2}{10 - Pr_a} \cdot (Pr_a - 0.71) + Nu_2, \tag{10.16}$$

$$Nu_1 = 2.80136 - 2.10514X^* + 0.411783X^{*2} + 4.11, \tag{10.17}$$

$$Nu_2 = 4.1880 - 3.14709X^* + 0.611075X^{*2} + 4.11, \tag{10.18}$$

$$X^* = \ln\left(\frac{L}{f_g} \cdot Re_{f_g} \cdot Pr_a\right), \tag{10.19}$$

where $Re_{f_g}$ and $Nu_{f_g}$ are the Reynolds number and the Nusselt number defined by the fin gap. $u$ is the air velocity through the fin gap.

Pressure drop $\Delta P$ of the conventional groove fins under a laminar flow regime is expressed as follows:

$$X = \frac{L_g}{Re_{D_e} \cdot D_e}, \tag{10.20}$$

$$\Delta P = \left(f \cdot \frac{4L_g}{D_e} + \zeta\right) \cdot \frac{1}{2}\rho \cdot u^2, \tag{10.21}$$

$$f = \frac{3.44}{X^{0.5} \cdot Re_{D_e}} + \frac{24 + \frac{0.674}{4 \cdot X} - \frac{3.44}{X^{0.5}}}{(1 + 0.000029 \cdot X^{-2}) \cdot Re_{D_e}}, \tag{10.22}$$

where $Re_{D_e}$ is the Reynolds number defined by hydraulic diameter $D_e$ $(= 4H_f f_g/(2H_f + 2f_g))$.

Chiba et al. measured the heat transfer coefficients and pressure drop using experimental apparatus of air-cooling heat sink [11]. The heat transfer coefficient $h_b$, based on the baseplate surface area defined by Eq. (10.3) for the experiment and Eq. (10.1) for the calculated results, respectively, is plotted as a function of the inlet velocity $u_0$ $(=U_t/A_w)$ in Fig. 10.6. The prediction for the lotus copper heat sink showed a good agreement with the experimental data within accuracy of $\pm 5$ %.

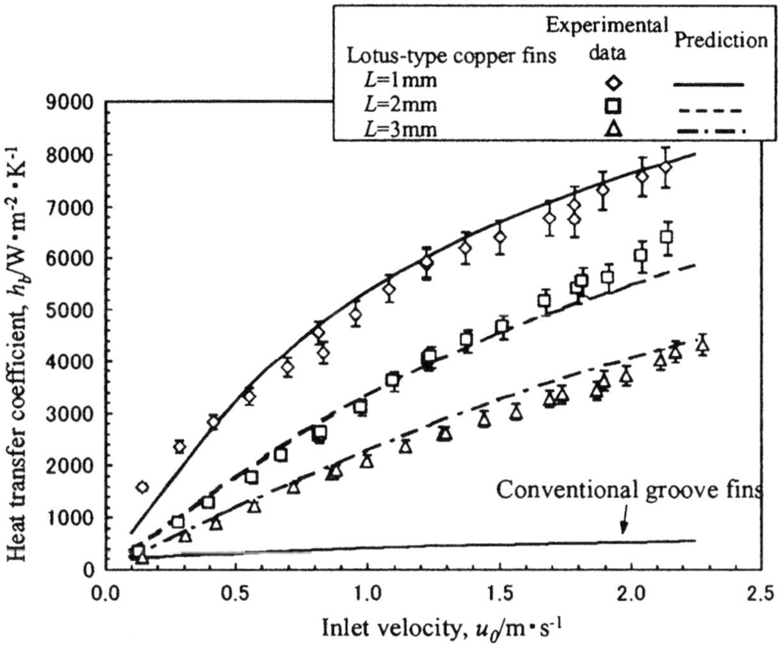


**Fig. 10.6** Comparison of heat transfer coefficient $h_b$ between experimental and predicted data (Reprinted with permission from [8] © 2008 Elsevier B.V.) (Reprinted with permission from [8] © 2011 American Society of Mechanical Engineering)

The experimental data for lotus copper heat sink with thickness $L = 1$ mm showed a very large-heat transfer coefficient of 5,000 W/(m$^2$ K) under an inlet velocity of 1.0 m/s, which is 13.2 times higher than that for the conventional groove fins.

The pressure drop $\Delta P$ in lotus copper heat sinks is plotted as a function of the inlet velocity to heat sink $u_0$ in Fig. 10.7. The predicted pressure drop of lotus heat sink, denoted by $3\Delta P_L$, shows a good agreement with the experimental data within accuracy of $\pm 10$ %. Comparing the pressure drop among all of heat sinks at the inlet velocity $u_0$ of 1.0 m/s, the experimental data show that the pressure drop of lotus copper heat sink with thickness $L = 3$ mm is 13.5 times higher than that of the conventional groove fins.

The heat transfer coefficient $h_b$ is shown as a function of the pumping power $\Delta P \cdot U_t$ that is defined by the product of the pressure drop $\Delta P$ and the total flow rate $U_t$ (Fig. 10.8). A comparison of the heat transfer coefficient among all lotus copper heat sinks under a pumping power of 0.02 W revealed that the experimental data for lotus copper heat sink with thickness of 1 mm are 11.3 times higher than those for the conventional groove fin.

In summary, the thinner the lotus copper fin is, the higher is the heat transfer capacity of a lotus copper heat sink under identical pressure drop and identical pumping power. It is preferable that the thickness of the lotus copper is small when a lotus copper fin is used for heat sink.

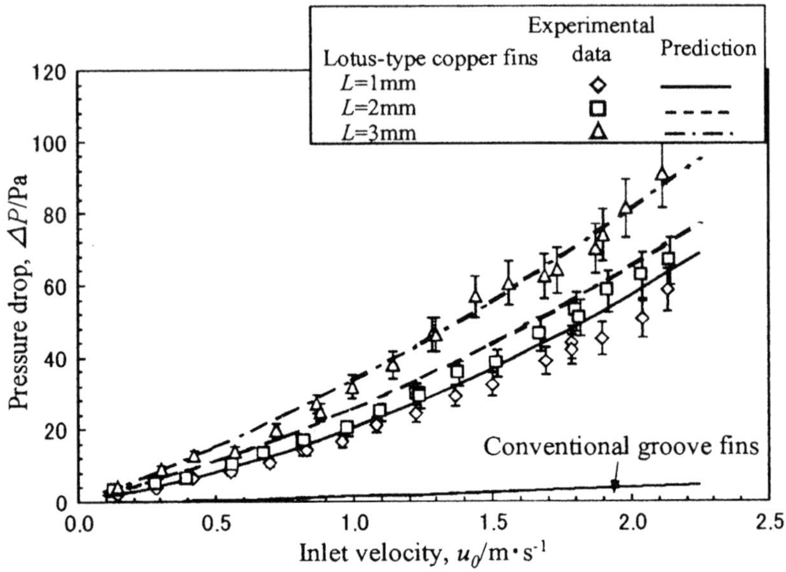


**Fig. 10.7** Comparison of pressure drop $\Delta P$ between experimental and predicted data (Reprinted with permission from [8] © 2011 American Society of Mechanical Engineering)

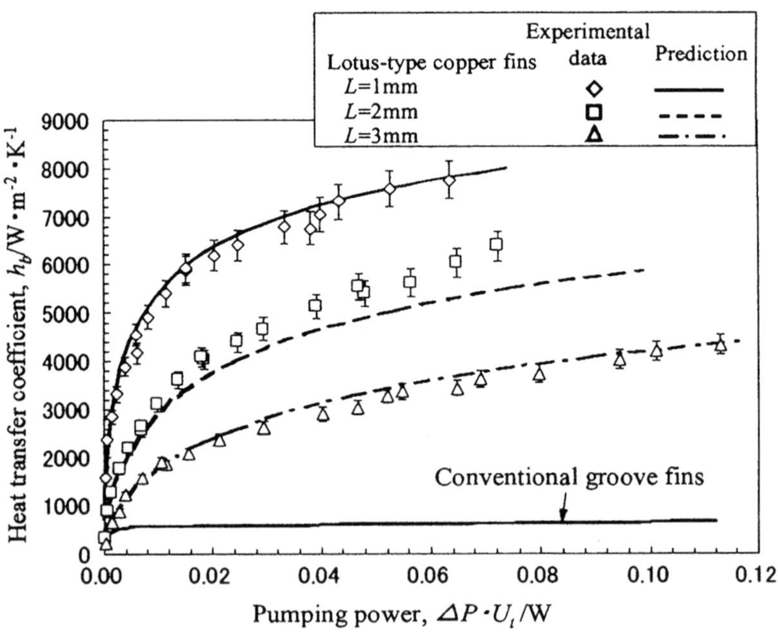


**Fig. 10.8** Comparison of heat transfer coefficient $h_b$ as a function of pumping power $\Delta P{\cdot}U_t$ (Reprinted with permission from [8] © 2011 American Society of Mechanical Engineering)

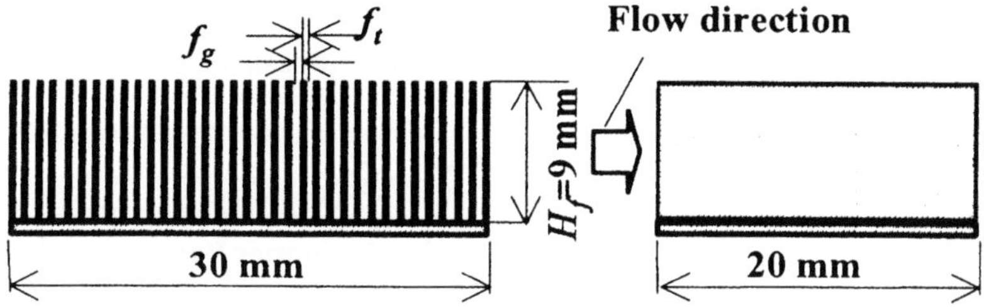


**Fig. 10.9** Configuration groove fins and microchannel copper heat sink (Reprinted with permission from [8] © 2011 American Society of Mechanical Engineering)

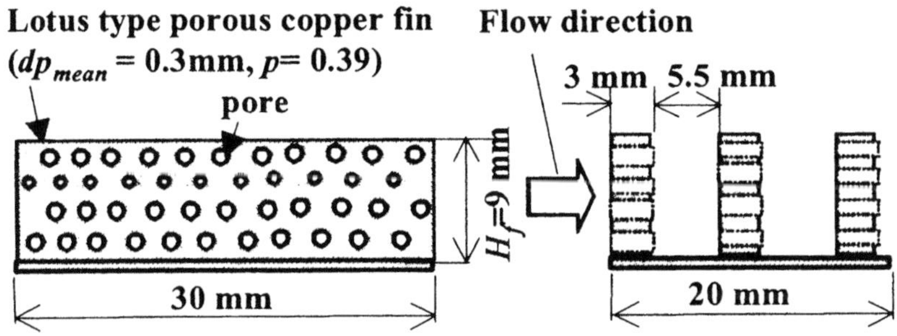


**Fig. 10.10** Lotus copper heat sink for water cooling (Reprinted with permission from [8] © 2011 American Society of Mechanical Engineering)

## *10.2.2 Heat Sink for Water Cooling*

The heat transfer capacity of three types of heat sinks for water cooling with conventional groove fins, groove fins with smaller fin gap (microchannel), and lotus copper were investigated. The configuration and the specifications of the conventional groove fins and microchannels are shown in Fig. 10.9. The conventional groove fins have fin gap of 3 mm and fin thickness of 1 mm. The microchannels have fin gap of 0.5 mm and fin thickness of 0.5 mm. The heat transfer capacity of the conventional groove fins is only for calculation. On the other hand, the configuration of lotus copper heat sink for water cooling is shown in Fig. 10.10. The heat sink has three pieces of lotus copper fins with the thickness of 3 mm along the flow direction. The lotus copper fins have pores with the mean diameter of 0.3 mm and the porosity of 39 %. Experimental apparatus for measuring the heat transfer capacity of heat sink is shown in Fig. 10.11. A cooling water was circulated through a filter and the test duct in which the heat sink is located. The circulator has a pump and water-cooling equipment. The heat sink consists of fins that are brazed on one side of a copper baseplate. The heat sink block with a heater was soldered on the other side of the baseplate. Inlet temperature of cooling

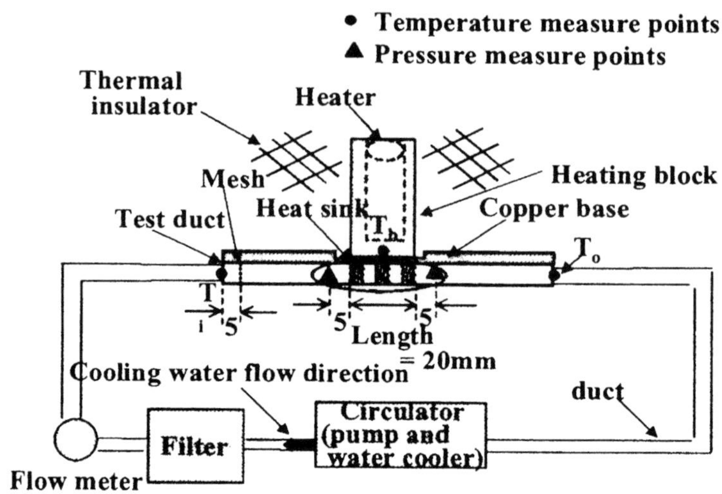


**Fig. 10.11** Experimental apparatus for measuring heat transfer capacity for water cooling (Reprinted with permission from [8] © 2011 American Society of Mechanical Engineering)

water $T_i$, a temperature of copper baseplate $T_{b'}$, and outlet temperature of cooling water $T_0$ were measured by K-type thermocouples. From the measurement result, heat transfer coefficient was calculated by Eq. (10.1). The pressure drop between the inlet and the outlet of the fins of every heat sink was measured.

The heat transfer coefficients based on baseplate surface area $A_b$ defined by Eq. (10.1) for experiment and Eq. (10.3) for prediction, respectively, are plotted for all of heat sinks as a function of the inlet velocity to the heat sinks $u_0$ in Fig. 10.12. The prediction for lotus copper heat sink showed good agreement with the experimental data within an accuracy of ±15 %. The experimental data of lotus copper heat sink showed very large-heat transfer coefficient of 80,000 W/(m$^2$ K) under the velocity $u_0$ of 0.2 m/s that is 1.7 times higher than that of the microchannels and 6.5 times higher than that of the conventional groove fins.

The value of pressure drop of all the heat sinks is compared in Fig. 10.13 as a function of $u_0$. The predicted pressure drop of lotus copper heat sink showed good agreement with the experimental data within an accuracy of ±5 %. Comparing the pressure drop among all of the heat sinks under the velocity $u_0$ of 0.2 m/s, the experimental data of the lotus copper heat sink is 2.5 times higher than that of the microchannels and 38 times higher than that of the conventional groove fins. The heat transfer coefficient $h_b$ is shown as a function of pumping power that is defined by a product of a flow rate $U_t$ and $\Delta P$ is shown in Fig. 10.14. Comparing the heat transfer coefficients among all of the heat sinks under the pumping power of 0.01 W, the experimental data of lotus copper heat sink is 1.3 times higher than that of the microchannels and 4 times higher than that of the conventional groove fins. Thus, the use of lotus metals is the most superior for the heat sinks of air cooling and water cooling.

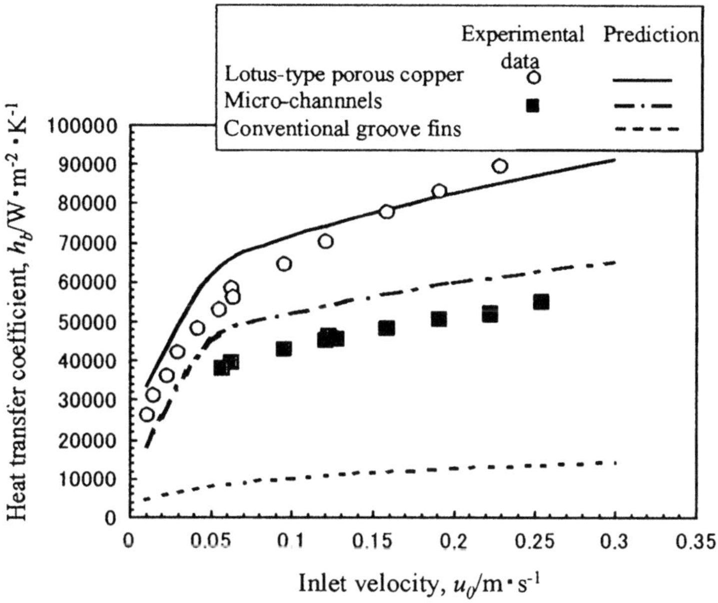


**Fig. 10.12** Comparison of heat transfer coefficient $h_b$ between experimental and predicted data (Reprinted with permission from [8] © 2011 American Society of Mechanical Engineering)

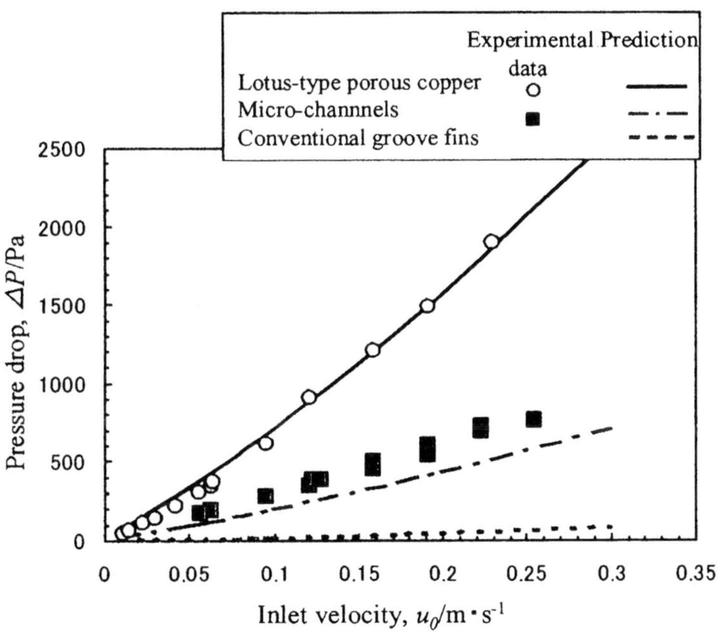


**Fig. 10.13** Comparison of pressure drop $\Delta P$ between experimental and predicted data (Reprinted with permission from [8] © 2011 American Society of Mechanical Engineering)

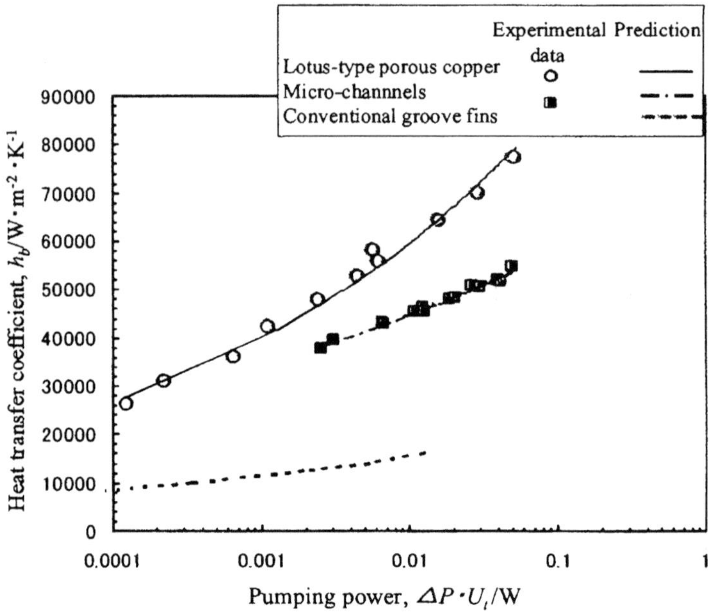


**Fig. 10.14** Comparison of heat transfer coefficient $h_b$ as a function of pumping power $\Delta P{\cdot}U_t$ (Reprinted with permission from [8] © 2011 American Society of Mechanical Engineering)

## 10.3 Vibration–Damping Materials

Nowadays, the demand for high-damping materials is increasing in various fields such as aerospace, transformation, and manufacturing industries. Also in an environmental problem, high-damping materials play an important role on the reduction of noise. Therefore, various damping materials have been produced and developed so far [12]. It is known that porous materials such as cellular materials and foamed metals show a high-damping capacity [13]. The combination of the two characteristics of lightness in weight and high-damping capacity makes porous metals more attractive. However, the mechanical strength of conventional porous materials is quite low due to their complicated porous structure, and this weak mechanical strength limits their applications. The strength of lotus metals is superior to that of conventional porous materials. If lotus metals show a high-damping capacity as well as the conventional porous materials, lotus metal can be a new high-damping material combining two characteristics of lightness in weight and high strength. Xie et al. investigated the damping capacity of lotus magnesium, which was measured by hammering–vibration–damping tests [14].

Figure 10.15 shows the schematic drawing of the apparatus for the hammering–vibration–damping method. The disc-shaped specimen is hanged with two strings, and the stainless steel sphere of 10 mm in diameter and 4 g in weight is also hanged with a string. They lift the sphere up to 20 mm in height and release it. The swung

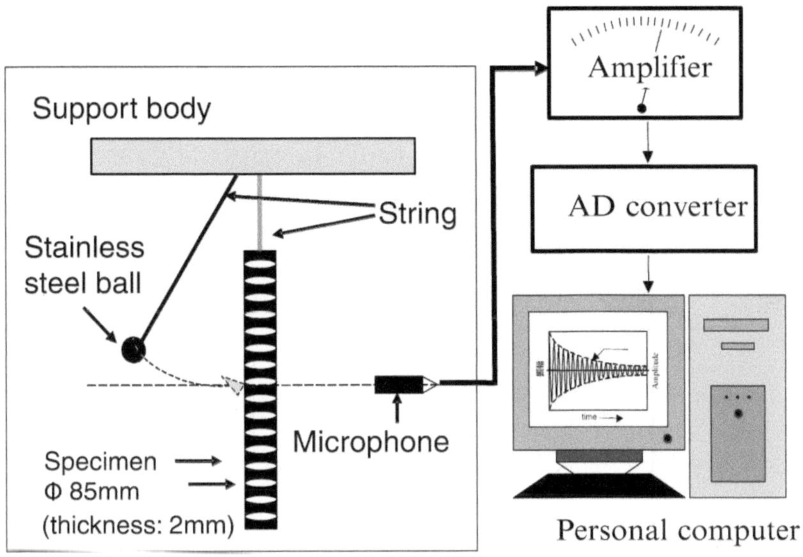


**Fig. 10.15** Schematic illustration showing the measurement system for the damping capacity (attenuation coefficient) by hammering–vibration–damping tests (Reprinted with permission from [14] © 2005 Elsevier B.V.)

sphere hits the center of the disc-shaped specimen and excites the free vibration. The acoustic microphone detected the amplitude of the specimen vibration through the acoustic waves. The AD converter transforms the analog signals obtained with the microphone into the digital signals, where the sampling frequency of the AD converter is 42 kHz. Figure 10.16 shows the vibration–damping curves of the nonporous and lotus magnesium with the porosity of 44.4 % after excitation of the vibration. While for nonporous magnesium, the amplitude of a free vibration gradually decreases with increase in time, the amplitude decreases steeply for lotus magnesium. Figure 10.17 shows the Fourier transforms of the vibration–damping curves of nonporous and lotus magnesium. The Fourier transforms of nonporous magnesium possess one large peak and two small peaks, which indicate that a hammering mainly excites the fundamental mode of resonance vibrations. On the other hand, many peaks observed for lotus magnesium indicate that a hammering excites various resonant-vibration modes.

Here, one considers a material vibration at a resonant frequency by cyclic external force. When the external force is removed, the resonant-vibration damps gradually. Then, the amplitude of a damping vibration, $A(t)$, can be expressed as

$$A(t) = A_0 \exp(-\alpha t + i2\pi f_r t), \tag{10.23}$$

where $t$ denotes the time after removal of an external force, $A_0$ denotes the amplitude at $t = 0$, and $f_r$ denotes a resonant frequency. $\alpha$ denotes the attenuation (damping) coefficient, which depends on the damping capacity of materials.

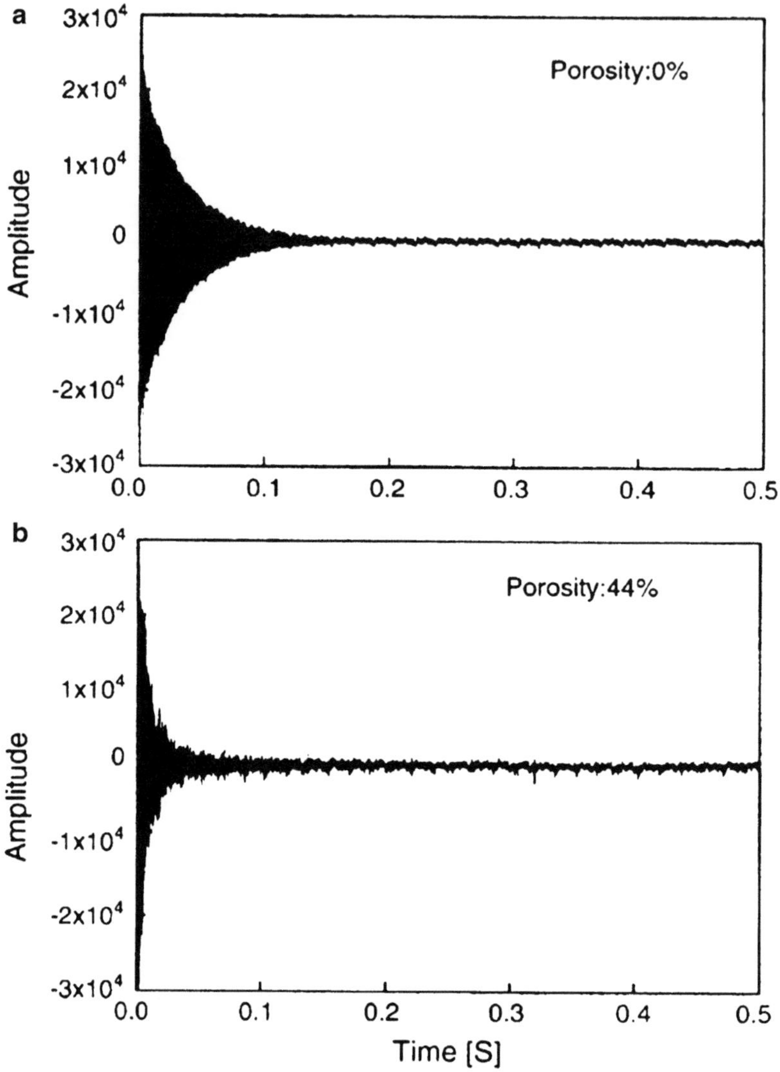


**Fig. 10.16** Vibration–damping curves of (**a**) nonporous magnesium (porosity 0 %) and (**b**) lotus magnesium with the porosity of 44.4 % (Reprinted with permission from [14] © 2005 Elsevier B.V.)

For lotus magnesium, various vibration modes are excited, and therefore they cannot obtain the attenuation coefficient by fitting Eq. (10.23) to the vibration–damping curve. To determine $\alpha$ in such a case, they picked up the maximum value of the positive amplitude of the damping curve and fitted the equation

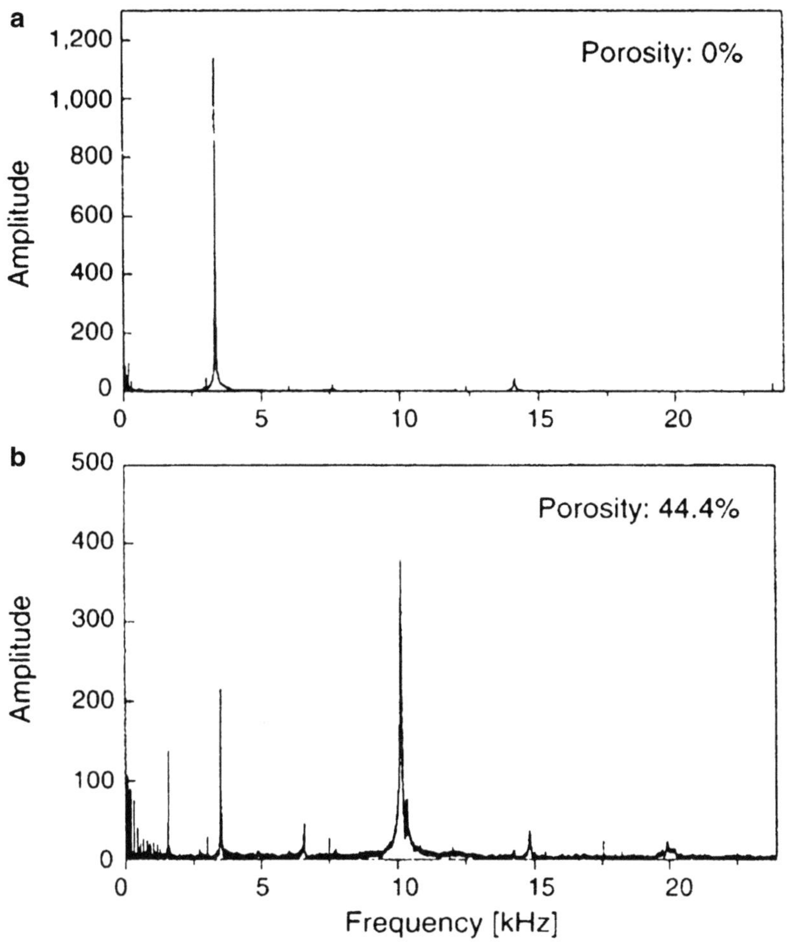


**Fig. 10.17** The Fourier transforms of the vibration–damping curves of (**a**) nonporous magnesium (porosity 0 %) and (**b**) lotus magnesium with the porosity of 44.4 % (Reprinted with permission from [14] © 2005 Elsevier B.V.)

$$A(t) = A_0 \exp(-\alpha t) + C \tag{10.24}$$

to the amplitude and obtained the apparent attenuation coefficient of lotus magnesium, where $C$ denotes a fitting coefficient. Figure 10.18 shows the vibration–damping curve of lotus magnesium of 44.4 % in porosity and the curves obtained by fitting Eq. (10.24) to the vibration–damping curves. Figure 10.19 exhibits the porosity dependence of the apparent attenuation coefficient of lotus magnesium. The apparent attenuation coefficient increases with increase in porosity.

When a resonant vibration of a material at a resonant frequency damps, the attenuation coefficient depends on the resonant frequency and intrinsic internal friction, $Q^{-1}$:

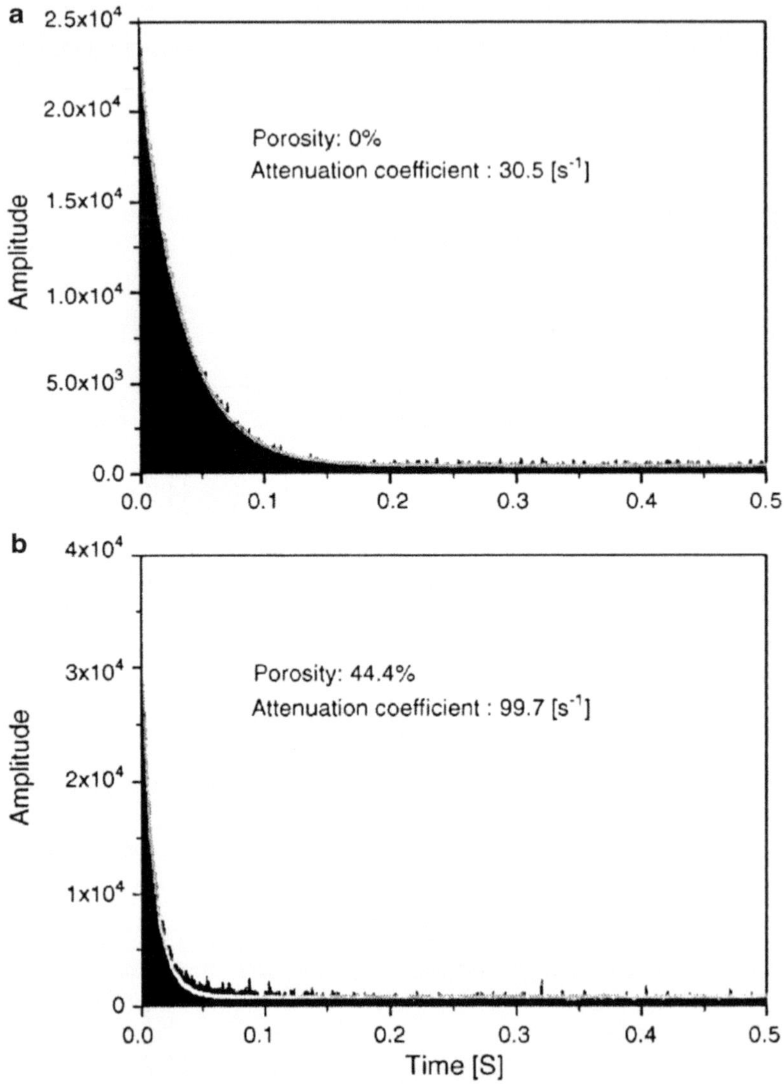


**Fig. 10.18** The vibration–damping curves of (**a**) nonporous magnesium and (**b**) lotus magnesium with the porosity of 44.4 %. The curves were obtained by fitting Eq. (10.24) to the vibration–damping curves (Reprinted with permission from [14] © 2005 Elsevier B.V.)

$$\alpha = \pi f_r Q^{-1}. \tag{10.25}$$

The measurement of the internal friction of lotus copper [15] revealed that the internal friction hardly depends on porosity at room temperature. Therefore, one can assume that $Q^{-1}$ of lotus magnesium is independent of its porosity and

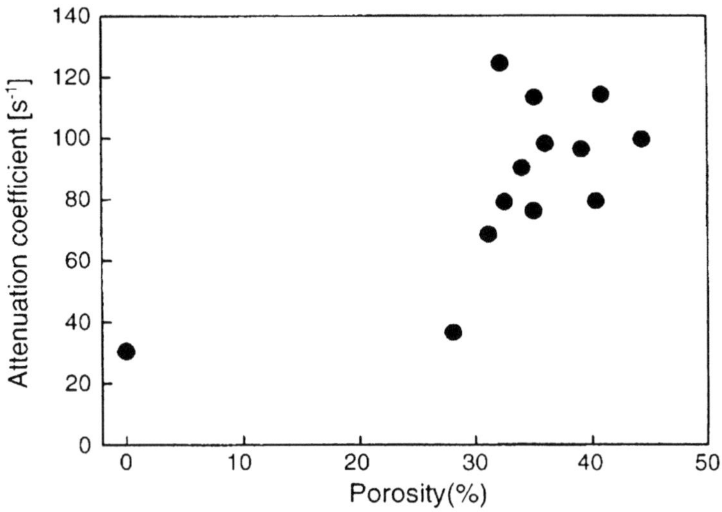


**Fig. 10.19** Porosity dependence of the attenuation coefficients of lotus magnesium, which was measured by hammering–vibration–damping tests (Reprinted with permission from [14] © 2005 Elsevier B.V.)

resonant-vibration mode, i.e., the attenuation coefficient depends only on the resonant frequency. The resonance frequency depends on the elastic constants, density, and dimension of the specimen. The presence of the pores decreases the frequency of the resonant-vibration modes, and therefore, the resonant frequency of the fundamental mode of the lotus magnesium of 44.4 % in porosity is lower than that of nonporous magnesium. If a hammering excites only fundamental mode, the attenuation coefficient of lotus magnesium is smaller than that of nonporous magnesium. However, in fact, various vibration modes are excited for lotus magnesium, which affects the apparent attenuation coefficient. In order to discuss the effect of such a mode excitation, they picked up the vibration–damping curves consisting of specific frequency using Fourier transforms technique. Figure 10.20 shows the vibration–damping curve of a specific vibration mode, where the curves containing the overall frequencies are shown for comparison. The vibration of nonporous magnesium mainly consists of one resonant-vibration mode, and the apparent attenuation coefficient mainly depends on the resonant frequency of 3.3 kHz. On the other hand, the vibration of the lotus magnesium of 44.4 % in porosity mainly consists of various resonance modes, and therefore, the amplitude of the overall vibration disperses to the various resonance modes. At $t = 0$, the amplitude of resonance mode of 10.2 kHz is larger than that of 1.6 kHz. However, the resonant vibration of 10.2 kHz damps faster than that of 1.6 kHz. For lotus magnesium, the frequencies of most of excited vibrations are higher than that of fundamental mode of nonporous magnesium, and the resonance modes of high frequencies are dominant in the apparent attenuation coefficient. Therefore, the attenuation coefficients of lotus magnesium are larger than that of nonporous magnesium.

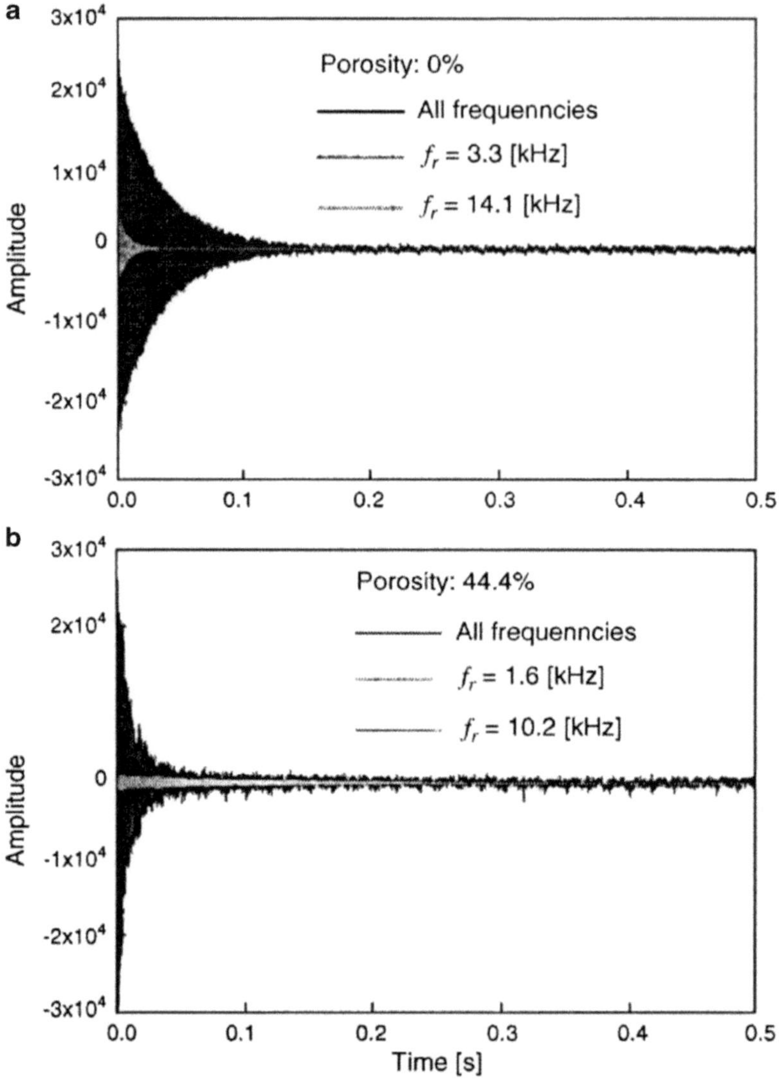


**Fig. 10.20** Vibration–damping curves of (**a**) nonporous magnesium and (**b**) lotus magnesium with the porosity of 44.4 %. (**a**) *Black line* denotes the vibration–damping curve consisting of the overall frequencies. *Dark-gray* and *light-gray lines* denote the damping curves of the resonant modes of 3.3 and 14.1 kHz, respectively. (**b**) *Black line* denotes the vibration–damping curve that consists of the overall frequencies. *Dark-gray* and *light-gray lines* denote the damping curves of the resonant modes of 10.2 and 1.6 kHz, respectively (Reprinted with permission from [14] © 2005 Elsevier B.V.)

In order to compare the vibration–damping capacity of lotus magnesium with those of various metals and alloys, Xie et al. measured the attenuation coefficients of various metals and alloys using the hammering–vibration–damping method.

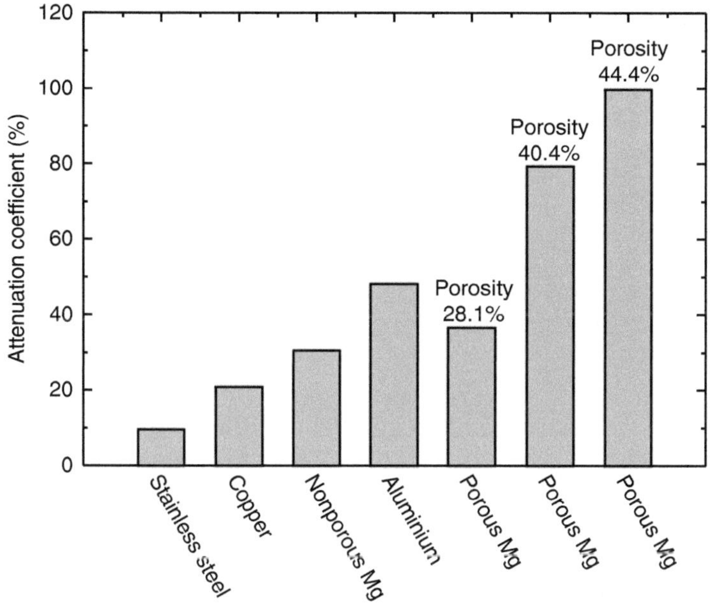


**Fig. 10.21** Attenuation coefficients of nonporous aluminum, copper, stainless steel magnesium, and lotus magnesium with different porosity (Reprinted with permission from [16] © 2006 Elsevier Ltd)

Figure 10.21 shows the attenuation coefficients of nonporous aluminum, copper, and stainless steel. The attenuation coefficients of lotus magnesium are larger than those of nonporous aluminum, copper, and stainless steel which indicates that lotus magnesium is newly considered as desirable candidate of high-damping material.

## 10.4 Golf Putter

Usually, new materials have been first adopted to produce sporting goods, weapons, space materials, etc., whose costs can be almost disregarded. Since the lotus metals are rather new materials, commercialization of lotus metals to sporting goods is reasonable. In 2002, lotus copper was for the first time commercialized as golf putter manufactured by Ryobi Corporation in Japan.

Nonporous copper, lotus copper, and resin were put into the putter flame as inserted materials. The vibration–damping capacity of the inserted materials such as nonporous copper, lotus copper, and resin was investigated [16]. Three damping curves were measured by hammering–vibration–damping method. Figure 10.22 shows the configuration of putter head, measurement apparatus of damping, the damping results, and the attenuation coefficients. The results

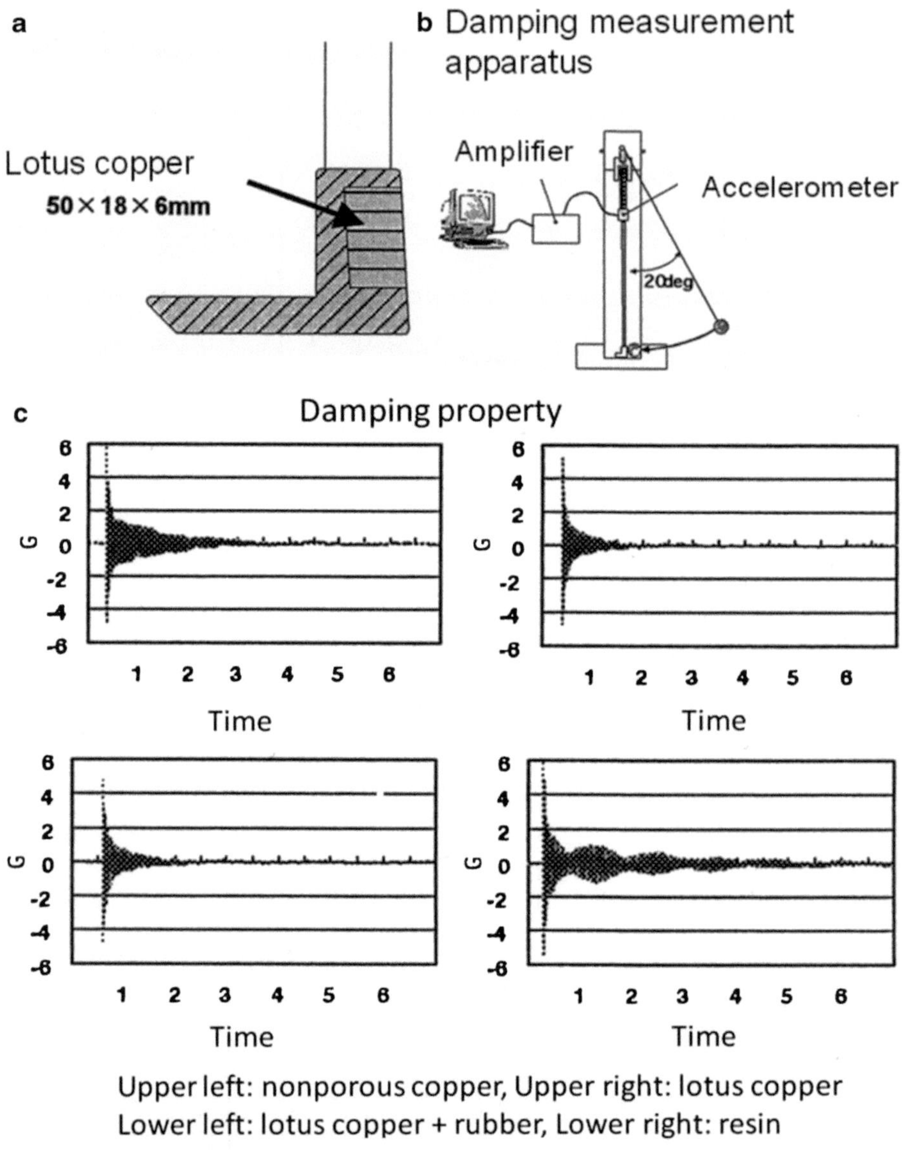


**Fig. 10.22** (**a**) Configuration of putter head, (**b**) damping measurement apparatus, (**c**) the damping property; the *vertical axis* shows the amplitude and the *transversal axis* shows the time (*upper left*, nonporous copper; *upper right*, lotus copper; *lower left*, lotus copper inserted by the rubber; *lower right*, resin)

exhibited that lotus copper had superior damping capacity, and thus, the company decided to commercialize the putter with lotus copper. It is said that feeling of a golf putter hitting a ball is mild. Figure 10.23 shows the golf putter whose central

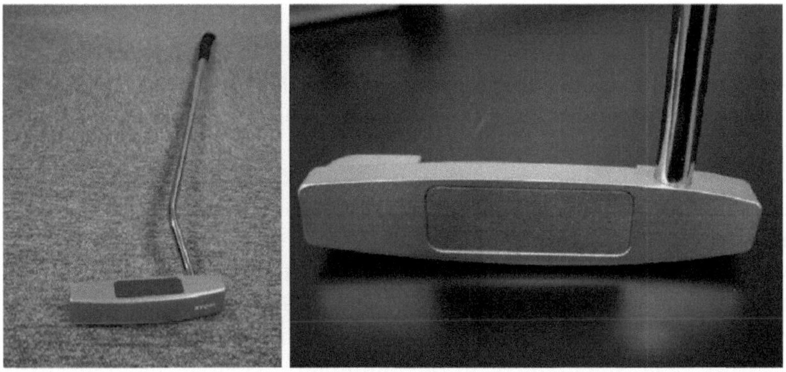

**Fig. 10.23** Golf putter whose central part is made of lotus copper. This putter exhibits superior vibration–damping capacity. Players feel soft to hit a ball (Reprinted with permission from [16] © 2006 Elsevier Ltd)

part is made of lotus copper: the porosity is 40 % and the pore size is about 100 μm in diameter.

## 10.5 Medical Devices

Porous metallic biomaterials have been developed for the reconstruction of hard tissues since those provide obvious advantages over typical monolithic implants. One of these is the possibility to match the mechanical properties to those of bone to prevent stress shielding and loosening of the implant. Metals and alloys processed to create a porous structure by means of powder metallurgy processes [17–19], solid free-form fabrication [20], or vapor deposition of the metal, i.e., tantalum on vitreous polymer network, have been shown to enhance the bone ingrowth and osseointegration [21–24]. In addition, a porous surface improves the mechanical interlocking between the host bone and the implant biomaterial [25], reinforcing the stability of the implant by biological fixation. Thus, application of the porous metallic biomaterials to medical devices has been increasingly expected.

### *10.5.1 In Vitro Cytocompatibility of Lotus Ni-Free Stainless Steel*

Three kinds of stainless steels containing different contents of chromium were employed in the investigation of in vitro cytocompatibility by Alvarez et al. [26]: Fe-25Cr-1N, Fe-23Cr-2Mo-1N, and AISI 446-1N [composition (wt.%)]. Beside

AISI 316L stainless steel was also tested for comparison. Lotus stainless steels were fabricated using continuous zone melting technique. In order to alloy the samples with nitrogen, a high temperature solution nitriding treatment was applied. The high temperature solution nitriding treatment was carried out by sealing the plate samples into a quartz tube of 20 mmϕ × 170 mm in a nitrogen atmosphere of 0.07 MPa. The encapsulated samples were heat-treated at 1,100 °C for 604.8 ks and then quenched into a water bath. After this heat treatment the stainless steel samples of originally ferritic structure become austenitic.

Cytoxicity of the solution nitride samples was evaluated using MC3T3-E1 murine calvaria osteoblasts-like cells. The cell line was acquired from an authorized cell and gene bank (Riken Cell Bank, Ibaraki, Japan). The cells were grown and maintained in α minimum essential medium (α-MEM) supplemented with 1 % kanamycin and 10 vol.% fetal bovine serum in a humidified incubator with 95 % air and 5 % $CO_2$ at 37 °C. Passaging and preparation of single ell suspensions for seeding on the metallic samples was achieved by enzymatic digestion using 0.25 vol.% trypsin, 0.02 % (w/v) EDTA solution in PBS at pH 7.4. Cells counts were performed using a Tatai-type hemacytometer under an inverted microscope. In vitro cytotoxicity was assessed by direct contact method. Disc-shaped specimens with approximate thickness of 1 mm and diameter of 10 mm were obtained from the nonporous alloys rods using electric discharge machining. Prior to cytotoxicity testing, each specimen was ground with 2000 grit SiC paper and ultrasonically cleaned with alcohol, and then rinsed in distilled water. Subsequently, the discs were autoclaved at 121 °C for 0.9 ks.

Cells were incubated at 37 °C in a humidified 5 % $CO_2$ atmosphere for 3- and 7-day intervals without renewal of the culture medium. The cells were examined under optical microscope and photographed with a Nikon DS-L2 color digital camera. The total cell number was analyzed after staining using digital image analysis. The measurements of the cell spreading area were analyzed at the different culture intervals and normalized to the maximum area available in the metallic surfaces. Cell proliferation and the assessment of extent of spreading and cell morphology assays were repeated four times. After 3 days of proliferation assays, media from the cells cultured with metallic materials were saved in quadruplicate for inductively coupled plasma (ICP) analysis.

Cytotoxicity testing of surgical implant materials is an important way to verify their biocompatibility. The quantitative results of the proliferation test obtained with the image analysis software are presented in Fig. 10.24 [26].

The cells cultured for 7 days multiplied actively on the four stainless steel surfaces compared to those cultured for 3 days. After each culture period, cells grown on stainless steel surfaces displayed a significantly lower proliferation rate than the cells cultured on type IV collagen-coated cell culture dish. One-way ANOVA showed no statistically significant difference for the cell numbers of four stainless steel samples after 3 and 7 days of culturing. In the case of pure Ni samples after 7 days in culture, the cell numbers were significantly smaller than that of four stainless steel samples and negative controls. While in the stainless steel

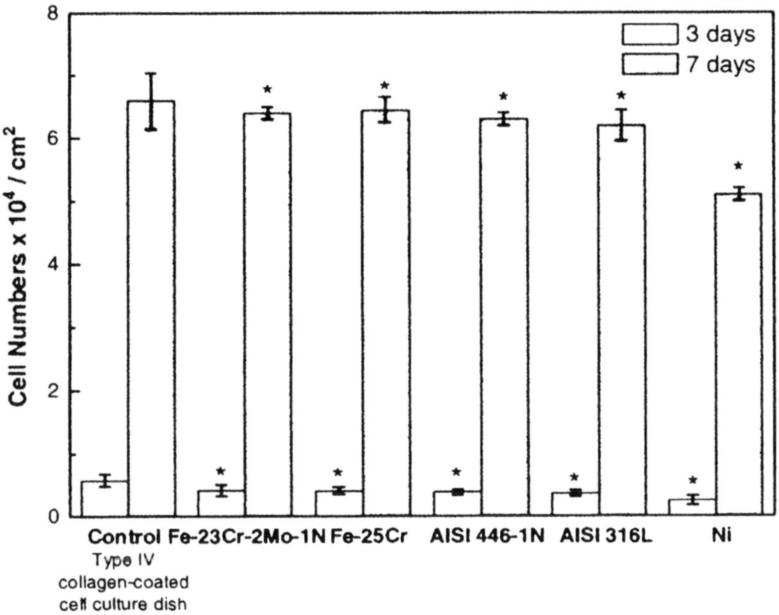


**Fig. 10.24** Proliferation rates after 3 and 7 days in direct contact with the evaluated metallic materials with respect to the control cultures grown on type IV collagen-coated cell culture dish (100 %). (Reprinted with permission from [26] © 2008 Springer Science + Business Media, LLC)

surface cells proliferated forming multilayers after 7 days of culture, in the pure Ni sample, the formation of randomly located multilayer cell nodules was observed, indicating that the spreading was scarce and not uniform.

Morphology of adherent cells after 3-day incubation period is shown in Fig. 10.25. Microscopic analysis of the cell monolayers revealed that there were no notable differences in cell morphology between the type IV collagen cell culture dish (Fig. 10.25c) and the stainless steels (Fig. 10.25a–h). The body of osteoblasts has a mean size of 15 μm and many of them developed armlike cytoplasmic extensions with lengths up to 25 μm. MC3T3-E1 cells seemed to adhere tightly to surface under high magnification (750×magnification) (Fig. 10.25e–h) in accordance with the flattened morphology surrounded by cytoplasmic prolongations. MC3T3-E1 osteoblasts cells proliferated uniformly on the surface of Ni-free stainless steels under optical microscope (100×magnification) (Fig. 10.25a–d). Microscopic analysis of the cell cultures on pure Ni samples revealed that the cell morphology was altered. Cells grown on pure Ni exhibited reduced cytoplasmic area and also abnormally large cells (giant cells) were observed in both cultivation periods (Fig. 10.26).

After 3 days of culture, the MC3T3-E1 cellular spreading was good in all the evaluated stainless steel surfaces but not as extensive as in the cell culture dish treated with type IV collagen, and after 7 days of culture, osteoblasts were evenly

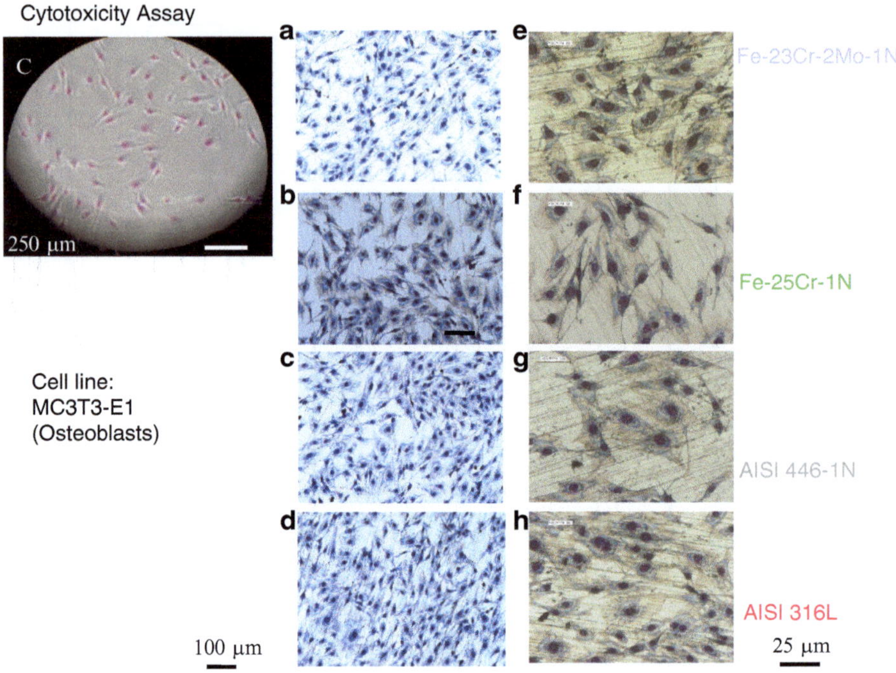


**Fig. 10.25** Representative optical microscopy images of spread MC3T3-E1 osteoblasts cells after 3 days of direct cultivation on (*C*) type IV collagen-coated cell culture dish (**a**) and (**e**) Fe-23Cr-2Mo-1N, (**b**) and (**f**) Fe-25Cr-1N, (**c**) and (**g**) AISI 446-1N, and (**d**) and (**h**) AISI 316L. MC3T3-E1 cells are approaching confluence and display normal morphology. Original magnification: (*C*)15×, (**a–d**) 100×, (**e–h**) 750× (Reprinted with permission from [26] © 2008 Springer Science + Business Media, LLC)

spread covering the entire available stainless steel surfaces. Statistically significant difference between the average area of cells cultivated on the Ni-free stainless steel samples and AISI 316L samples was not observed.

## *10.5.2 In Vivo Osteocompatibility of Lotus Ni-Free Stainless Steel*

A new generation of stainless steels, called nickel-free austenitic stainless steels, does not contain any nickel or the amount of nickel is very reduced (<0.5 wt%) [27], and those have been developed to be used in biological systems. The high nitrogen content is the reason of their superior corrosion resistance and elevated mechanical properties, and these advantages have been widely reported in the literatures [28]. As for the mechanical properties are concerned, the stiffness of

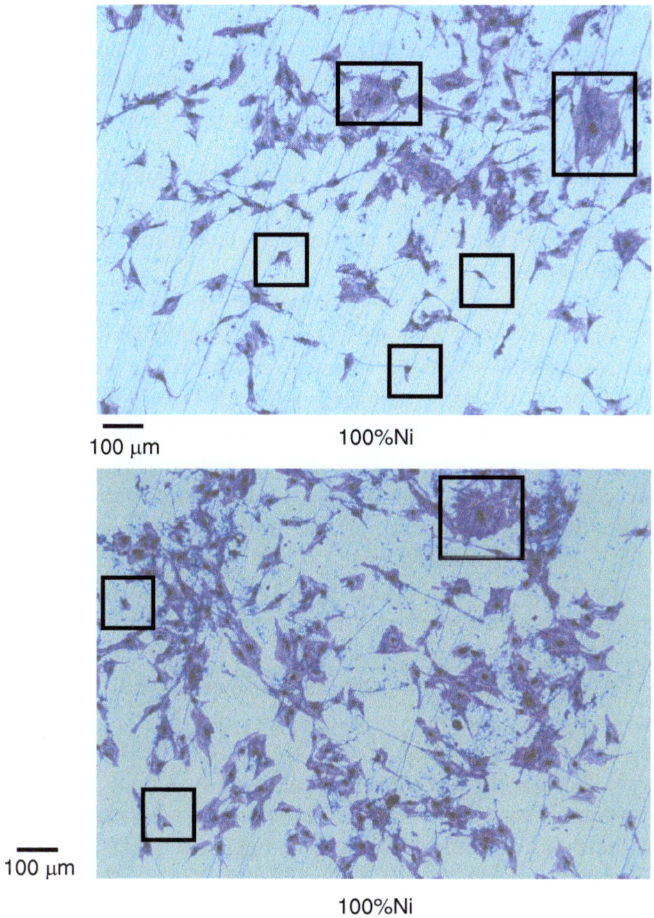


**Fig. 10.26** Optical microscopic images of MC3T3-E1 osteoblast cells grown on pure Ni samples. A higher frequency of reduced cytoplasmic area cells and giant cells was observed (Reprinted with permission from [26] © 2008 Springer Science + Business Media, LLC)

stainless steels is far higher than that of compact bone, and this can lead to stress shielding to the host bone, which may result in detrimental resorptive bone remodeling. Moreover, high elastic moduli of stainless steel implants can cause delayed callus formation and bone healing. However, in order to solve this problem, a number of fabrication processes to produce porous bulk metals and alloys, including stainless steels, have been developed.

The introduction of pores into a structural material will drastically change its mechanical properties of stiffness, strength, and fatigue resistance. Moreover, the geometric microstructure of the porosity is able to change the physical properties such as specific surface area, apparent density, and degree of anisotropy. Based on clinical outcomes and histological evidence, the advantages of porous biomaterials for bone–implant applications are multifold. Bone ingrowth is possible, which could greatly improve the bone–implant interface and may allow for efficient soft tissue attachment supplementing the stability of the implant by biological fixation. A strong bone–implant interface reduces harmful tissue strain and micromotion and therefore promotes the osseointegration of the implant, which is a basic requirement for its successful functioning as host organism. Additionally, by increasing the porosity of the implant material, its stiffness decreases correspondingly, reducing the stress-shielding effect, which can lead to resorption of the bone and the loosening of the implant.

For many clinical situations, a metallic porous graft offers structural integrity. These porous structures must provide temporary support and bone ingrowth. The greatest application fields are in the osseous reconstructive surgery, such as treatment of small bone defects, cranio-maxillofacial reconstruction, tooth root replacements, and vertebral body replacement. In applications in which a porous structure bridges gaps in tissues, the success of bone ingrowth is affected by the pore structure characteristics, such as pore density; pore orientation along length, width, or radial direction; pore size; and pore spacing. New porous metallic structures have to be developed to help advance the current generation of metallic grafts and address unresolved clinical applications. The evaluation of the bone apposition rates within differently oriented pores is also necessary in order to accomplish an efficient implant design.

Alvarez et al. [29] produced a lotus nickel-free austenitic stainless steel with longitudinally oriented pores by continuous zone-melting technique. In the former section, lotus structure was tested in vivo in a canine model achieving excellent bone ingrowth. This section describes further and detailed evaluation of lotus nickel-free austenitic stainless steel as a biomaterial candidate for the treatment of osseous defects and osseous reconstructive repairs in persons with strong nickel allergy predisposition.

The lotus rods of AISI 316L and AISI 446 were fabricated using the continuous zone-melting technique. AISI 316L is commonly used in biomedical applications and was selected as a reference material. AISI 446 stainless steel is a ferritic-type stainless steel that contains 0.3 wt% Ni as impurity. Sixty male, 8 weeks of age Sprague–Dawley rats, received bilateral femoral and trial transcortical implants. In each femur and tibia, one sterile implants of the same alloy was carefully placed into the defect with the pores parallel or perpendicular to the long axis of the rat bone using a sterile forceps (Fig. 10.27). Animals were placed in pairs in cages and were given free access to food and water. The rats were sacrificed by ether suffocation. The left and right tibia bones were retrieved at 4, 8, and 12 weeks after implantation for histology analysis. The femurs also retrieved at 4, 8, and 12 weeks were reserved for mechanical testing. Twenty rats were used for each implantation period.

Push-out testing was performed after 4, 8, and 12 weeks postoperatively by applying a longitudinal compressive force to the cylinder implants. Schematic diagram for testing shear bonding strength between bone tissues and the implant is shown in Fig. 10.28. Push-out testing of the bone–implant samples was

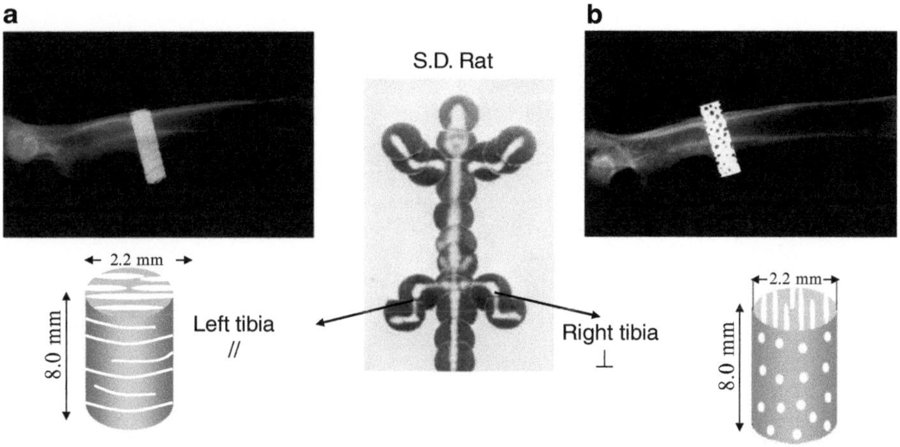


**Fig. 10.27** Radiographs showing the implant placement configuration. (**a**) The elongated direction of the pores of the implant was parallel to the long axis of the rat bone. (**b**) The elongated direction of the pores of the implant was perpendicular to the long axis of the rat bone

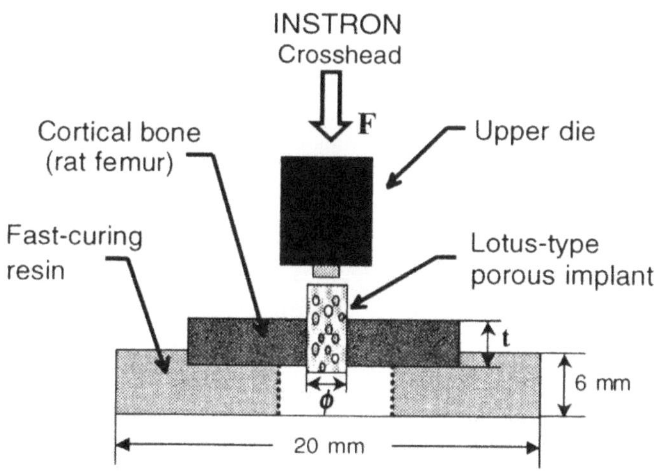


**Fig. 10.28** Schematic diagram of the arrangement used for the mechanical push-out tests. *F* force applied on implant, *t* thickness of the cortical bone surrounding the implant, $\phi$ implant diameter (Reprinted with permission from [29] © 2008 Elsevier B.V.)

performed by using an INSTRON 4482 universal testing machine. The crosshead speed was set at 0.5 mm min$^{-1}$ and the cylinder implants were tested to failure.

Signs of almost complete repair were observed with both types of implants. Histological evaluations revealed bone ingrowth even in the most central part of the lotus implants. Photomicrographs of representative histologic sections after 12 weeks implantation are shown in Fig. 10.29. At 12 weeks the histological

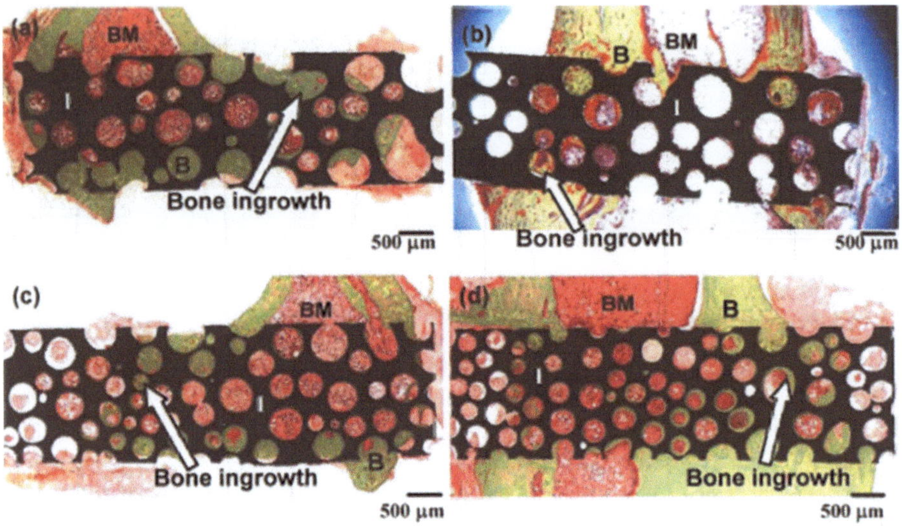


**Fig. 10.29** (**a**) Lotus-type porous AISI 316L implant inserted with the elongated direction of the pores parallel (//) to the tibia bone long axis direction. (**b**) Lotus-type porous AISI 316L implant inserted with the elongated direction of the pores perpendicular (⊥) to the tibia bone long axis direction. (**c**) Lotus-type porous AISI 446-IN implant inserted with the elongated direction of the pores parallel (//) to the tibia bone long axis direction. (**d**) Lotus-type porous AISI 446-IN implant inserted with the elongated direction of the pores perpendicular (⊥) to the tibia bone long axis direction. Light microscopy histological sections after 12 weeks of implantation stained with Villanueva-Goldner's-Trichrome. (Reprinted with permission from [29] © 2008 Elsevier B.V.)

sections displayed an increase of bone contact and bone ingrowth. Indeed, it was found in many histological sections that the area of bone ingrowth within the implant surface was almost complete at 12 weeks. At higher magnifications, the fabricated implants demonstrated healthy cortical bone directly interfacing with the metallic materials (Fig. 10.30). The concept of bone–implant contact (osseointegration) is based on the light microscopic appearance of the biological side of the interface. Osseointegration implies an interface with remodeled, viable bone in direct contact with the implant surface, with no interposed fibrous tissue membrane. This study showed that osseointegration can be attained regularly after 12 weeks implantation time. In the lotus nickel-free implants, the mean bone ingrowth distance through the porous graft was 1,000 μm at 12 weeks, and the appearance of the host cortical bone located far from the implants was similar to the newly ingrown bone inside the pores.

Differences in bone ingrowth between AISI 316L and AISI 446-1N implants are presented in Fig. 10.31. In general, the amount of newly mineralized bone and the bone ingrowth distance into the implants progressively increased with time and maximum bone ingrowth was noted by 12 weeks. The results for both implant types showed that the alignment of the pores parallel (||) to the bone long axis permitted a higher percentage of bone ingrowth. Histological analysis and computer-assisted

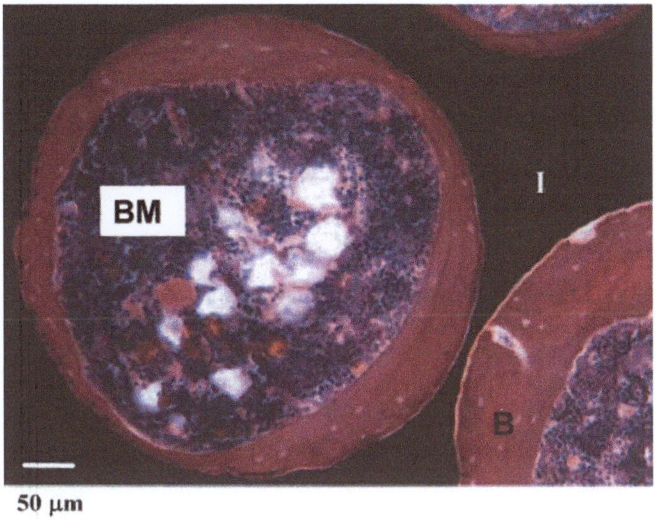


**Fig. 10.30** Higher magnification light micrograph showing bone formation in apparent close contact with lotus AISI 446-1N Ni-free stainless steel at the internal part of the implant. Bone is *stained pink*, bone marrow *purple*, and Ni-free stainless steel is observed in *black*. Bar, 50 μm (Reprinted with permission from [29] © 2008 Elsevier B.V.)

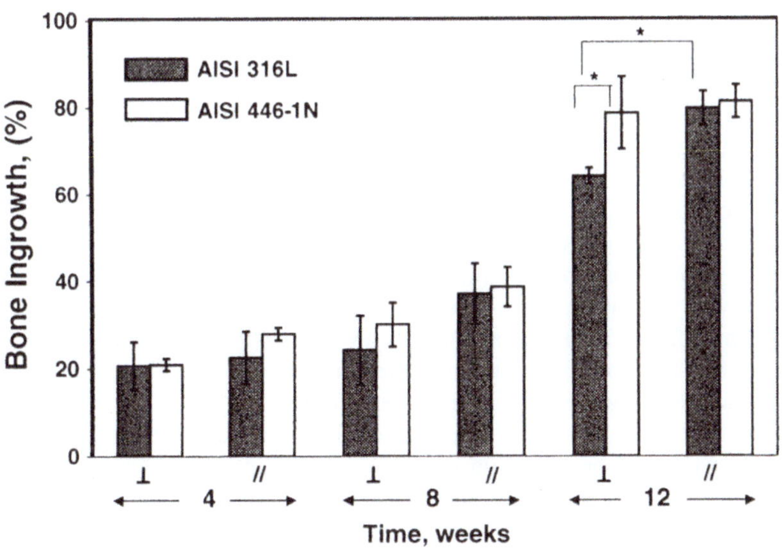


**Fig. 10.31** Percentage of the newly formed bone in the total available pore space in the region of interest for lotus AISI 446-1N Ni-free stainless steel and lotus AISI 316L implants. // and ⊥ indicate parallel and perpendicular pore orientation with respect to the tibia long axis, respectively (Reprinted with permission from [29] © 2008 Elsevier B.V.)

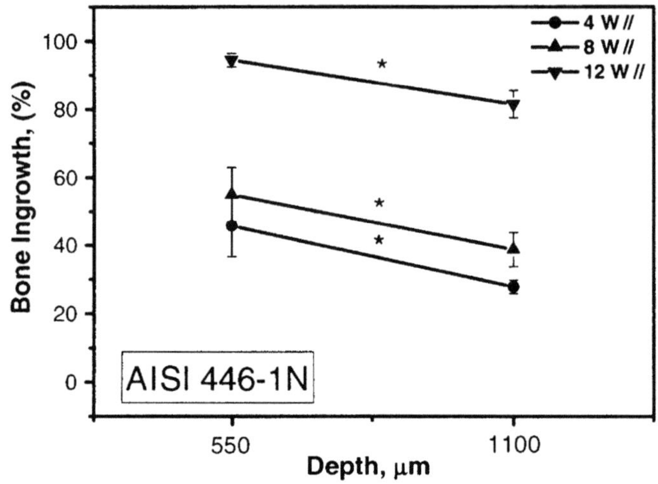


**Fig. 10.32** Changes in the bone ingrowth as a function of the depth for each implantation period for lotus AISI-1N Ni-free stainless steel implants inserted with the elongated direction of the pores parallel with respect to the tibia long axis (Reprinted with permission from [29] © 2008 Elsevier B.V.)

quantification of the bone growth formed revealed that the bone formation within lotus AISI 446-1N had already reached in average 27.9 and 20.9 % after 4 weeks in the implants inserted parallel (||) and perpendicular (⊥) to the bone axis direction, respectively. On the other hand, for lotus AISI 316L by 4 weeks, newly formed bone was 22.5 and 20.6 % in the implants inserted parallel and perpendicular to the bone long axis direction, respectively. Between the two types of materials, no statistically significant differences were noted with respect bone ingrowth.

To determine how deep the bone grew inside the lotus implants, bone ingrowth was determined from two consecutive histological sections. The bone ingrowth parameter was then examined as a function of depth into the implant. For lotus AIAI 446-1N implants, these results are presented in Fig. 10.32. The bore ingrowth decreases with increasing depth into lotus implants. After 12 weeks in the middle of the implants, 82 and 79 % of bone ingrowth were obtained when the implants were inserted with the elongated direction of the pores oriented parallel or perpendicular to the tibia bone long axis direction, respectively.

Mechanical testing was performed since higher bone apposition does not necessarily imply greater bone-bonding strength. The applications of compressive shear forces at the bone–implant interface are a useful means of distinguishing bone-bonding quality and bone contact at the interface. Changes of bonding strength between bone tissues and the implants with the AISI 316L stainless steels are shown in Fig. 10.33. During the push-out tests, all the implants failed through the interface of implant and bone, and two specimens were broken on the bone side. For both alloys, the shear strength of the interface between bone and implants increased with an increase of the implantation time during the first 8 weeks, and this could be

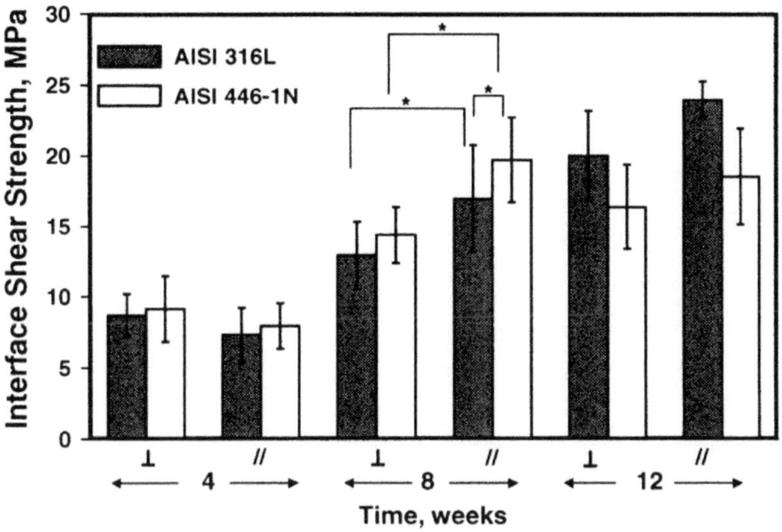


**Fig. 10.33** Changes in the interface shear strength between the rat femur bone and the lotus implants with the implantation time. // and ⊥ indicate parallel and perpendicular pore orientation with respect to the tibia long axis, respectively (Reprinted with permission from [29] © 2008 Elsevier B.V.)

attributed to the increasing with time of the extent of penetration of new bone across the full diameter of the implants.

Although there were no significant differences among the 12 weeks strength values, maximum interfaces shear strength of approximately $22 \pm 3$ and $24 \pm 1$ MPa for lotus AISI 446-1N and lotus AISI 316L, respectively, were obtained at 12 weeks, and these values differed significantly from those of the 4 weeks group. When comparing in the literatures the failure loads of nonporous (smooth or rough surface) and porous-coated metallic implants with the failure loads of lotus implants, it turned out that the strength of bone–implant bonding of the fabricated lotus implants is about ten times higher than in porous-coated or nonporous metallic implants. According to the high shear forces needed to loose the implants at the interface, it is concluded that fabricated lotus implants provide tight fixation offering adequate pore volume for bone ingrowth.

### *10.5.3 Biocompatibility of Lotus Stainless Steel and Titanium in Alveolar Bone*

Metallic materials for artificial hip joints, femoral heads, and metal plates to fix fragments are too strong and heavy for elderly patients, causing wear of the patients' bones, damaging and fracturing metals. In dental implantology,

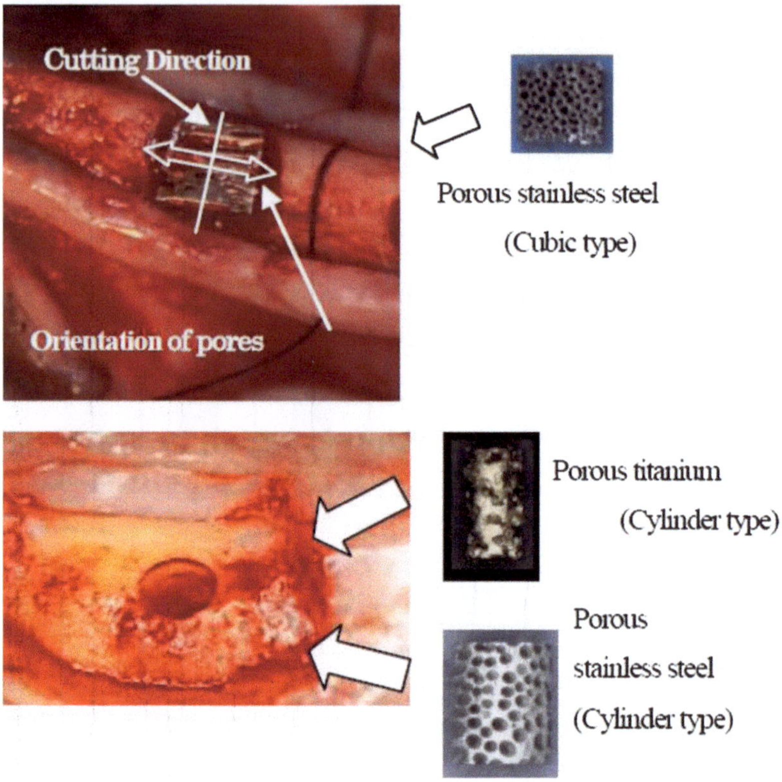


**Fig. 10.34** Insertion of lotus metals to alveolar bone (Reprinted with permission from [32] © 2006 WILEY-Verlag GmbH & Co.)

insufficient trabecular bones contribute to poorer clinical results. It is due to insufficient bone–metal contact. One of the solutions to this phenomenon is the use of porous metallic materials. In natural teeth, Sharpey's fibers make bundles to go into cementum. Minute pores of the cementum work as scaffold to another fiber bundles. The fiber bundles then resist vertical, lateral, and torsional stresses applied on the teeth [30]. The biomaterials with 150–200 μm pores are confirmed to facilitate bone ingrowth [31] and there exists a possibility to further improve functionality of the biomaterials. The conventional manufacturing technology, however, did not control shape and size of the pores so as to maintain certain mechanical strength. Higuchi et al. paid special attention to lotus metals whose pore size and orientation are controlled by utilizing gas pressure to investigate the biocompatibility of lotus stainless steel and titanium in alveolar bone in order to investigate the possibility of an implant by animal experiments [32].

Ten 5-year-old beagle dogs were used in this animal experiment. The dogs were under general anesthesia with intravenously injected pentobarbital sodium. The mandible premolars were extracted. When bone healing was obtained after three months, epithelial mucosa was stripped at the mandible premolars to expose

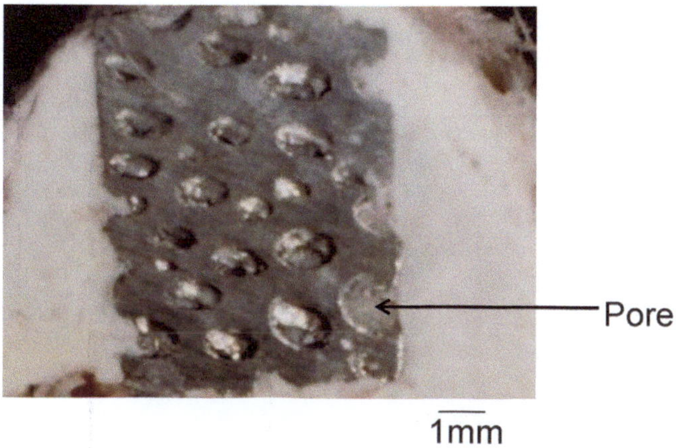


**Fig. 10.35** Osteogenesis in the lotus stainless steel. Observed with actual condition by optical microscope after 4 weeks (Reprinted with permission from [32] © 2006 WILEY-Verlag GmbH & Co.)

mandibular bone. The cortical bone was drilled with a round bur under saline irrigation to create the implantation holes as shown in Fig. 10.34. The lotus metal samples were then implanted and the tissue was sutured. The samples were placed so that the pore direction is mesio-lateral to the mandibular bone. If the pore direction is buccolingual, the openings of the pores will contact the cortical bone, making it difficult to have active tissue ingrowth into the pores. The dogs were fed with water and pellet feeds for 2, 4, and 8 weeks before being sacrificed. At the end of the study period, the dogs were sacrificed with overdose of anesthetics. The specimens were observed with actual condition microscope, optical microscope, and SEM.

As a result, newly formed bones were not observed in lotus stainless steel SUS 304L after 2 weeks. In 4 weeks the active osteogenesis was observed on the surface of the pores as shown in Fig. 10.35, and it advanced into the depth of the pores in 8 weeks. Angiogenesis was rapid in cancellous bones and slower in cortical bones as shown in Fig. 10.36. After 4 weeks the sinusoidal vessels in porous dehydrogenated titanium formed dense vascular plexus advancing into depth of the pores (Fig. 10.37). Osteogenesis was identified in as early as 2 weeks. Osteogenesis was confirmed in the pores both after 4 and 8 weeks as shown in Fig. 10.38.

There is a technical problem to fabricate lotus titanium with uniform porosity and pore size like stainless steel. Living tissues, however, actively grew into the pores. In 2 weeks, active formation of vessels was observed around the metal with invasion into the pores, which was not observed in lotus stainless steel. In 4 weeks the sinusoidal vessels were developed into dense plexus, entering into the deep area of the pores. It is the same with osteogenesis. Newly formed bones were observable in 2 weeks. Pores were filled with new bones in 4 weeks. High biocompatibility of titanium seemed to include new bone formation without induction materials. Biocompatibility of titanium includes excellent cell adhesion property and

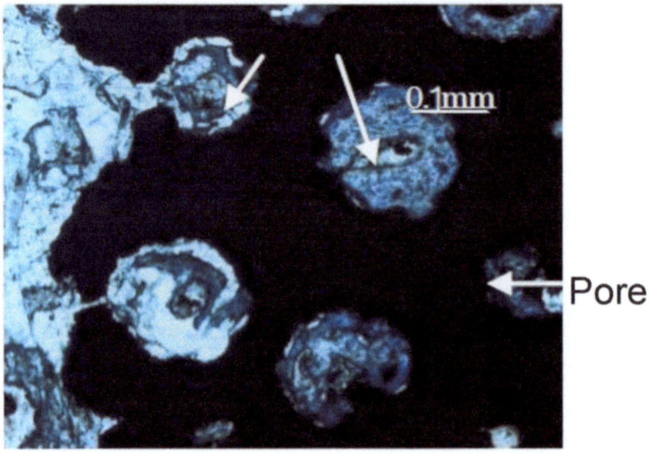


**Fig. 10.36** Osteogenesis in the lotus stainless steel. Staining with truiduin blue after 8 weeks (Reprinted with permission from [32] © 2006 WILEY-Verlag GmbH & Co.)

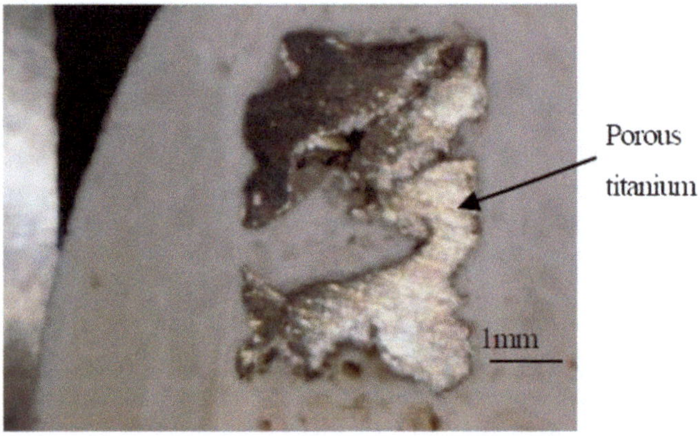


**Fig. 10.37** Osteogenesis in porous titanium. Observed with actual condition by optical microscope after 4 weeks (Reprinted with permission from [32] © 2006 WILEY-Verlag GmbH & Co.)

biomineralization or the ability to deposit hydroxyapatite-like calcium phosphate on the surface under the presence of body fluid. The cells thus strongly adhered to the titanium surface release extracellular matrix to form scaffold for tissue formation. The vessels go into the pores along the scaffold, forming network earlier than lotus stainless steel. Early formation of capillary vessels allows early migration of osteoblasts in the porous titanium. Combination of osteoblasts and biomineralization [33] provides titanium better osteogenic environment than stainless steel. Though titanium is a metal, it has tissue compatibility equal to that of ceramics. One of the reasons for this is that titanium is easily oxidized and covered with an oxide layer. Exposure of titanium to body fluid through this oxide layer was

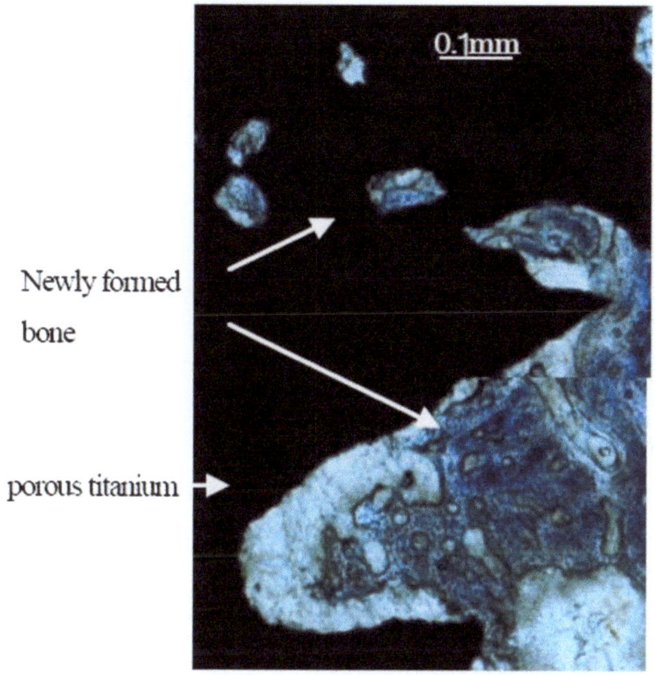


**Fig. 10.38** Osteogenesis in porous titanium. Staining with truiduin blue after 8 weeks (Reprinted with permission from [32] © 2006 WILEY-Verlag GmbH & Co.)

considered to be causing ceramic-like behavior. Recently, however, it was found that immersion of titanium in body fluid-like solution produced the layer of hydroxyapatite. This phenomenon was not observed in gold alloy, gold–silver–palladium alloy, cobalt–chromium–nickel alloy, nickel–chromium alloy, or in stainless steel alloys. Furthermore, deposit on the Ti-6Al-4V or on Ti-50Ni alloy is not apatite. Higuchi et al. [33] conducted in vitro study that extracellular matrix became more absorbable to titanium through Ca ions. The favorable environment for extracellular matrix adsorption via Ca ion is thus created around the implant surface. Extracellular matrix formed in the tissue around titanium makes barrier to through extracellular matrix increases physical resistance, contributing to the initial fixation of titanium implants.

Regenerative medicine using cell proliferation has been the topic of research recently. It takes long time, however, before this technique becomes a mainstream due to such problems as carcinogenicity, ethical issues, complicated culture operation, and high cost. Scaffold for cell proliferation is indispensable for living tissue architecture in the regenerative medicine. The materials currently used are conventional biomaterials. Such hard tissue as bones and teeth needs to have stress resistance. It is highly expected to develop better combination of stiff biomaterials such as lotus metals and regenerative medicine for tissue repair. Development of artificial bones (Fig. 10.39) and joints of metals and ceramics has been active reconstruct structure supporting body and interlocking with joints.

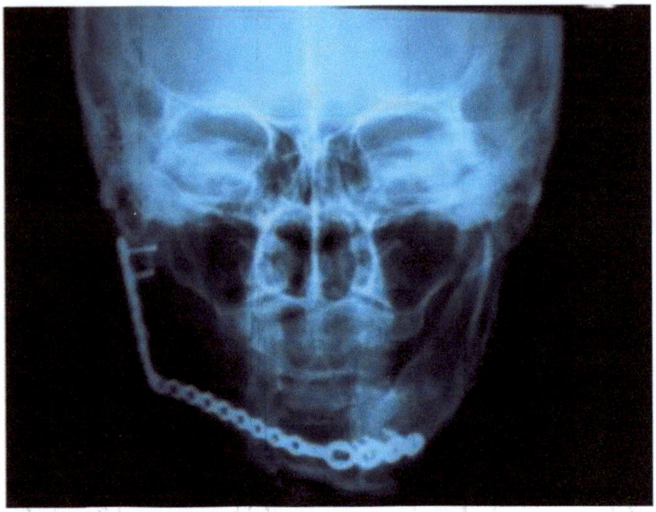

**Fig. 10.39** Parenchymatous defect of the mandible and biomaterial (Reprinted with permission from [32] © 2006 WILEY-Verlag GmbH & Co.)

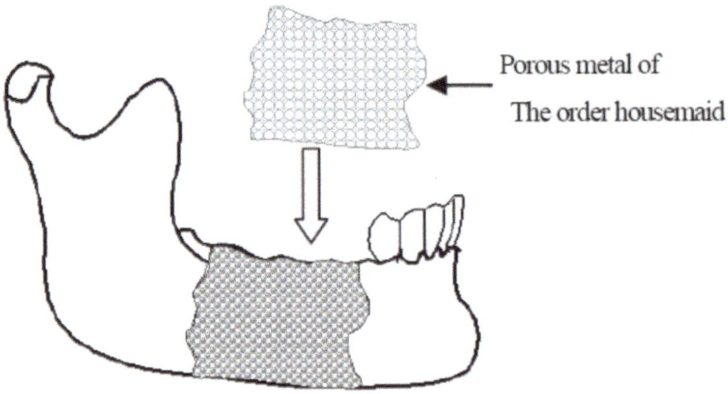


**Fig. 10.40** Application of lotus metal for parenchymatous defect of the mandible (Reprinted with permission from [32] © 2006 WILEY-Verlag GmbH & Co.)

Metals have been used as biomaterials since it has sufficient stiffness to resist stress concentration fracture. Bones bear body weight, the stress of which acts as stimulus for repetitive remodeling. Remodeling is important in maintaining trabecular structures. Long implantation of stiff metals in the bone reduces stimulus to the adjunct bone, causing disuse atrophy. The challenge for the future is to make metal stiffness close to that of the bone (elastic modulus: 7–30 GPa). The lotus materials allow porosity control, giving metals elasticity. It is one of the ways to provide artificial materials with elastic modulus similar to bone modulus (Fig. 10.40). Porous metals with pore of 150–200 μm are said to be ideal for

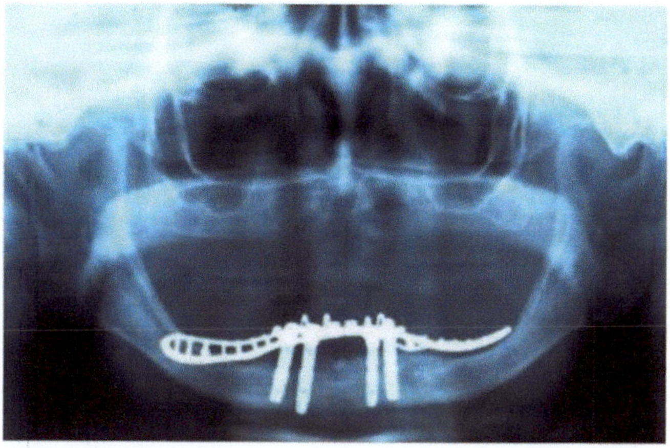

**Fig. 10.41** In the implant case, the bone resorption in maxilla mandible posterior is remarkable (Reprinted with permission from [32] © 2006 WILEY-Verlag GmbH & Co.)

artificial bones. The pores allow ingrowth of vessels and bones in the pores. They have a potential to develop into artificial bones with bone marrow, which is the final goal.

The implants allow contact between the metal and the gingiva. Absence of fibrous connection makes it more probable that bacteria and food residues enter from peri-implant space. Without more careful cleaning after implantology treatment, patients will experience peri-implant inflammation, which may be the cause of implant removal. Implants are composed of three components: super structures, fixtures, and abutment going through mucosa. Greatest difference between dental implants and other biomaterials is that implants communicate with ex vivo environment. Biomaterials other than dental implants are implanted inside the body, giving little opportunity for external bacteria coming into the body. Dental implants, on the other hand, are easily exposed by external bacteria. It is the reason why we have to make thorough oral care.

It has been impossible or less successful to use conventional dental implants for bones with compromised volume or osteoconduction (Fig. 10.41). Lack of fibrous connection gives no buffer against occlusive impacts, reducing abutment–mucous membrane connection. Porous implants should be developed to solve issues we currently face. There are several ways to manufacture porous implants: metal deposition onto fixture surface under high temperature and deposition through melting metal beads and fixture surface. The manufacturing process of porous implants with high functionality will allow to create pores in fixtures without depositing metal beads.

Lotus stainless steel and titanium can be utilized to create pores of 150–200 μm. It will allow growth of bony tissues alone in the pores to enhance holding force. Pore size can be controlled to 10 to 20 μm to allow fibers alone to come into the pores to add buffer against stress. Tailor-made many different types of implant can

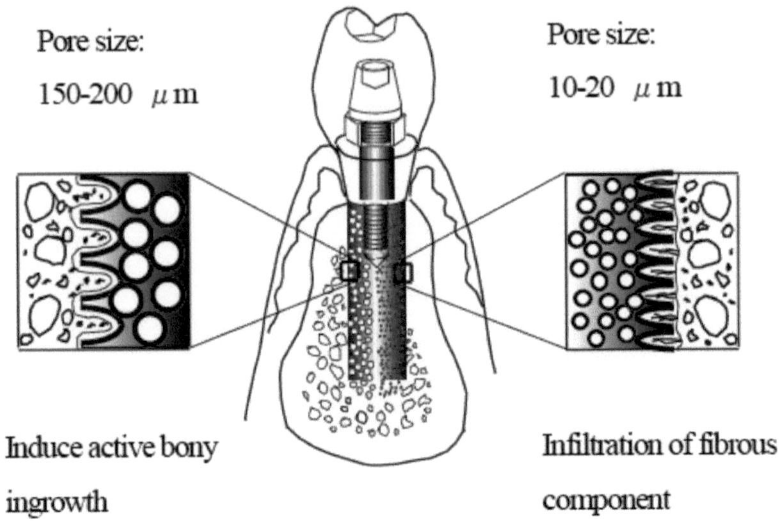


**Fig. 10.42** Serve as basis for bone and fibrous tissue regeneration (Reprinted with permission from [32] © 2006 WILEY-Verlag GmbH & Co.)

be produced. Some can have strongly physical connection by fibers at the site where body communicates with external environment. The lotus metals can be better controlled to improve implant functions (Fig. 10.42). Fixture pore size can be controlled to 150–200 μm to allow only bony tissue to come into the pores and to enhance holding force. The size can be decreased to 10–20 μm to allow only fibrous tissue to come into pores, thus adding buffer to the stress. The fibrous connection is applied at the site of communication between implants and the external environment, enhancing physical binding strength. In the future the porous metals with porosity of 80 % can be used to tailor-made artificial bones.

## References

1. Nishio S (2001) J Thermal Eng. Jpn Soc Mech Eng 617–622
2. Wei X, Joshi Y (2004) J Electron Packaging 126:60–66
3. Bastawros AF, Evans AG (1999) ASME EEP (Am Soc Mech Eng Electronic Packaging Div) 1:733–736
4. Hunt ML, Tien CL (1988) Int J Heat Mass Trans 31:301–309
5. Kamiuto K, Yee SS (2005) Int Comm Heat Mass Trans 32:947–953
6. Zhang HY, Pinjala D, Joshi YK, Wong TN, Toh KC, Ilyer MK (2004) Proceedings of IEEE ITherm 2004, pp 640–647
7. Chiba H, Ogushi T, Nakajima H, Ikeda T (2003) JSME J Series B—Fluids Thermal Eng 47:5843–5847
8. Chiba H, Ogushi T, Nakajima H (2011) Proceedings of ASME/JME 2011 8th thermal engineering joint conference, 13–17 March, Honolulu, ATTEC2011-44108

9. Shah RK, London AL (1978) Advances in heat transfer series, Suppl 1. Academic, New York, pp 192–197
10. Shah RK, London AL (1978) J Fluids Eng 100:177–179
11. Chiba H, Ogushi T, Nakajima H (2010) J Thermal Sci Tech 5:222–237
12. Yoshida I, Monma D, Iino K, Otsuka K, Asai M, Tsuzuki H (2003) J Alloy Comp 355:79–84
13. Golovin IS, Sinning HR (2003) J Alloy Comp 355:2–9
14. Xie ZK, Tane M, Hyun SK, Okuda Y, Nakajima H (2006) Mater Sci Eng A 417:129–133
15. Ota K, Ohashi K, Nakajima H (2003) Mater Sci Eng A341:139–143
16. Nakajima H (2007) Prog Mater Sci 52:1091–1173
17. Greiner C, Oppenheimer SM, Dunand DC (2005) Acta Biomater 1:705–716
18. Jee CSY, Ozguven N, Guo ZX, Evans JRG (2000) Metall Mater Trans B 31:1345–1352
19. Oh IH, Nomura N, Masahi N, Hanada S (2003) Scr Mater 49:1197–1202
20. Bram M (2000) Adv Eng Mater 2:196–199
21. Cameron HU, Lilliar RM, Macnab I (1976) J Biomed Mater Res 10:295–302
22. Bobyn JD, Toh KK, Hacking SA, Tanzer M, Krigier JJ (1999) J Arthroplasty 14:347–354
23. Galante J, Rostoker W, Lueck R, Ray RD (1971) J Bone Jt Surg 53A:101–114
24. Zhang X, Ayers RA, Thorne K, Moore JJ, Schowengerdt F (2001) Biomed Sci Instrum 37:463–468
25. Story BJ, Wagner WR, Gaisser DM, Cook SD, Rust-Dawicki AM (1998) Int J Oral Maxillofac Implants 13:749–756
26. Alvarez K, Hyun SK, Fujimoto S, Nakajima H (2008) J Mater Sci Mater Med 19:3385–3397
27. Speidel MO (2004). US Patent 6,682,582
28. Gavriljuk VG, Berns H (1999) High nitrogen steels. Springer, Berlin, pp 267–268
29. Alvarez K, Hyun SK, Nakano T, Umakoshi Y, Nakajima H (2009) Mater Sci Eng C 29:1182–1190
30. Sims MR (1984) Connec Tissue Res 13:59–67
31. Hulbert SF, Cooke FW, Klawitter JJ, Leonard RB, Sauer BW, Moyle DD (1973) J Biomed Mater Res 4:1–23
32. Higuchi Y, Ohashi Y, Nakajima H (2006) Adv Eng Mater 8:907–912
33. Higuchi Y, Takahashi K, Asai T, Murata Y (2001) J Osaka Dent Univ 35:83–86

# Chapter 11
# Summary

This overview summarized the present status of the research of porous materials with directionally elongated pores. In comparison with the conventional porous and foamed materials with isotropically almost spherical pores, the lotus materials exhibit peculiar features derived from alignment of long pores such as anisotropy of various properties. Besides, these materials possess lightweight and integrated mechanical properties. Thus, these materials should be considered as one of the advanced engineering materials from the point of view of not only materials science but also applications for the practical use. In this review, the author mentioned only a few kinds of applications, but there are a variety of possible prospective applications such as air bearings, filters, mechanical parts of airplanes, and automobiles. We hope the further research and development of this kind of materials will surely be expanded to widen a path to porous materials science and technology.

H. Nakajima, *Porous Metals with Directional Pores*,
DOI 10.1007/978-4-431-54017-5_11, © Springer Japan 2013

# About the Author

**Hideo Nakajima** received his Bachelor's degree in 1971 and Ph.D. from Tohoku University in 1977. Then, he was a postdoctoral associate at Rensselaer Polytechnic Institute, USA, until 1980. From 1980 to 1992, he was an assistant and associate professor at Institute for Materials Research, Tohoku University. In 1992 he moved to Iwate University as a professor. Since 1996, he had been a professor of Institute of Scientific and Industrial Research, Osaka University. He has been the director of the Wakasa Wan Energy Research Center at Tsuruga in Fukui Prefecture since 2012 and is an emeritus professor of Osaka University.

His research interests focus on fabrication, properties, and application of porous metals. He was awarded the Medal with Purple Ribbon in 2009 by the Emperor of Japan.

H. Nakajima, *Porous Metals with Directional Pores*,
DOI 10.1007/978-4-431-54017-5, © Springer Japan 2013

# Index



H. Nakajima, *Porous Metals with Directional Pores*,
DOI 10.1007/978-4-431-54017-5, © Springer Japan 2013

**T**

MIX
Papier aus verantwortungsvollen Quellen
Paper from responsible sources
FSC® C105338

If you have any concerns about our products,
you can contact us on
**ProductSafety@springernature.com**

In case Publisher is established outside the EU,
the EU authorized representative is:
**Springer Nature Customer Service Center GmbH**
**Europaplatz 3, 69115 Heidelberg, Germany**

Printed by Libri Plureos GmbH
in Hamburg, Germany